Computational Techniques in Transient and Turbulent Flow

Vol. 2

Computational Techniques in Transient and Turbulent Flow

Vol. 2 in Series
Recent Advances in Numerical Methods in Fluids

Edited by:

C. Taylor and K. Morgan
University College of Swansea, Swansea, U.K.

1981

PINERIDGE PRESS LIMITED

Swansea, U.K.

First Published in 1981 by
Pineridge Press Limited
91, West Cross Lane, West Cross, Swansea, U.K.

Computational techniques in transient and turbulent fluid flow
   1. Fluid dynamics — Mathematics
   2. Electronic digital computers — Scientific applications
   I. Taylor, C.   II. Morgan, K.
   532'.051028'5   QA911
   ISBN 0–906674–17–4

Printed in Great Britain by
Robert MacLehose and Co. Ltd.
Printers to the University of Glasgow

ADVISORY PANEL

CONTENTS

Page

PREFACE

1. A Survey of Finite Differences with Upwinding 1
   for Numerical Modelling of the Incompressible
   Convective Diffusion Equation.
   B.P. LEONARD

2. A Finite Element Method for the Numerical Solution 37
   of Transient Inviscid Incompressible Flows.
   M. BERCOVIER, G. BERWALD and O. PIRONNEAU

3. Differential Analysis of Strong Discontinuities 59
   in One Dimensional Gas Flow.
   N.N. YANENKO and E.V. VOROZHTSOV

4. Solution of the Unsteady Navier-Stokes Equations 97
   by a Finite-Element Projection Method.
   J. DONEA, S. GIULIANI, H. LAVAL and L.QUARTAPELLE

5. Large Scale Weather Forecasting Using Finite 133
   Element Methods.
   M.J.P. CULLEN

6. Variational Formulation of Free Surface Gravity 165
   Flow Including Gated Flow Using Adaptive Finite
   Elements.
   W.D. LIAM FINN and E. VAROGLU

7. Iterative Procedure for Simultaneous Calculation 191
   of a Stress Transport Turbulence Model in Two
   Dimensional Eliptic Fields
   D. NOAT

8. Computer Analysis of Fluid Flow and Heat Transfer 223
   S.V. PATANKER

9. Approaches to the Finite Element Solution of 253
   Dimensional Turbulent Flows.
   B.E. LAROCK and D.R. SCHAMBER

10. Analysis of Turbulent Flow with Separation Using 283
    the Finite Element Method.
    C. TAYLOR, C.E. THOMAS and K. MORGAN

SUBJECT INDEX

PREFACE

This second volume in the series of texts devoted to recent advances in the numerical solution of flow problems follows on from the first successful text printed in 1980. Again each chapter is devoted to a specific area of research and the contributors have developed and adequately referenced recent aspects of the topic being considered. It is evident that considerable advances are apparent in all areas of research associated with flow problems and, hopefully, further developments will evolve from ideas projected in this book.

In each chapter the authors introduce each topic, providing adequate referencing, and develop the application of the proposed technique to specific problems. In common with the first volume, these range from potential flow to those where turbulence phenomena dominate. The advocated techniques are demonstrated by specific application, and, where necessary, most authors indicate the limitations of these by direct comparison with other numerical models or experimental results.

The editors wish to express their thanks to the authors and the advisory panel in making this publication possible.

Finally, the editors wish to state that views expressed and conclusions drawn in the text are solely those of the contributing authors.

C. TAYLOR

K. MORGAN

CHAPTER 1

A SURVEY OF FINITE DIFFERENCES WITH UPWINDING FOR NUMERICAL
MODELLING OF THE INCOMPRESSIBLE CONVECTIVE DIFFUSION EQUATION

B.P. Leonard

City University of New York, College of Staten Island

ABSTRACT

Central difference methods, although excellent for problems
dominated by even-ordered spatial derivatives, have been
plagued by unphysical oscillations or computational non-
convergence when applied to the first-derivative convection
term in fluid mechanics. This has led to the development of
a variety of non-centered upstream-shifted convective differencing schemes, the general philosophy of which is often
known as "upwinding." Although smooth and computationally
docile solutions can be achieved by upwinding, it is often
at the direct expense of accuracy. This is true even in
the case of so-called "optimally weighted" upwind schemes
unless they have been carefully generalized in a consistent
manner. As the present analysis will show, a misinformed
application of finite differences with (certain types of)
upwinding can lead to unacceptable inaccuracies without any
indication that gross errors have been committed. It will
also be shown, however, that a straight-forward upstream-
shifted third-order convective differencing scheme
automatically combines inherent stability and accuracy, and
is algorithmically consistent with (standard) second-order
diffusive differencing. Slight problems in thin boundary
layers can be resolved (accurately) by introducing alternate
interpolation functions into the basic third-order technique.

INTRODUCTION

Finite difference methods have been very successful when
applied to physical problems such as diffusion and wave
motion, for which the governing differential equations
involve dominant second-order spatial derivatives. The
same is true for solid mechanics problems in which spatial
fourth derivatives are dominant. For all these problems,

classical (second-order) central differencing is entirely
adequate in terms of accuracy, stability and algorithmic
simplicity. Thus it seems natural to try to apply similar
methods to the first derivative convection term in fluid
mechanics problems. Unfortunately, central-difference
methods for convection are usually plagued by unphysical
oscillations or explosive nonconvergence -- a well known (but
poorly understood) phenomenon. A popular "remedy" for this
has been the use of first-order upstream-shifted convective
differencing (Roache, 1976), which gives very smooth and
computationally docile results. It is now generally
recognized, however, that the "artificial diffusion"
associated with this so-called "full upwinding" method
severely corrupts the solution (Leonard, 1978). Recently,
"partial upwinding" techniques (based on a weighted combination of first- and second-order differences, or the
equivalent in terms of a finite element formulation) have
been strongly advocated (Heinrich & Zienkiewicz, 1979), the
degree of upwinding being based on the exact solution of a
specialized model problem (Raithby & Torrance, 1974; Heinrich
et al., 1977). There has been a tendency to apply these so-
called optimal upwinding methods to more general problems
with the assumption (or hope!) that high accuracy is being
achieved. However, under even only moderately high convection conditions, optimal upwinding reverts to full (first-
order) upwinding, and to the extent that the particular
problem differs from the model problem on which the
technique is based, accuracy is severely impaired by the
artificial diffusion thereby introduced. Optimal upwinding
methods can be generalized, and this has recently been done
within the framework of finite element techniques (Hughes &
Brooks, 1979; Brooks & Hughes, 1980). It is clear that
similar techniques could be developed in terms of finite
differences; however, this type of generalization is not
pursued here. Rather, an algorithmically straight-forward
third-order finite difference method is introduced which
possesses inherent stability and is of high accuracy. Sharp
boundary-layer regions can then be accurately resolved
without numerical oscillations by introducing an adjusted
interpolation scheme -- specifically, the three-point
quadratic interpolation of the standard third-order method
is partially replaced by (an approximation to) a three-point
exponential interpolation in appropriate regions. The
relative weighting between quadratic and exponential interpolation is determined by local shape characteristics of the
solution itself, leading to excellent accuracy throughout
all flow regimes.

The model incompressible convective diffusion equation

$$\frac{\partial \phi}{\partial t} = - u \frac{\partial \phi}{\partial x} + \Gamma \frac{\partial^2 \phi}{\partial x^2} + S \tag{1}$$

is useful for the study of computational methods in fluid mechanics. The equation is nominally one-dimensional; however, it is important to recognize that it is appropriate to a much wider class of flow situations because other transport terms not written explicitly in Equation (1) can be considered to be contained in S (in addition to explicit source terms). The necessity of including an effective source term in a model equation cannot be over-emphasized. One of the problems with so-called optimal upwinding is that, until recently (Brooks & Hughes, 1980) it was based on an exact *steady-state* solution of Equation (1) with $S \equiv 0$. It is not surprising that early forms of optimal upwinding break down under unsteady conditions (Gresho & Lee) or when there is an effective source term due either to actual sources or to cross-grid transport (Hughes & Brooks, 1979) in multi-dimensional flow.

For studying the relative merits of spatial differencing techniques, Equation (1) can be rewritten in a nominally steady (one-dimensional) form as

$$u \frac{\partial \phi}{\partial x} = \Gamma \frac{\partial^2 \phi}{\partial x^2} + S^* \tag{2}$$

where $S^* = S - \partial \phi / \partial t$. In a real (unsteady, multi-dimensional, compressible, turbulent) flow simulation, at any instant of time and at a given (y,z) location (line), all effects not represented in the first two terms of Equation (2) can be lumped into $S^*(x)$. Although the study of variable coefficients is obviously important, the *essential* qualities of spatial finite difference methods can be studied by assuming u and Γ to be constants. This is the philosophy behind the model test problem studied here, using a simple specified source term $S^*(x)$.

In the next section, a number of spatial finite difference operators are analyzed in terms of their Taylor series expansions (as an indication of accuracy) and with respect to a property called "feedback sensitivity" (which determines whether an algorithm has inherent damping, is neutral, or is actually unstable). It will be shown, for example, that one of the main problems with central differencing (of *any* order) is a total lack of feedback sensitivity when applied to derivatives of *odd* order. Thus, there is no inherent damping of oscillatory perturbations. By contrast, non-centered convective differencing schemes always have a non-zero feedback sensitivity, and this can be made negative (thereby guaranteeing inherent damping) by choosing an upwind bias. Until recently, most upwinding strategies have been based on first-order difference schemes or modifications thereof, with some preliminary flirtations with second-order upwinding (Hodge *et al*, 1979). Unfortunately, such schemes tend to have low accuracy; but, because they produce

4

plausible looking (i.e. "wiggle"-free) results, they have
remained very popular in spite of their known (and often
ignored) shortcomings. An alternative philosophy, based on
a simple third-order upwind scheme, is advocated in the
present analysis. As will be seen, the third-order convective
differencing scheme is a logical extension of second-order
central differencing for diffusion -- both are exact for a
cubic polynomial. For reference, a centered fourth-order
scheme is also briefly discussed.

The remainder of the paper forms a comparative review of
several spatial differencing schemes as applied to a model
test problem for which exact solutions are also given. The
oscillatory nature of second-order central differencing is
made quite apparent. The intolerable artificial diffusion of
both first-order (hybrid) *and* so-called optimal schemes is
demonstrated. The high accuracy and strong inherent damping
characteristics of the third-order scheme are immediately
clear, although the polynomial interpolation scheme is still
unable to resolve sharp boundary layers without (a few)
wiggles. Finally, the "exponentially adjusted" third-order
scheme is seen to give results which are graphically almost
indistinguishable from the exact solution over the complete
range of Péclet numbers, from pure diffusion to pure convection.

ANALYSIS OF FINITE-DIFFERENCE OPERATORS

The diffusion term

By far the most popular finite-difference approximation to
the diffusion term in Equation (1) involves the second
central difference. Using classical Taylor series analysis,
this can be written

$$\Gamma \left[\frac{\phi_{i+1} - 2\phi_i + \phi_{i-1}}{\Delta x^2} \right] = \Gamma \left[\left(\frac{\partial^2 \phi}{\partial x^2} \right)_i + \frac{1}{12} \phi_i^{(iv)} \Delta x^2 + \ldots \right] \qquad (3)$$

The discretization error is proportional to Δx^2; thus this
operator is known as being "second-order accurate." Note
that the coefficient of Δx^2 involves a fourth-derivative
term, so that the discretization is an exact representation
of the second derivative (at the point i) for a *cubic*
polynomial. This means, in particular, that in a solution
using this operator (assuming all other terms are at least
as accurate), local truncation error is proportional to Δx^4.
This operator combines high formal accuracy with algorithmic
simplicity. It also possesses another attractive property:
negative feedback sensitivity -- as will now be demonstrated.

<u>Feedback sensitivity</u>

In order to understand the concept of feedback sensitivity,
assume that a numerical simulation of Equation (1) has been
made and consider the evolution of the central node value,
ϕ_i:

$$\frac{\partial \phi_i}{\partial t} = \text{RHS} \tag{4}$$

where RHS represents the numerically modelled terms on the
right-hand side. In general, RHS will involve some dependence
on ϕ_i; thus, the evolution of perturbations in ϕ_i can be
studied by taking the variation of Equation (4) with respect
to ϕ_i, giving

$$\frac{\partial \delta \phi_i}{\partial t} = \frac{\partial \text{RHS}}{\partial \phi_i} \delta \phi_i \tag{5}$$

which has a formal solution

$$\delta \phi_i = \exp\left[\int \Sigma \, \partial t\right] \tag{6}$$

where the *feedback sensitivity* is given by

$$\Sigma = \frac{\partial \text{RHS}}{\partial \phi_i} \tag{7}$$

Clearly, a positive feedback sensitivity would lead to
exponential growth of any perturbations and is therefore
undesirable. If $\Sigma = 0$, there is no inherent feedback, the
algorithm has neutral sensitivity and (certain types of)
perturbations can be superimposed on the solution without
affecting the RHS -- the algorithm is insensitive to these
errors and no automatic corrective action is taken. This
type of neutral sensitivity is often associated with
temporal and spatial oscillations (i.e. wiggles!). It is
directly analogous to the oscillatory nature of marginal
stability in dynamic systems. Negative feedback sensitivity,
on the other hand, assures that corrective action will
automatically be taken by the algorithm to damp out random
fluctuations. This is clearly a highly desirable property of
any numerical algorithm.

In the case of the diffusion operator given by Equation (3)
it is clear that

$$\Sigma = -2\Gamma/\Delta x^2 \tag{8}$$

and since Γ is physically always positive, the operator has
negative feedback sensitivity giving it good inherent damping

6

properties. Note that if a *computed* diffusion coefficient
itself takes on an erroneous negative value, the consequent
positive feedback then indicated by Equation (8) usually
guarantees that computational diaster will follow rapidly.

Other diffusion operators
It is interesting to compare other second-order finite-
difference operators for diffusion which may not be quite so
well known (for reasons that will become obvious). Consider,
for example

$$\Gamma \left[\frac{\phi_{i+2} - \phi_{i+1} - \phi_{i-1} + \phi_{i-2}}{3\Delta x^2} \right] = \Gamma \left[\left(\frac{\partial^2 \phi}{\partial x^2} \right)_i + \frac{5}{12} \phi_i^{(iv)} \Delta x^2 + \ldots \right] \quad (9)$$

Superficially, the Taylor expansion is not unlike that of
Equation (3): the second-order discretization error has a
fourth-derivative coefficient (which is zero for a cubic);
the only obvious difference is the numerical factor. However,
in this case, ϕ_i does not appear in the operator and so the
feedback sensitivity is

$$\Sigma \equiv 0 \quad (10)$$

-- there is no corrective feedback! In fact, solutions using
this operator are susceptible to unphysical parasitic
oscillations (of wavelength $3\Delta x$), clearly an undesirable
characteristic.

An even less desirable diffusion operator can be formed as
follows

$$\Gamma \left[\frac{\phi_{i+2} - 2\phi_{i+1} + 2\phi_i - 2\phi_{i-1} + \phi_{i-2}}{2\Delta x^2} \right] =$$

$$\Gamma \left[\left(\frac{\partial^2 \phi}{\partial x^2} \right)_i + \frac{7}{12} \phi_i^{(iv)} \Delta x^2 + \ldots \right] \quad (11)$$

which again has a Taylor series expansion superficially
similar to Equations (9) or (3). But, for this operator, the
feedback sensitivity is *positive*:

$$\Sigma = \Gamma/\Delta x^2 \quad (12)$$

-- an algorithm using this operator would be inherently
unstable!

There are many other practical finite-difference operators
for diffusion. For obvious reasons, the fourth-order

central-difference form

$$\Gamma \left[\frac{-\phi_{i+2} + 16\,\phi_{i+1} - 30\,\phi_i + 16\,\phi_{i-1} - \phi_{i-2}}{12\,\Delta x^2} \right]$$

$$= \Gamma \left[\left(\frac{\partial^2 \phi}{\partial x^2} \right)_i - \frac{1}{90}\,\phi_i^{(vi)} \Delta x^4 + \ldots \right] \tag{13}$$

is very popular. Note that the operator is exact for a quintic polynomial; thus, local truncation error is $O(\Delta x^6)$. The feedback sensitivity is negative:

$$\Sigma = -5\Gamma/2\Delta x^2 \tag{14}$$

and similar to (slightly stronger than) that of the standard second-order operator given by Equation (8).

Modelling the convection term
When one considers the success of second-order central differencing for diffusion, it is natural to consider using a similar operator for the first-derivative convection term in Equation (1). This would be

$$- u \left[\frac{\phi_{i+1} - \phi_{i-1}}{2\Delta x} \right] = - u \left[\left(\frac{\partial \phi}{\partial x} \right)_i + \frac{1}{6}\,\phi_i''' \Delta x^2 + \ldots \right] \tag{15}$$

which has the same formal order of discretization error as that of Equation (3); note, however, that this convective operator is exact only for a quadratic polynomial, giving a formal local truncation error $O(\Delta x^3)$. As should by now be immediately obvious, this convective operator has *neutral* feedback sensitivity, $\Sigma \equiv 0$. And this is essentially the major problem with this operator. As will be seen, neutral feedback sensitivity is a characteristic common to all *central* difference methods (of any order) when applied to *odd*-order derivatives.

When second-order central differencing is used for both convection and diffusion in modelling Equation (1), negative feedback sensitivity is dependent entirely on the diffusion term. Under high convection conditions, this is relatively very weak and the solution is susceptible to both temporal and spatial unphysical oscillations. The nondimensional parameter which represents the relative ratio of modelled convection and diffusion terms is the generalized grid Péclet number

$$P_\Delta = u\Delta x/\Gamma \tag{16}$$

Typically, for second-order central differencing, when $P_\Delta \lesssim 2$, the negative feedback sensitivity of the diffusion operator is strong enough to damp out potential wiggles. But to the extent that P_Δ exceeds 2, the damping becomes relatively weaker and weaker, and in the high-convection regime the wiggles may corrupt large regions of the flow domain.

The problems associated with central-differencing's neutral feedback sensitivity are not cured by going to higher-order *central* methods. For example, the fourth-order central-difference operator for convection is

$$- u \left[\frac{-\phi_{i+2} + 8\phi_{i+1} - 8\phi_{i-1} + \phi_{i-2}}{12\Delta x} \right]$$

$$= - u \left[\left(\frac{\partial \phi}{\partial x} \right)_i - \frac{1}{4} \phi_i^{(v)} \Delta x^4 + \ldots \right] \tag{17}$$

Although this has high formal accuracy, there is no ϕ_i term in the operator, and $\Sigma \equiv 0$. Once again there is the possibility of wiggles under high-convection conditions, although the amount of physical diffusion needed to dampen wiggles is smaller than that for the second-order case.

Note that the corresponding operator for the third derivative

$$\frac{\phi_{i+2} - 2\phi_{i+1} + 2\phi_{i-1} - \phi_{i-2}}{2\Delta x^3} = \left(\frac{\partial^3 \phi}{\partial x^3} \right)_i + \frac{1}{4}\phi_i^{(v)} \Delta x^2 + \ldots \tag{18}$$

also has neutral feedback sensitivity. A little reflection will reveal that this must be a common property of all *central* difference methods when modelling *odd*-order derivatives.

The upwinding strategy

<u>First-order upwinding</u> Early "remedies" for central-differencing's problems with convection involved a strategy of using the one-sided first difference, usually with "physical plausibility" arguments to determine which direction should be favoured for taking the difference (e.g. see Roache, 1976). Instead of making *a priori* judgements as to the appropriateness of the direction, consider both first differences of the convection term:

$$- u \left[\frac{\phi_i - \phi_{i-1}}{\Delta x} \right] = - u \left[\left(\frac{\partial \phi}{\partial x} \right)_i - \frac{1}{2} \phi_i'' \Delta x + \ldots \right] \tag{19}$$

and

$$- u \left[\frac{\phi_{i+1} - \phi_i}{\Delta x} \right] = - u \left[\left(\frac{\partial \phi}{\partial x} \right)_i + \frac{1}{2} \phi_i'' \Delta x + \dots \right] \qquad (20)$$

Note that the leading discretization error is equivalent to a physical diffusion term (i.e. a second derivative) with an effective diffusion coefficient of

$$\Gamma_{num} = \pm u \Delta x / 2 \qquad (21)$$

Thus, by choosing Equation (19) when $u > 0$ and (20) when $u < 0$, the discretization error terms are always stabilizing. This, of course, is the "upwinding" strategy. The same conclusion is reached by studying the feedback sensitivity,

$$\Sigma = \mp u / \Delta x \qquad (22)$$

which is always negative if upwinding is used.

In fact, by combining Equations (19) and (20), first-order upwinding can be written in terms of central-difference operators (regardless of the sign of u) as

$$- u \left[\frac{\phi_{i+1} - \phi_{i-1}}{2\Delta x} \right] + |u| \left[\frac{\phi_{i+1} - 2\phi_i + \phi_{i-1}}{2\Delta x} \right]$$

$$= - u \left[\left(\frac{\partial \phi}{\partial x} \right)_i + \frac{1}{2} |u| \phi_i'' \Delta x + \dots \right] \qquad (23)$$

from which it is immediately clear that

$$\Gamma_{num} = |u| \Delta x / 2 \qquad (24)$$

and

$$\Sigma = - |u| / \Delta x \qquad (25)$$

Although at first sight the negative feedback sensitivity of first-order upwinding seems quite attractive in comparison with central differencing's total lack, it must be remembered that the equation being modelled is the convective *diffusion* equation and that by using first-order upwinding for convection a "false diffusion" term is automatically added which will corrupt the modelling of the physical diffusion term unless

$$\Gamma_{num} \ll \Gamma \qquad (26)$$

which, from Equations (16) and (21), is the same as requiring

$$P_\Delta \ll 1 \tag{27}$$

-- a much more stringent requirement than the $P_\Delta \lesssim 2$ condition needed for (non-wiggly) computations using second-order central differencing. There have been arguments in the literature to the effect that first-order upwinding actually becomes more accurate for very *large* values of P_Δ (Spalding, 1972; Raithby, 1976). However, as is shown quite dramatically in the next section, such arguments are entirely fallacial, being based on the results of a degenerate test problem which has virtually no relevance to most practical flow situations.

It is now generally acknowledged among most professional computational fluid dynamicists that the false diffusion introduced by first-order upwinding renders this method unacceptable for practical purposes. To overcome these difficulties, essentially two different types of philosophy have developed, which might be categorized as "partial first-order upwinding" (a weighted mixture with second-order central differencing) and "higher-order upwinding." Unfortunately, most partial upwinding methods revert to full first-order upwinding under (even only moderately) high convection conditions (e.g. $P_\Delta \gtrsim 5$). The real problem is that such methods have been widely advertised as being highly accurate under such conditions, whereas they certainly are not! Once again, the fallacy stems from the degenerate test problem mentioned earlier. One would have to take great care to generalize a finite-difference method based on partial first-order upwinding to take account of the unsteady effects, cross-grid transport, and explicit source terms, all of which are missing in currently popular techniques. So far, this has apparently not been done, although progress is being made in the corresponding development of finite-*element* methods (Brooks & Hughes, 1980).

<u>Second-order upwinding</u> With respect to higher-order upwind schemes, there has been tentative interest in a second-order full upwind scheme, for which the convective term is written (Hodge *et al.*, 1979)

$$- u \left[\frac{3\phi_i - 4\phi_{i-1} + \phi_{i-2}}{2\Delta x} \right]$$

$$= - u \left[\left(\frac{\partial \phi}{\partial x} \right)_i - \frac{1}{3} \phi_i''' \Delta x^2 + \frac{1}{4} \phi_i^{(iv)} \Delta x^3 + \ldots \right] \tag{27}$$

for $u > 0$, and

$$- u \left[\frac{-\phi_{i+2} + 4\phi_{i+1} - 3\phi_i}{2\Delta x} \right]$$

$$= - u \left[\left(\frac{\partial \phi}{\partial x} \right)_i - \frac{1}{3} \phi_i''' \Delta x^2 - \frac{1}{4} \phi_i^{(iv)} \Delta x^3 + \dots \right] \tag{28}$$

for $u < 0$. The feedback sensitivity is $\Sigma = - 3|u|/2\Delta x$, which of course is stabilizing. Note that the discretization error of the combined operator can be written

$$\frac{1}{3} u \, \phi_i''' \Delta x^2 - \frac{1}{4}|u| \phi_i^{(iv)} \Delta x^3 + \dots \tag{29}$$

While the second term in this expression tends to have a damping effect, the first term is potentially oscillatory. The overall benefit is rather disappointing. Some improvement can be achieved by averaging this operator with second order central differencing (Fromm, 1968), thereby somewhat reducing the size of the ϕ_i''' term. Of course, by judicious weighting, this potentially wiggly term could be eliminated entirely -- but this in fact just leads exactly to the upstream-shifted *third*-order method (Leonard, 1979).

<u>Third-order upwinding</u> The third-order upwind convection operator can be written in symmetrical form (for $u \gtrless 0$) as

$$- u \left[\frac{- \phi_{i+2} + 8\phi_{i+1} - 8\phi_{i-1} + \phi_{i-2}}{12 \, \Delta x} \right]$$

$$- |u| \left[\frac{\phi_{i+2} - 4\phi_{i+1} + 6\phi_i - 4\phi_{i-1} + \phi_{i-2}}{12 \, \Delta x} \right]$$

$$= - u \left(\frac{\partial \phi}{\partial x} \right)_i - \frac{1}{12}|u| \phi_i^{(iv)} \Delta x^3 + \frac{1}{4} u \phi_i^{(v)} \Delta x^4 + \dots \tag{30}$$

which could be interpreted as a fourth-order central-difference convective term with a fourth-difference term "arbitrarily" added in order to dampen potential wiggles. Of course, the numerical coefficient is not arbitrary, but is in fact consistent with *cubic* interpolation through the four points (i-2), (i-1), i, and (i+1) when $u > 0$, or through (i-1), i, (i+1), and (i+2) when $u < 0$.

As was previously noted, for this cubic interpolation the second derivative (at point i) is exactly the second central difference at i. In this interpretation, the third-order upwind convection operator is *consistent* with second-order central differencing for diffusion. Referring to Equations (3) and (30), the combined operator for convection and diffusion has a Taylor expansion which can be written

$$-u\left(\frac{\partial \phi}{\partial x}\right)_i \; + \; \Gamma \left(\frac{\partial^2 \phi}{\partial x^2}\right)_i \; - \; \frac{|u|}{12}\left[1 - \frac{1}{P_\Delta}\right]\phi_i^{(iv)}\Delta x^3 + \dots \qquad (31)$$

which is the appropriate form for high convection conditions (large P_Δ). The corresponding feedback sensitivity is

$$\Sigma \; = \; - \; |u|/2\Delta x \; - \; 2\Gamma/\Delta x^2 \qquad (32)$$

which, being unconditionally negative for all (physical) values of u and Γ, represents very good damping qualities. As will be seen in the next section, in practical applications the performance of this Uniformly Third-Order Polynomial Interpolation Algorithm (UTOPIA) in terms of accuracy and inherent stability is quite impressive.

MODEL TEST PROBLEM

For comparative studies of the performance of different spatial differencing methods, a simple model test problem can be based on the nominally steady one-dimensional convective diffusion equation as given by Equation (2). As explained in the introduction, the inclusion of a non-zero S^* term is an essential part of the model. One could study the effects of a number of different forms of $S^*(x)$; for the test problem used here, a relatively simple form is used as shown in Figure 1, which also shows the imposed boundary conditions. The problem is summarized as follows

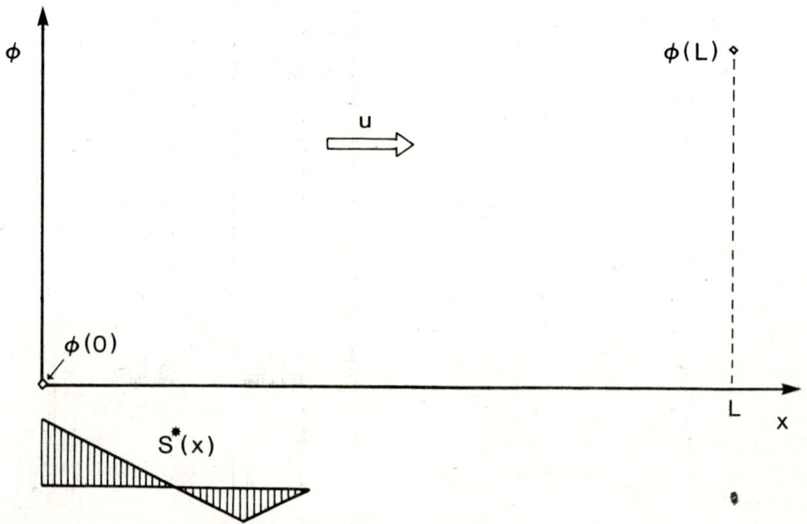

Figure 1. Formulation of model problem.

$$u \frac{d\phi}{dx} = \Gamma \frac{d^2\phi}{dx^2} - S^*(x), \quad 0 \le x \le L \tag{33}$$

$$\phi(0) = 0, \quad \phi(L) = \text{const} \tag{34}$$

where u and Γ are positive constants.

Second-order central differencing

When second-order central differencing is used for both the convection and diffusion terms in Equation (33), the finite difference equation becomes

$$u \left[\frac{\phi_{i+1} - \phi_{i-1}}{2\Delta x} \right] = \Gamma \left[\frac{\phi_{i+1} - 2\phi_i + \phi_{i-1}}{\Delta x^2} \right] + S^*(x) \tag{35}$$

which can be rearranged into a standardized form as

$$\frac{1}{2} \left[\phi_{i+1} - \phi_{i-1} \right] = \frac{1}{P_\Delta^*} \left[\phi_{i+1} - 2\phi_i + \phi_{i-1} \right] + \bar{S}^* \tag{36}$$

where $\bar{S}^* = S^* \Delta x/u$ and P_Δ^* is the effective grid Péclet number -- in this case, $P_\Delta^* \equiv P_\Delta$, as emphasized in Figure 2, for later comparison with other methods. As noted, this method is relatively well-behaved for $P_\Delta < 2$, but may lead to oscillatory results when $P_\Delta > 2$.

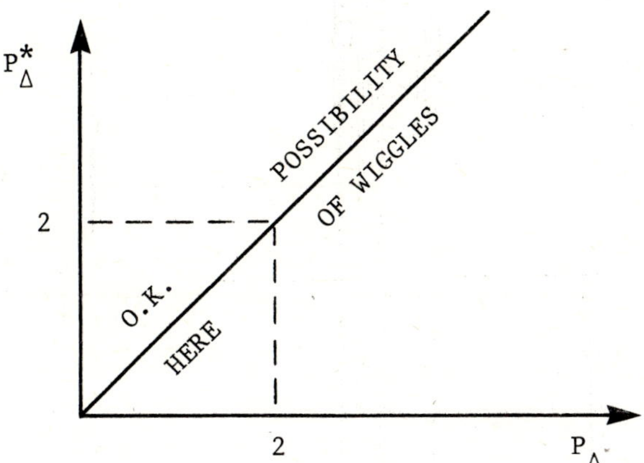

Figure 2. Effective grid Péclet number for central differencing.

Figure 3 shows the results for $P_\Delta = 1$, in comparison with the exact solution (solid curve). Note that the numerical results

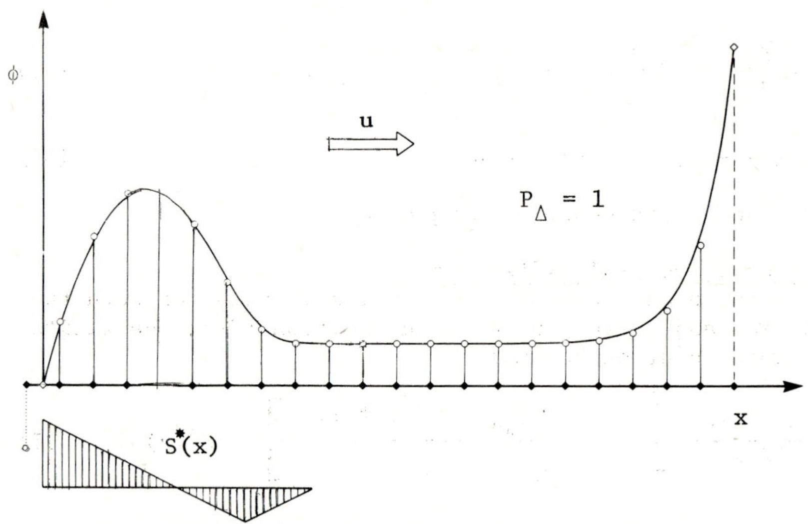

Figure 3. Second-order central differencing, $P_\Delta = 1$.

tend to be less accurate in regions of strong curvature --
the computed points generally lie slightly away from the
convex side of curves in the exact solution. This effect is
even more apparent in Figure 4, for $P_\Delta = 2$. In this case,

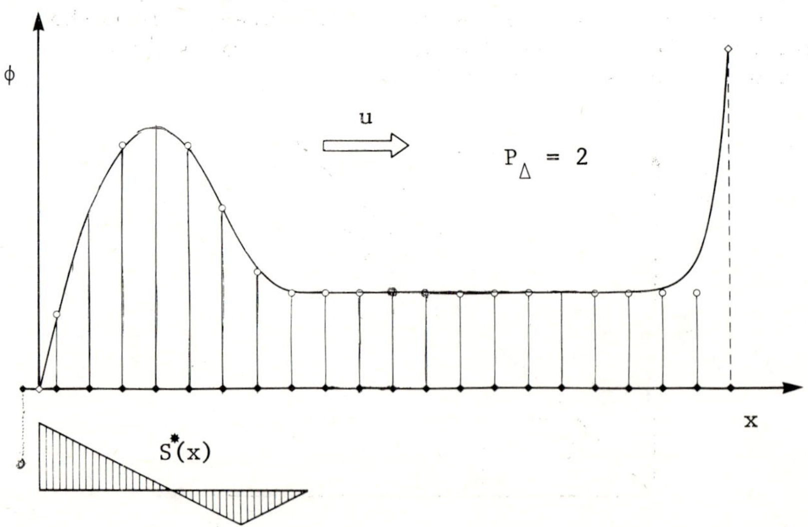

Figure 4. Second-order central differencing, $P_\Delta = 2$.

computed results in the downstream boundary layer have
reached a critical condition. Further increase in P_Δ produces
wiggles in this region, as shown in Figure 5 for $P_\Delta = 5$.

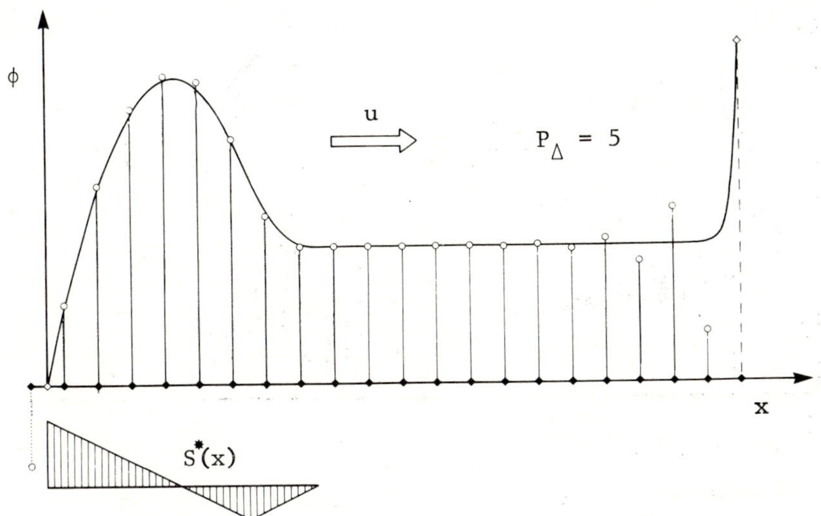

Figure 5. Second-order central differencing, $P_\Delta = 5$.

<u>Wiggle penetration distance</u> The extent of the penetration of
the unphysical oscillations can be estimated from the exact
solution to the difference equation in this region (i.e.
without a source term) (Gresho, 1979). If one asks how many
grid-point steps, n, from the downstream boundary are
necessary for the solution to settle down to within $\varepsilon\Delta\phi$ of
the exact solution (where $\Delta\phi$ is the jump across the boundary
layer), the result is

$$n \;\approx\; P_\Delta \, \ln(\varepsilon^{-1/4}) \tag{37}$$

so that if $\varepsilon = 2\%$, for example, $n \approx P_\Delta$, which is quite
obviously demonstrated in Figures 5 ($P_\Delta = 5$) and 6 ($P_\Delta = 10$).

This direct proportionality between wiggle penetration
distance and grid Péclet number is the most serious short-
coming of second-order central differencing, and can lead
to total disruption under high-convection conditions, as
seen in Figure 7 ($P_\Delta = \infty$).

<u>First-order upwinding and hybrid methods</u>
If full first-order upwinding is used for the convection
term, the consistent approximation for modelling the physical
diffusion term is to delete it altogether (i.e. first-order
upwinding is consistent with linear interpolation, for which
the second derivative is, of course, identically zero). Thus,
the model equation becomes (for $u > 0$)

$$u\left[\frac{\phi_i - \phi_{i-1}}{\Delta x}\right] \;=\; 0 + S^* \tag{38}$$

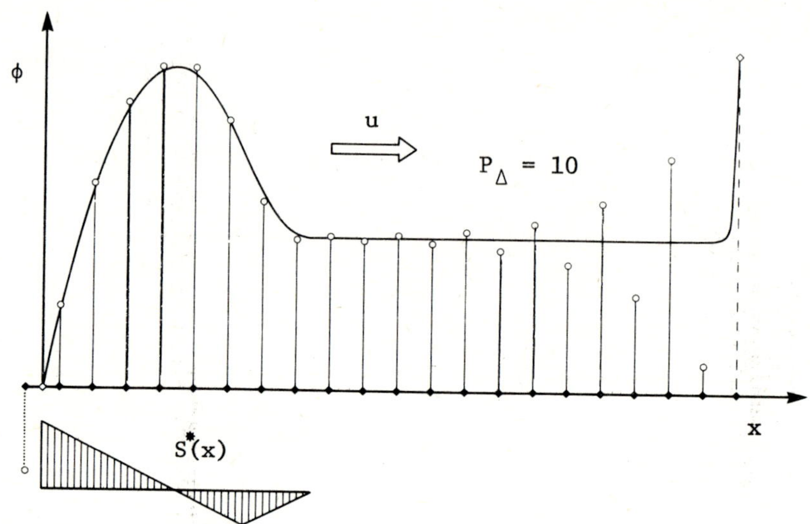

Figure 6.   Second-order central differencing, $P_\Delta = 10$.

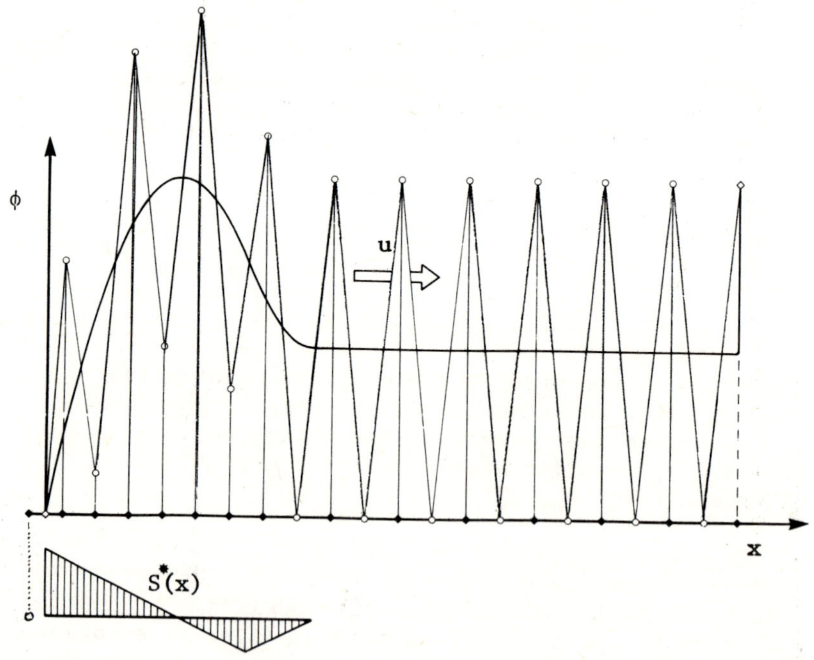

Figure 7.   Second-order central differencing, $P_\Delta = \infty$.

which can be rewritten in the standard form as

$$\frac{1}{2}\left[\phi_{i+1} - \phi_{i-1}\right] = \frac{1}{P_\Delta^*}\left[\phi_{i+1} - 2\phi_i + \phi_{i-1}\right] + \bar{S}^* \qquad (39)$$

where $P_\Delta^* \equiv 2$. Thus, full upwinding is equivalent to second-order central differencing *at a single* P_Δ *value*. Computed solutions are *independent* of the actual physical P_Δ value. For example, all solutions of the present test problem for any P_Δ are identical to the computed solution shown in Figure 4, clearly an unacceptable situation.

This defect is partially corrected by a hybrid technique in which second-order central differencing is used for both convection and diffusion when $P_\Delta \leq 2$, and full upwinding (neglecting physical diffusion) when $P_\Delta \geq 2$. The governing difference equation can still be written in the form of Equation (39), except that now

$$P_\Delta^* \equiv P_\Delta \qquad \text{for } P_\Delta \leq 2 \qquad (40)$$

and

$$P_\Delta^* \equiv 2 \qquad \text{for } P_\Delta \geq 2 \qquad (41)$$

as shown in Figure 8. This means that for $P_\Delta \leq 2$, results

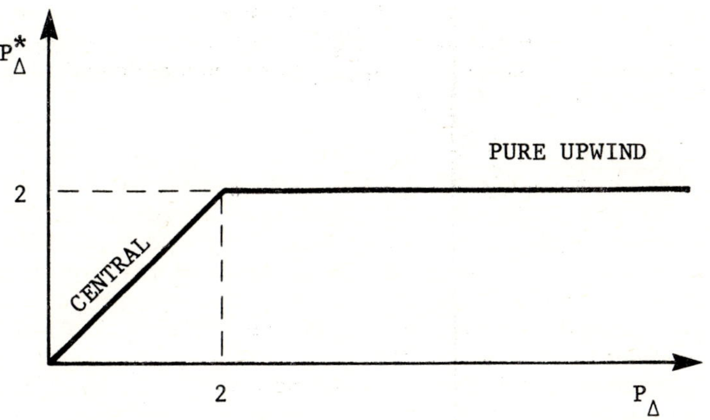

Figure 8. Effective grid Péclet number for the hybrid method.

are the same as for central differencing. However, the computed solution still "saturates" at a single form for $P_\Delta > 2$, as emphasized in Figure 9. One can see then that the whole concept of the hybrid method is clearly ill-founded:

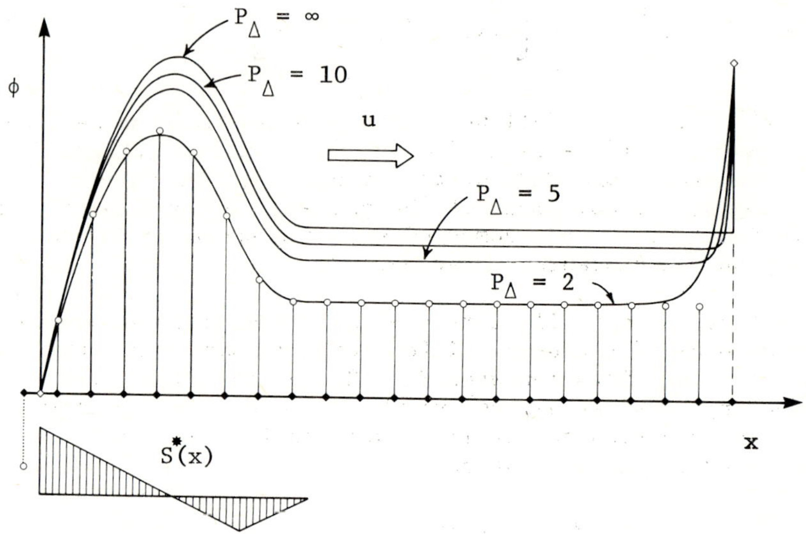

Figure 9. Hybrid method, P_Δ = 2,5,10 and ∞; for $P_\Delta \leq 2$, results are the same as for second-order central differencing.

for $P_\Delta \leq 2$, there is no improvement over central differencing, and for $P_\Delta > 2$, the method is so inaccurate that it is useless for any practical purposes. It is indeed rather surprising that it is still so widely used. [The real reason for its popularity is a totally unscientific one: it is very easy to work with, being algorithmically simple and giving plausible-looking trouble-free results without the obviously unphysical wiggles of central differencing or the convergence problems to which these often lead!]

"Optimal" upwinding

The hybrid scheme just described is an example of a class of finite-difference methods which combine features of both second-order central differencing and first-order upwinding with some weighting strategy which depends on the local grid Péclet number. In the case of the hybrid scheme, the weighting strategy consists of a switch at $P_\Delta = 2$. Clearly, continuous change-over strategies are also possible. For example, the convective term could be modelled by a weighted combination such as

$$u \left\{ \alpha \left[\frac{\phi_i - \phi_{i-1}}{\Delta x} \right] + (1-\alpha) \left[\frac{\phi_{i+1} - \phi_{i-1}}{2\Delta x} \right] \right\} =$$

$$\Gamma \left[\frac{\phi_{i+1} - 2\phi_i + \phi_{i-1}}{\Delta x^2} \right] + S^* \tag{42}$$

where the diffusion term is modelled by second-order central differencing (Raithby & Torrance, 1974). The weighting factor α is related to P_Δ. The philosophy behind the so-called "optimal" weighting strategy is to choose α to that the computed solution agrees with the exact solution of a specific (*source-free*!) model problem for which P_Δ is a shape parameter. If Equation (42) is written in the standard form of Equation (39), "optimal" upwinding is equivalent to second-order central differencing using an effective grid Péclet number given by (Gresho & Lee, 1979)

$$P_\Delta^* = 2 \tanh(P_\Delta/2) \tag{43}$$

which is shown in Figure 10. This should be compared with

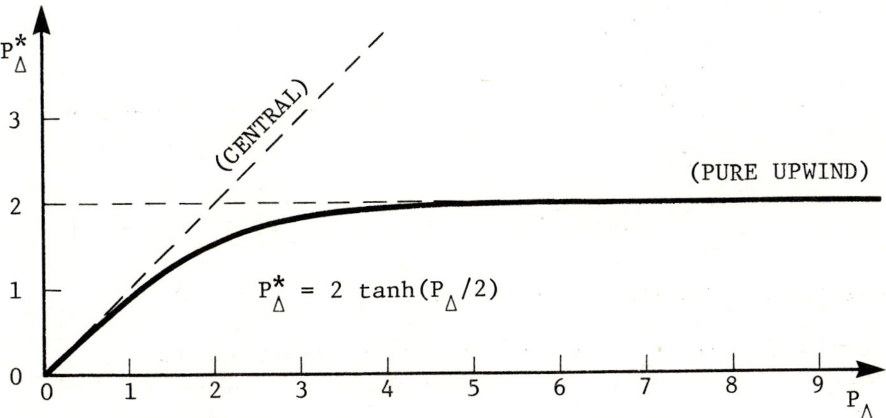

Figure 10.   Effective grid Péclet number for "optimal" upwinding.

Figures 2 and 8. Since the effective diffusion coefficient is

$$\Gamma_{\text{eff}} = u\Delta x/P_\Delta^* \tag{44}$$

it can be seen from the behaviour of the hyperbolic tangent function that

$$\Gamma_{\text{eff}} \to \Gamma \quad \text{for} \quad P_\Delta \lesssim 0.5 \tag{45}$$

and

$$\Gamma_{\text{eff}} \to u\Delta x/2 \quad \text{for } P_\Delta \gtrsim 5 \tag{46}$$

Otherwise, Γ_{eff} is always greater than that of either central differencing or full upwinding.

Figure 11 shows that the additional artificial diffusion is already apparent at $P_\Delta = 1$ ($P_\Delta^* = 0.924$).

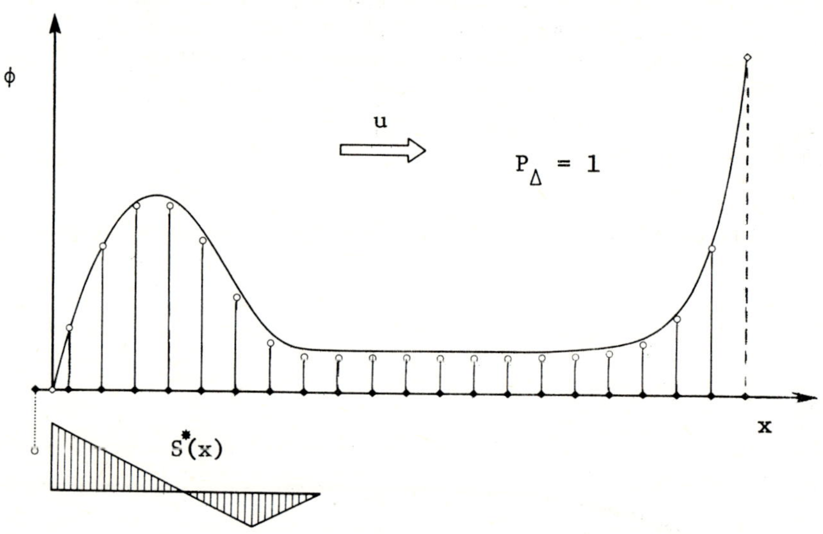

Figure 11. "Optimal" upwinding, $P_\Delta = 1$.

As P_Δ is increased, the artificial diffusion becomes relatively more dominant as shown in Figure 12, for $P_\Delta = 2$

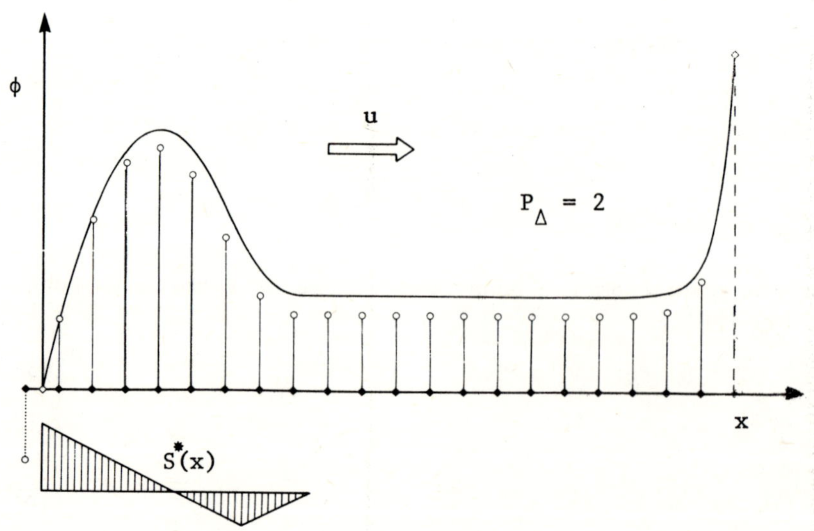

Figure 12. "Optimal" upwinding, $P_\Delta = 2$.

$(P^*_\Delta = 1.52)$, until the solution effectively saturates at $P_\Delta = 5$ $(P^*_\Delta = 1.97)$ as seen in Figures 13, and in 14 $(P_\Delta = 10, P^*_\Delta = 2.00)$ and 15 $(P_\Delta = \infty, P^*_\Delta \equiv 2)$.

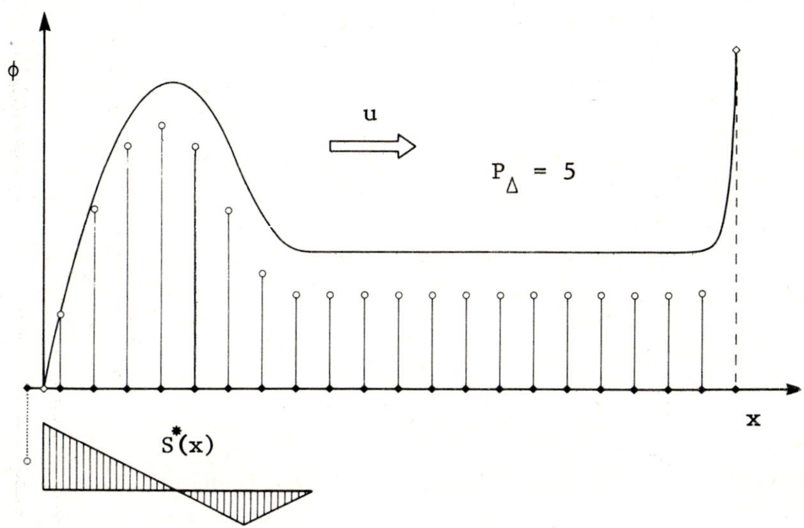

Figure 13. "Optimal" unwinding, $P_\Delta = 5$.

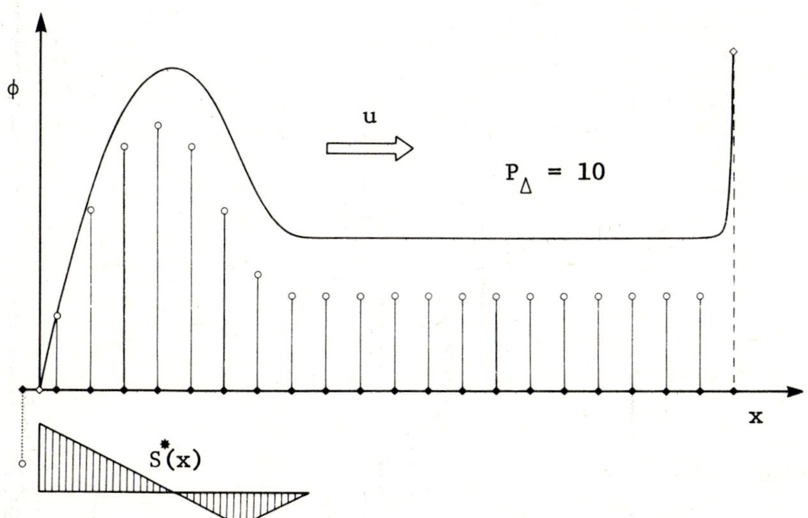

Figure 14. "Optimal" upwinding, $P_\Delta = 10$.

As can be seen from these results, "optimal" upwinding is actually *worse* than the (already unacceptable) hybrid scheme over the complete range of Péclet numbers. There appear to be two possible courses of action with respect to first/

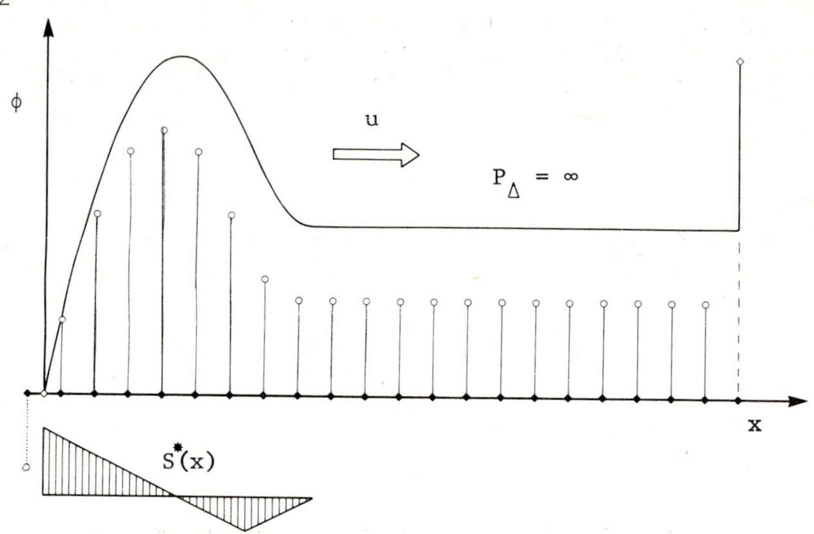

Figure 15. "Optimal" upwinding, $P_\Delta = \infty$.

second-order weighting schemes: (i) either abandon the idea
completely or (ii) suitably generalize the method to take
into account the effects of S^* (i.e. unsteady terms, cross-
grid transport, and actual source terms). The first course
will be followed in the remainder of this article.

The QUICK method
A control-volume formulation of the UTOPIA scheme results in
Quadratic Upstream Interpolation for Convective Kinematics
(Leonard, 1979). The interpolation for the control-
volume right-face value and gradient is shown in Figure 16

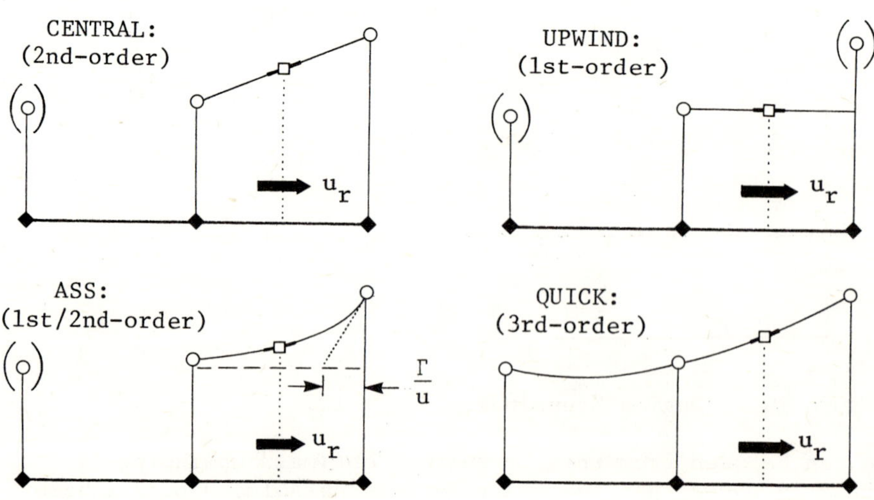

Figure 16. Control-volume interpolation schemes.

in comparison with that for second-order central, first-order upwind, and the Allen-Southwell scheme (Allen & Southwell, 1955) (which is the basis for "optimal" upwinding). Note that central differencing uses linear (first-order) interpolation between two points; pure upwinding used "zeroth-order interpolation" through a single (upstream) value; the Allen-Southwell scheme is based on an exponential interpolation through two points with a *specified* length constant (= Γ/u); the QUICK method uses quadratic interpolation through three points: two points straddling the control-volume face together with an additional adjacent *upstream* point.

If the right-face velocity were to the left in Figure 16, only the central-difference interpolation would remain unchanged; first-order upwinding would be based on the right node value; the ASS interpolation would involve an exponential term decreasing to the right rather than increasing; and the QUICK method would use a quadratic interpolation through the two node values straddling the face together with that of the next (upstream) node to the right (not shown). Note the interpretation of the right-face gradient in Figure 16 (shown by the short heavy dashes on the interpolation curves). In particular, the consistent first-order modelling of gradients assumes them to be identically zero. Because of a geometrical property of the parabola (that the gradient half-way between two points is equal to that of the chord joining the points), the QUICK gradients are identical to those of central differencing.

For the QUICK method, the right-face value is

$$\phi_r = \frac{1}{2}(\phi_i + \phi_{i+1}) - \frac{1}{8}\,\text{CURV} \tag{47}$$

where the curvature term is given by

$$\phi_{i-1} - 2\phi_i + \phi_{i+1} \quad \text{if} \quad u > 0 \tag{48}$$

or

$$\phi_i - 2\phi_{i+1} + \phi_{i+2} \quad \text{if} \quad u < 0 \tag{49}$$

As mentioned, the consistent modelling of the gradient is given by

$$\left(\frac{\partial \phi}{\partial x}\right)_r = \frac{\phi_{i+1} - \phi_i}{\Delta x} \tag{50}$$

regardless of the sign of u. Of course, analogous formulas for the left face can be obtained by decreasing all indexes by 1. This means that the method is "value conservative": $\phi_\ell(i) = \phi_r(i-1)$; and the analogous formula for the gradient shows that it is also "gradient conservative."

24

For constant u (>0) and Γ, the control-volume QUICK formulation of the model equation can be written

$$\frac{1}{2}\left[\phi_{i+1} - \phi_{i-1}\right] - \frac{1}{8}\left[\phi_{i+1} - 3\phi_i + 3\phi_{i-1} - \phi_{i-2}\right]$$

$$= \frac{1}{P_\Delta^*}\left[\phi_{i+1} - 2\phi_i + \phi_{i-1}\right] + \bar{S}^* \tag{51}$$

in which $P_\Delta^* \equiv P_\Delta$. The high accuracy of this method is immediately apparent in Figure 17 for $P_\Delta = 1$, where computed

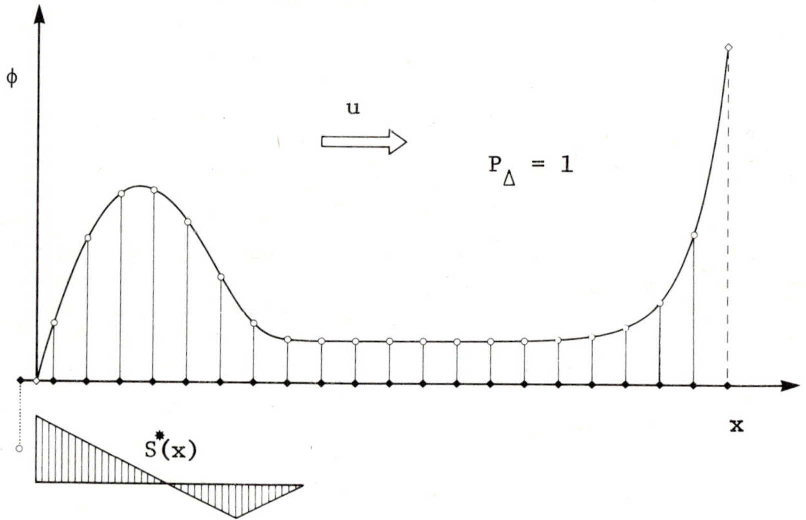

Figure 17. QUICK solution, $P_\Delta = 1$.

points are graphically indistinguishable from the exact solution. As the convection is made relatively stronger, the QUICK method remains highly accurate except near the downstream boundary layer where a slight discrepancy is just noticeable in Figure 18, for $P_\Delta = 2$.

Although still very accurate over most of the region for $P_\Delta = 5$, Figure 19 shows that some wiggles are occurring in the region of sudden adjustment to the downstream boundary condition. As P_Δ is increased further, the extent of the wiggly region also increases as seen in Figure 20, for $P_\Delta = 10$.

But, in strong contrast to the behaviour of central differencing, the QUICK wiggles have a limiting form as P_Δ is increased indefinitely as seen in Figure 21. Counting back from the downstream boundary node, the successive relative

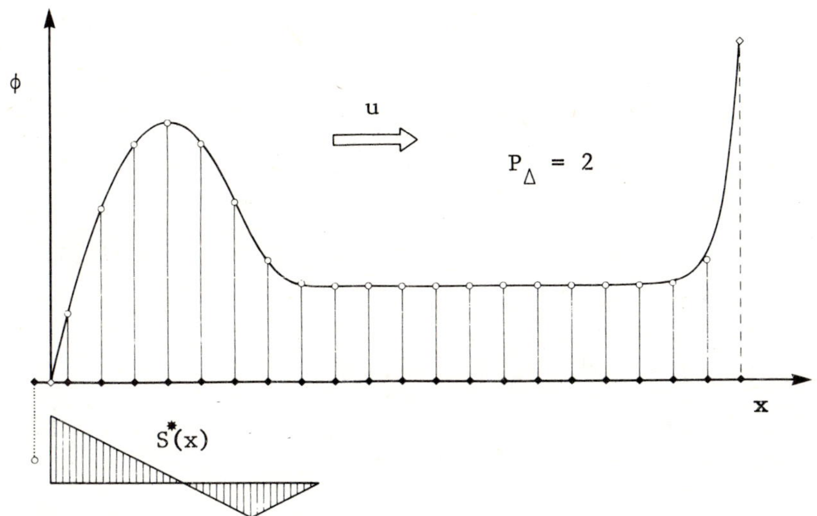

Figure 18. QUICK solution, $P_\Delta = 2$.

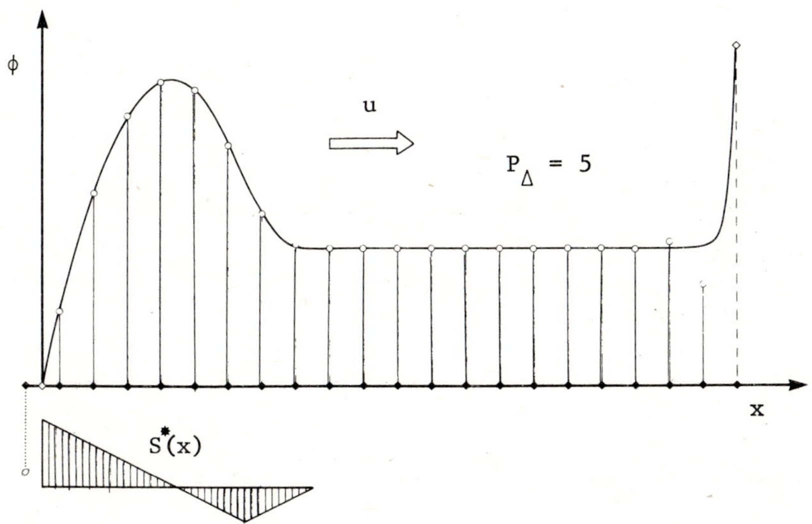

Figure 19. QUICK solution, $P_\Delta = 5$.

amplitude of adjacent wiggles in the limiting form is approximately $10^{-1/3}$, so that at three nodes in from the boundary the discrepancy is 10% of the overall jump whereas at six nodes it is only 1%. Of course these wiggles would not be acceptable in a practical computation, and further steps need to be taken to resolve the difficulty -- e.g. grid refinement, alternate interpolation, or a re-evaluation of the imposed boundary condition. However, even under conditions of pure convection ($P_\Delta = \infty$) the extent of any

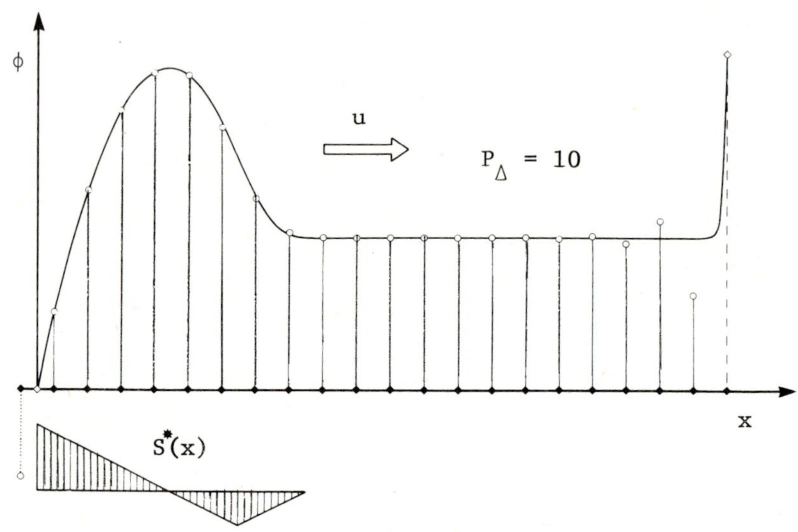

Figure 20. QUICK solution, $P_\Delta = 10$.

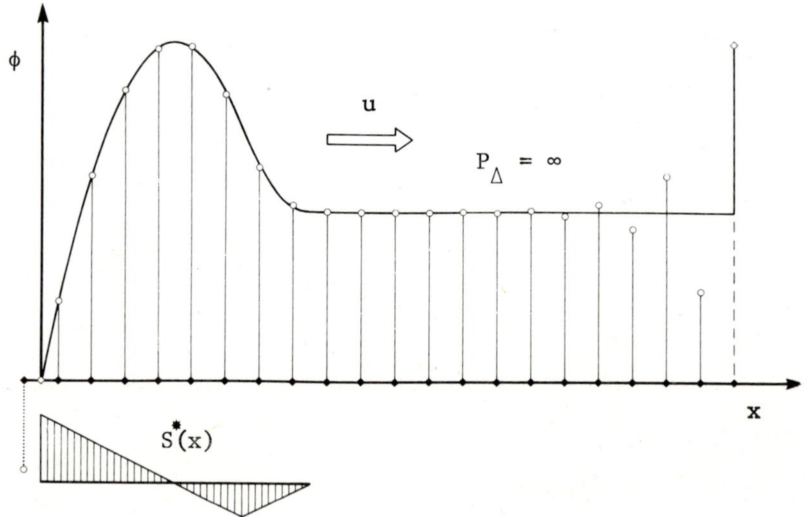

Figure 21. QUICK solution, $P_\Delta = \infty$.

anomolous behaviour is confined to within a few grid points --
unlike central differencing, for which the extent of the
corruption is proportional to P_Δ. Thus, in simulations of
practical flow problems the basic QUICK method is extremely
"managable" from the user's point of view, generating
solutions of high accuracy over most of the flow domain in
a stable and algorithmically efficient manner.

The EXQUISITE Method

The problem of resolving a sudden jump in value in the flow direction over a few grid points can be solved by using a more appropriate interpolation function than the quadratic polynomial used in the QUICK method. One needs an interpolation function which can change rapidly and monotonically (when necessary). For this, the exponential function is quite appropriate. Figure 22 shows how the same three

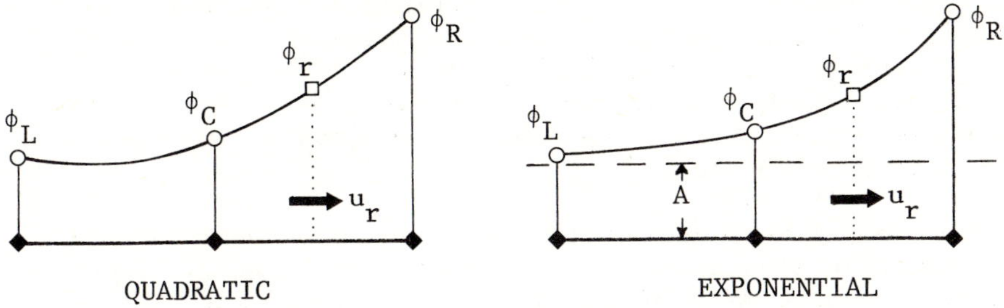

Figure 22.   Quadratic and exponential three-point interpolation.

(monotonic) node values can be interpolated by either a quadratic function or an exponential.

For the right-face value, quadratic interpolation gives (as before)

$$\phi_r = \frac{1}{2}(\phi_C + \phi_R) - \frac{1}{8}\,\mathrm{CURV} \tag{52}$$

where, for $u > 0$,

$$\mathrm{CURV} = \phi_R - 2\phi_C + \phi_L \tag{53}$$

whereas a three-parameter exponential of the form

$$\phi = A + B\,e^{C\xi} \tag{54}$$

where $\xi = x - x_i$, leads to

$$\phi_r = A \pm \sqrt{(\phi_R - A)(\phi_C - A)} \tag{55}$$

where the asymptote is given by

$$A = \frac{\phi_R \phi_L - \phi_C^2}{\phi_R - 2\phi_C + \phi_L} \tag{56}$$

The plus sign is taken in Equation (55) when the asymptote is below the smallest value (as in Figure 22), and the minus sign if it happens to lie above the largest value (i.e. if ϕ_C were closer in value to ϕ_R than to ϕ_L). It is important to stress that there is no *a priori* length scale built into this type of three-point exponential interpolation; the shape is determined by the actual solution itself -- not by some local solution to a hypothetical (source-free) problem (such as that assumed in the two-point ASS interpolation involving the length scale Γ/u).

Of course one must decide when it is more appropriate to use quadratic interpolation and when to use exponential. Some progress can be made in this regard by introducing the normalized variable

$$\tilde{\phi} = \frac{\phi - \phi_L}{\phi_R - \phi_L} \tag{57}$$

[Note that $\tilde{\phi}_R = 1$ and $\tilde{\phi}_L = 0$.] In particular,

$$\tilde{\phi}_C = \frac{\phi_C - \phi_L}{\phi_R - \phi_L} \tag{58}$$

then, for $u > 0$, the interpolated right-face value for the QUICK algorithm becomes

$$\tilde{\phi}_r = \frac{3}{8} + \frac{3}{4}\tilde{\phi}_C \tag{59}$$

and for exponential interpolation

$$\tilde{\phi}_r = \frac{\sqrt{\tilde{\phi}_C(1 - \tilde{\phi}_C)^3} - \tilde{\phi}_C^2}{(1 - 2\tilde{\phi}_C)} \tag{60}$$

which is, of course, valid only in the monotonic range : $0 \leq \tilde{\phi}_C \leq 1$.

Equations (59) and (60) are summarized graphically in Figure 23, in which the exponential relationship in the

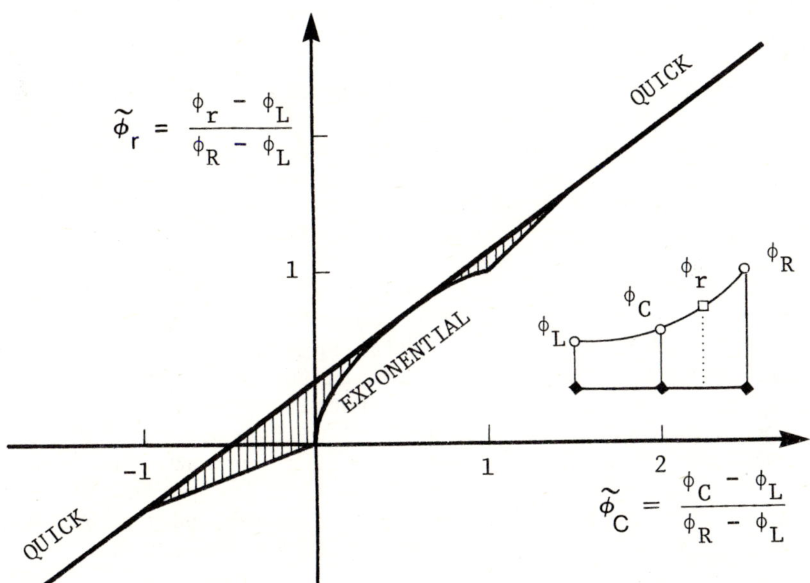

Figure 23.   Normalized face value for the EXQUISITE method

monotonic region has been extended by *ad hoc* "wings" connecting the ends of the exponential curve to the QUICK line in a simple and convenient manner, while keeping $\partial\tilde{\phi}_r/\partial\tilde{\phi}_C > 0$, a necessary condition for stability. Depending on the value of $\tilde{\phi}_C$, the interpolated right-face value may lie on the QUICK line or in the shaded region. Just how close to the lower curve it will lie is determined by weighting factor which depends on adjacent local curvature criteria. Essentially, if the curvature of the ϕ-function at the right face is similar to that at the adjacent upstream face, the weighting is chosen close to the QUICK value; alternatively, if the curvature is changing rapidly a weighting closer to the lower boundary in Figure 23 is more appropriate.

The right-face gradient can be treated in a similar manner by defining

$$\left(\frac{\partial\phi}{\partial x}\right)_r = \frac{\phi_{RG} - \phi_C}{\Delta x} \tag{61}$$

For quadratic interpolation, of course, $\phi_{RG} \equiv \phi_R$; but for exponential interpolation, the normalized value is

$$\tilde{\phi}_{RG} = \tilde{\phi}_C + \frac{\sqrt{\tilde{\phi}_C(1 - \tilde{\phi}_C)^3}}{(1 - 2\tilde{\phi}_C)} \ell_n\left(\frac{1 - \tilde{\phi}_C}{\tilde{\phi}_C}\right) \tag{62}$$

for $0 < \tilde{\phi}_C \leq 0.5$.   A weighting procedure is used similar to that used for ϕ_r.

In a practical implementation of a working algorithm, Equations (60) and (62) can be approximated quite adequately without resorting to computationally expensive functions. The resulting method is naturally referred to as EXponential or Quadratic Upstream Interpolation for Solution of the Incompressible Transport Equation (EXQUISITE).

The results of the model test problem certainly justify this acronym, as seen in Figures 24 through 28.   The computed points are graphically almost indistinguishable from the exact solution over the complete flow domain for any value of P_Δ. Clearly, this method can handle (effective) source terms and boundary jumps with high accuracy and with never any hint of unphysical wiggliness.   The corresponding *unsteady* Adjusted Quadratic Upstream Algorithm for Transient Incompressible Convection (AQUATIC) is similarly impressive, and the basic idea is easily generalized to two and three dimensions.

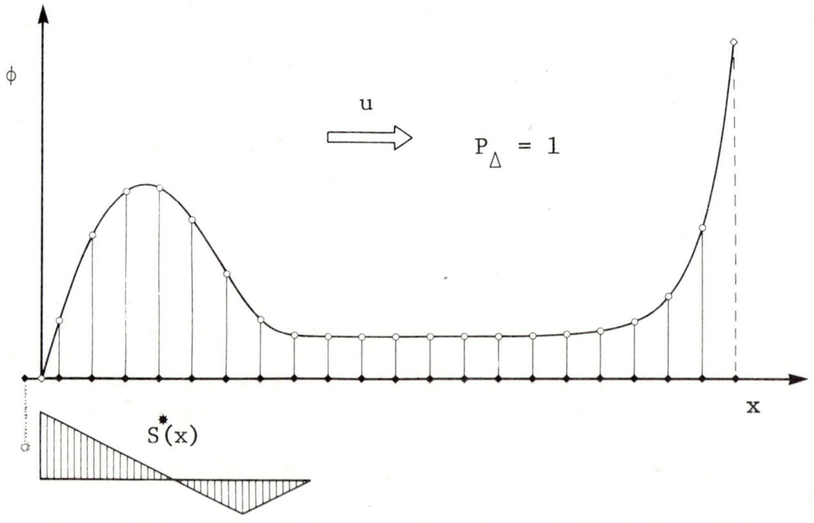

Figure 24.   EXQUISITE method, $P_\Delta = 1$.

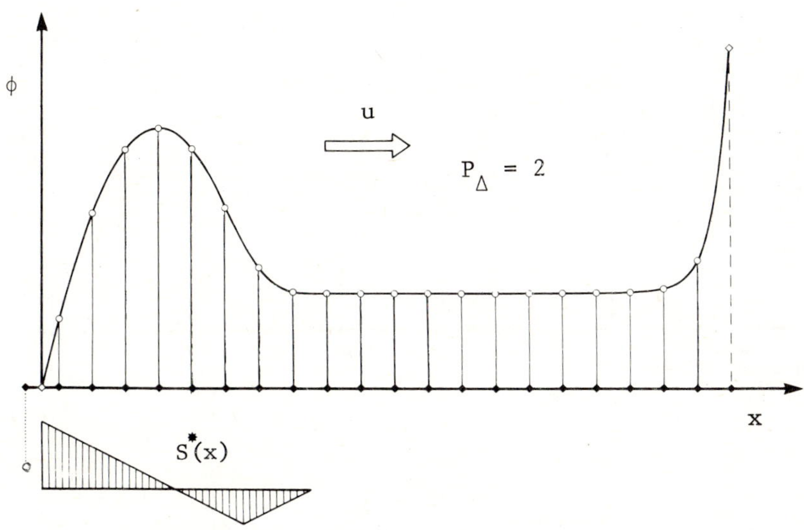

Figure 25.   EXQUISITE method, $P_\Delta = 2$.

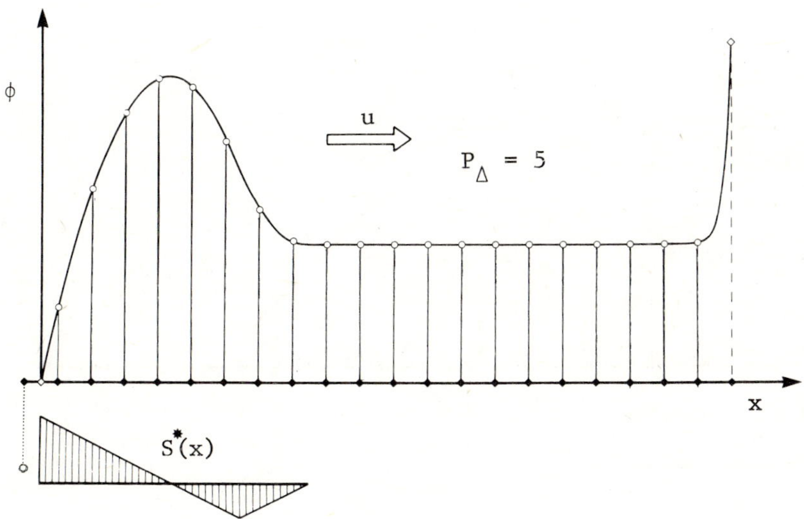

Figure 26.   EXQUISITE method, $P_\Delta = 5$.

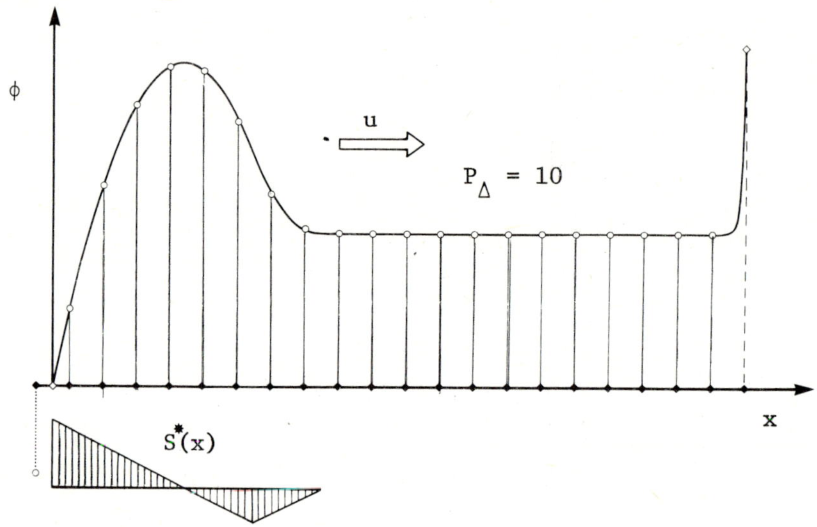

Figure 27. EXQUISITE method, $P_\Delta = 10$.

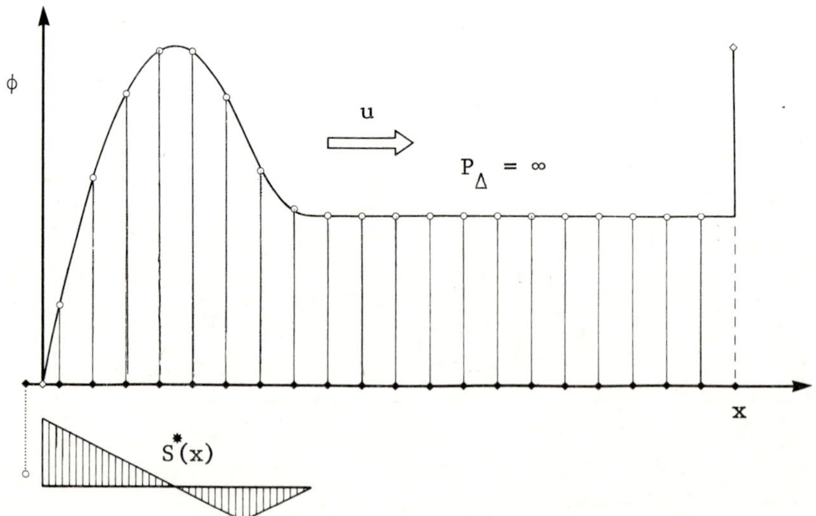

Figure 28. EXQUISITE method, $P_\Delta = \infty$.

SUMMARY

It has been shown that standard central-difference methods are
entirely adequate for modelling physical mechanisms involving
even-ordered spatial derivatives, such as diffusion, wave
motion, and elastic flexure. However, for odd-ordered
derivatives, central-difference methods of any order lack
inherent feedback damping qualities and may lead to problems
involving unphysical oscillations or computational nonconvergence. This, of course, is particularly relevant to
computational fluid dynamics which is distinguished from the
other branches of computational physics by the importance of
the *first*-derivative convection term.

The earliest "remedy" for modelling strongly convective flows,
viz.: first-order upwinding, is now generally recognized to
be totally *in*adequate as a predictive tool for modelling
practical engineering or geophysical flows. Unfortunately,
there has not been widespread agreement on the similar
inadequacy of the related methods involving "hybrid" or
"optimal" upwinding techniques based on weighted combinations
of first- and second-order convective differencing. But, as
the results of the model problem studied here have quite
forcefully shown, these methods offer no improvement in
accuracy in the "difficult" high-convection regime (say
$P_\Delta > 2$). No doubt truly optimal techniques could be developed
by taking into account the effects of unsteady terms,
transverse cross-grid transport, and source terms -- all of
which are absent from currently popular "optimal" weighting
methods. However, as has been demonstrated here, it is
relatively easy to develop a *third*-order upwind scheme which
possesses high accuracy and inherent damping qualities.

Results of the model test problem show the basic power of the
third-order upwind method. In the one-dimensional control-
volume formulation, cell face values and gradients are based
on local (upstream-weighted) three-point polynomial interpolation. Under most circumstances the interpolating parabolas
are adequate for handling variations in the transported
variable. But, obviously, a parabola is not the most
appropriate interpolant for sudden step-like behaviour; in
this case, a three-point exponential interpolation is more
appropriate from both the physical and the computational
points of view. Since both the quadratic polynomial and the
exponential interpolations use the same three node values, it
is not difficult to devise an algorithm which is a weighted
combination of both interpolants, with a weighting criterion
based on the (slightly extended) local spatial behavior of
the solution as it evolves.

Under high convection conditions, second-order central
differencing suffers from severe wiggle problems whenever

there is required to be a sudden change in the transported variable along the flow direction. The spatial extent of the wiggles is directly proportional to the generalized (longitudinal) grid Péclet number, P_Δ; so that under practically interesting cases of very high P_Δ, central-differencing's wiggles may corrupt much of the flow domain. By contrast, the basic (polynomial) third-order upwind method involves inherent numerical damping so that even under conditions of pure convection ($P_\Delta \to \infty$), wiggle problems are limited in extent to a few grid points near sudden changes in value. And even these minor annoyances can be entirely eliminated (without sacrificing accuracy) by using the exponential-quadratic weighting technique.

Although the present analysis has concentrated on a steady one-dimensional formulation, there is nothing in the third-order upwind scheme which restricts its being applied to multi-dimensional unsteady flow situations. For full formal third-order accuracy some additional small terms (such as transverse curvature, for example) must be included; however, any viable (albeit convectively "unstable") second-order central-difference code can be QUICKened in a brute force manner simply by adding the appropriate upstream-weighted normal curvature terms, thus extending its predictive capabilities to the high convection regime. This strategy is, of course, particularly suited to codes based on quasi-one-dimensional techniques such as time-splitting and alternating direction (explicit or implicit) spatial solution methods. Fully third-order upwind techniques for unsteady one-dimensional and steady two- and three-dimensional flows have been developed, and work is proceeding on the unsteady multi-dimensional version.

REFERENCES

Allen, D.N. de G. and R.V. Southwell (1955) Relaxation Methods Applied to Determine the Motion in Two Dimensions of a Viscous Fluid past a Fixed Cylinder, *Quarterly Journal of Mechanics and Applied Mathematics* <u>8</u>, 129.

Brooks, A. and T.J.R. Hughes (1980) Streamline-Upwind/Petrov-Galerkin Methods for Advection Dominated Flows, California Institute of Technology Preprint (personal communication).

Fromm, J.E. (1968) A Method for Reducing Dispersion in Convective Difference Schemes, *Journal of Computational Physics* <u>3</u>, 176.

Gresho, P. and R.L. Lee (1979) Don't Suppress the Wiggles — They're Telling You Something, in AMD-34, *Finite Element Methods for Convection Dominated Flows*, T.J.R. Hughes (ed.), ASME, New York, 37.

Heinrich, J.C., P.S. Huyakorn, O.C. Zienkiewicz and A.R. Mitchell (1977) An "Upwind" Finite Element Scheme for Two-Dimensional Convective Transport Equation, *International Journal for Numerical Methods in Engineering* 11, 131.

Heinrich, J.C. and O.C. Zienkiewicz (1979) The Finite Element Method and 'Upwinding' Techniques in the Numerical Solution of Convection Dominated Flow Problems, in AMD-34, *Finite Element Methods for Convection Dominated Flows*, T.J.R. Hughes (ed.), ASME, New York, 105.

Hughes, T.J.R. and A. Brooks (1979) A Multidimensional Upwind Scheme with no Crosswind Diffusion, in AMD-34, *Finite Element Methods for Convection Dominated Flows*, T.J.R. Hughes (ed.), ASME, New York, 19.

Leonard, B.P. (1978) A Consistency Check for Estimating Truncation Error Due to Upstream Differencing, *Applied Methematical Modelling* 2, 239.

Leonard, B.P. (1979) A Stable and Accurate Convective Modelling Procedure Based on Quadratic Upstream Interpolation, *Computer Methods in Applied Mechanics and Engineering* 19, 59.

Raithby, G.D. (1976) A Critical Evaluation of Upstream Differencing Applied to Problems Involving Fluid Flow, *Computer Methods in Applied Mechanics and Engineering* 9, 75.

Raithby, G.D. and K.E. Torrance (1974) Upstream-Weighted Schemes and their Application to Elliptic Problems Involving Fluid Flow, *Computers and Fluids* 2, 191.

Roache, P.J. (1976) *Computational Fluid Dynamics*, Hermosa Publishers, Albuquerque, New Mexico.

Spalding, D.B. (1972) A Novel Finite Difference Formulation for Differential Expressions Involving Both First and Second Derivatives, *International Journal for Numerical Methods in Engineering* 4, 551.

CHAPTER 2

A FINITE ELEMENT METHOD FOR THE NUMERICAL SOLUTION OF
TRANSIENT INVISCID INCOMPRESSIBLE FLOWS.
M. Bercovier[*], G. Berwald[*], O. Pironneau[**]

(*) Hebrew University of Jerusalem
(**) University of Paris-Nord and INRIA.

ABSTRACT

This paper describes an efficient adaptation of the vortex method to compute solutions of the Euler system of equations for
incompressible inviscid fluids.

The method is applied to a problem of aeronautical engineering :
the vortex alleviation problem. The method is also generalized
to 3-D problems.

INTRODUCTION

In [1] we have introduced a finite element version of the Vortex
Method [2], [3]. For two dimensional non viscous flows, this method is based on an old concept of fluid mechanics which says
that the vorticity in the fluid is transported by the flow. Thus
if we know an initial distribution of a finite number of point
vortices the flow at later time is found by transport of these
point vortices along the stream line of the flow that they create.

In order to present our method let us recall the basic equations

Let Ω be the region occupied by the flow ; let Γ be the boundary of Ω. Let $u(x,t)$, $p(x,t)$ denote the velocity and pressure of
the flow at $x \in \Omega$ at time t.
Then (u,p) is the solution of the Euler system of equations :

$$\frac{\partial u}{\partial t} + u\nabla u = -\nabla p \ ; \ \nabla.u = 0 \text{ in } \Omega \times \ ]0,T[\tag{1}$$

$$u.n = g \text{ on } \Gamma \times \ ]0,T[\ ; \ \nabla \times u = \omega_\Gamma \text{ on } \Sigma^- = \{(x,t):y(x).n(x)<0, x\in\Gamma\}$$

$$u(t=0) = u^o$$

For 2-D flows (1) can be rewritten in term of the vorticity

$$\omega = \nabla \times u \text{ in } \Omega \times \ ]0,T[\tag{2}$$

and the stream function Ψ

$$\nabla \times \Psi = u \text{ in } \Omega \times \ ]0,T[. \tag{3}$$

Hence (1) is equivalent to

$$\frac{\partial \omega}{\partial t} + u\nabla\omega = 0 \quad \text{in} \quad \Omega \times \ ]0,T[$$

$$\omega = \omega_\Gamma \quad \text{on} \quad \Sigma^- \tag{4}$$

$$\omega(t=0) = \nabla \times u^o$$

$$- \Delta\Psi(x,t) = \omega(x,t) \text{ in } \Omega \times \ ]0,T[$$

$$\Psi|_\Gamma(s) = \int_{s^o}^{s} g(\sigma)d\sigma \tag{5}$$

$$u(x,t) = \nabla \times \Psi(x,t) \text{ in } \Omega \times \ ]0,T[. \tag{6}$$

If $\omega|_\Gamma = 0$ and if $\omega^0 = \nabla \times u^0$ is a sum of Dirac functions centered a x_i^0 :

$$\omega^0(x) = \Sigma \, \omega_i^0 \, \delta(x - x_i^0)$$

then (4) - (6) has a semi-explicit solution :

$$\frac{dx_i}{dt} = \begin{bmatrix} u(x_i,t) \quad ; \quad x_i(t=0) = x_i^0 \\ 0 \; \text{ if } \; x_i(t) \notin \Omega \end{bmatrix} \tag{7}$$

$$\omega(x,t) = \Sigma \, \omega_i^0 \, \delta(x - x_i(t)) \tag{8}$$

$$- \Delta\Psi(x,t) = \omega(x,t)$$
$$\Psi|_\Gamma = \int^s g \, d\sigma \tag{9}$$

For complex geometries one would like to use the finite element method to solve (9). However as explained in [1] it is then much easier to replace Dirac functions by step functions or any finite element approximation of the Sobolev space $L^2(\Omega)$.

Our aim in this paper is to present one finite element implementation of the vortex method and its applications on a problem of aeronautical engineering. By mean of this example we illustrate and confirm the theoretical results obtained in [1] for the stability and error estimates of the method. Moreover we shall discuss the importance of vortex conservation and how to implement it. Then we shall present further extentions of the method to 3-D flows.

1. PRESENTATION OF THE METHOD

Computation of the characteristics

Assume for the time being that $u(x,t)$ is known at all $(x,t) \in \Omega \times]0,T[$ and satisfies the continuity equation everywhere :

$$\nabla . u = 0 \tag{1.1}$$

Then consider equation (4) :

$$\frac{\partial\omega}{\partial t} + u \, \nabla\omega = 0 \quad \text{in } \Omega \times]0,T[$$
$$\omega = \omega_\Gamma \text{ on } \Sigma^- = \{(x,t):u(x,t).n(x) < 0, \; x \in \Gamma\} \tag{1.2}$$

The exact solution of this equation is given first by computing the characteristics backward :

$$\frac{dX}{d\tau}(\tau) = \begin{bmatrix} u(X(\tau),\tau) & \text{if } X(\tau) \in \Omega \\ 0 & \text{if } X(\tau) \in \Gamma \end{bmatrix} \qquad \forall \tau \in \,]0,t[\qquad (1.3)$$

$$X(t) = x.$$

Define $X^{x,t}(0) = X(\tau=0)$

and then set

$$\omega(x,t) = \tilde{\omega}(X^{x,t}(0)) \qquad (1.4)$$

where $\tilde{\omega}$ is defined as

$$\begin{aligned} \tilde{\omega}(x) &= \omega^0(x) \quad \text{if} \quad x \in \overset{\circ}{\Omega} \\ &= \omega_\Gamma(x) \quad \text{if} \quad x \in \Gamma . \end{aligned} \qquad (1.5)$$

In physical terms one says that the vorticity is transported by the flow ; therefore to compute the vorticity at (x,t) one should first compute $X^{x,t}(0)$ the position at initial time of the molecule which is in x at time t. If one hits a boundary then one stays there (second condition in (1.3)) and then it is not the initial condition of ω which is transported but the boundary condition ω_Γ .

Now let $\mathcal{J}_h$ be a triangulation of Ω, made of non overlaping triangles T_i ; h is as usual the "mesh size". Let Δt be a constant time step and denote by P_{im} the prism of $\Omega \times \,]0,T[$:

$$P_{im} = T_i \times \,]m\Delta t,(m+1)\Delta t[. \qquad (1.6)$$

Suppose now that $u(x,t)$ is constant on each triangle and constant on each time interval $]m\Delta t,(m+1)\Delta t[$. Then (1.3) can be integrated explicitly by the following procedure :

Procedure 1 : (Computation of characteristics)

Step 1 : find the prism P_{im} which contains (x,t) and
$(x- \varepsilon u(x,t),t-\varepsilon)$, for some ε positive arbitrarly small.

Step 2 : find the intersection of the half-line
$\{x-\mu u(x,t), \ t-\mu : \forall \mu > 0\}$ with the boundary of P_{im}
(see figure 2).

<u>Step 3</u> : If $t-\mu = 0$ or if $x-\mu u(x,t) \in \Gamma$ then

$$\text{set } X^{x,t}(0) = x-\mu u(x,t)$$

$$\text{else set } t = t-\mu, \ x = x-\mu u(x,t) \text{ and go back to 1.}$$

<u>Remark 1</u>

If u is not divergence free in the sens that the flux through any close curve is not null then there may not be a P_{im} which satisfies the condition of Step 1. Indeed one may get "stuck" at the intersection of two prisms on which the velocities point each one to the interior of its own prism. This cannot happen if $u = \nabla \times \Psi$ and Ψ is piecewise linear continuous in x and piecewise constant in t.

<u>Remark 2</u>

If u is approximated by a higher order finite element method, say piecewise linear or quadratic or more, then one can replace step 3 by a Runge-Kutta method of suitable order k (it seems that if $u|_{T_i} \in P^\ell$ then a good choice is $k = \ell+1$).

<u>Approximation of the vorticity</u>

Let L_h be a finite dimensional space approximating $L^2(\Omega)$. For example the simplest choice is

$$L_h = \{\omega_h : \omega_h|_{T_i} \text{ constant } \forall T_i\} \tag{1.7}$$

Another choices could be

$$L_{h,k} = \{\omega_h \quad \text{continuous} : \omega_h|_{T_i} \in P^k \quad \forall T_i\} \tag{1.8}$$

$$L'_{h,k} = \{\omega_h : \omega_h|_{T_i} \in P^k \quad \forall T_i\} \tag{1.9}$$

(where P^k is the space of polynomials of degree k at most).

If (1.7) is selected and if we choose to take ω_h also piecewise constant in time then to compute ω_h we have the following scheme :

<u>Procedure 2</u> : (computation of ω_h)

<u>Step 1</u> : for each prism P_{im} choose an arbitrary point $(\xi,\theta) \in P_{im}$

<u>Step 2</u> : compute $X^{\xi,\theta}(0)$ by Procedure 1.

<u>Step 3</u> :

$$\omega_h(x,t) = \tilde{\omega}(X^{\xi,\theta}(0)) \quad \forall(x,t) \in P_{im} \tag{1.10}$$

Remark 3

If N is the number of triangles and M the number of time intervals, one has $N \times M$ characteristics to compute : this is a costly step ; therefore we suggest an alternative which uses the fact that (ξ,θ) can be properly chosen.

Procedure 3 : (Forward-backward computations of ω_h)

Step 0 : choose a set of points $\xi_i^o \in T_i$. Set $m = 0$.

Step 1 : compute the characteristics forward in time starting
at $(\xi_i^m, m\Delta t)$; i.e. compute $\xi_i^{m+1} = X((m+1)\Delta t)$ where

$$\frac{dX}{d\tau} = \begin{cases} u(X,\xi) \quad ; \quad X(m\Delta t) = \xi_i^m \\ 0 \quad \text{if} \quad X \in \Gamma . \end{cases} \tag{1.11}$$

Step 2 : if (x,t) and $(\xi_i^{m+1},(m+1)\Delta t)$ both belong to $P_{j,(m+1)}$
set :
$$\omega_h(x,t) = \omega_h(\xi_i^m,m\Delta t)$$

Step 3 : for all triangles T_j which have no ξ_i^{m+1} in them choose
ξ_i^{m+1} arbitrarily and compute $X^{\xi_i^{m+1}(m+1)\Delta t}(0)$ by Procedure 1 and define ω_h by (1.10).

Step 4 : go back to Step 1 with $m = m+1$.

Comments

Since at a given time several characteristics, as computed in step 2, can arrive in this same triangle T_j we have the following variants :

1. Pick any (first, last,...) $\xi_i^{(m+1)\Delta t}$ to define ω_h and keep
this ξ_i only for the next loops. All other characteristics that ends in T_j die there.

2. Proceed as above but keep all points $\xi_i^{(m+1)\Delta t}$ as starting
points of characteristics in the next loop, except those introduced in step 3.

Approximation of the stream function

As usual (9) is approximated in variational form by

$$\int_{\Omega} \nabla\Psi_h \, \nabla\varphi_h \, dx = \int_{\Omega} \omega_h \, \varphi_h \, dx \qquad \forall \varphi_h \in H_{oh} \tag{1.12}$$

$$\Psi_h - \Psi_\Gamma \in H_{oh}$$

where H_h is an approximation of $H^1(\Omega)$ and

$$H_{oh} = \{\varphi_h \in H_h : \varphi_h|_\Gamma = 0\}$$

For example to get a piecewise constant velocity

$$u_h = \nabla \times \Psi_h \tag{1.13}$$

we choose

$$H_h = \{\varphi_h \text{ continuous} : \varphi_h|_{T_i} \text{ linear } \forall T_i\} \tag{1.14}$$

Note that (1.12) yields a linear system independant of t, therefore we can factorize the matrix once and for all.
Note also that the choice of approximation of ω_h allows exact integrations of the load vector (right member in (1.12)).

<u>Coupling the procedures</u>
In theory the ω-system and the Ψ-system are coupled. Therefore we have to solve a non linear system and each iteration will involve the computation of characteristics all the way back to the origin of time.

However if u is piecewise constant in time and procedure 3 is selected then the systems are decoupled.

<u>Algorithm 1</u>
Indeed assume that on $]0,\Delta t[$ u is equal to

$\nabla \times \Psi$ where Ψ is the solution of

$$-\Delta\Psi = \omega^o \quad ; \quad \Psi|_\Gamma = \Psi_\Gamma \ .$$

Then ω_h can be computed on $]0,\Delta t[$ by procedure 3. Next on $]\Delta t, 2\Delta t[$ set u equal to its values at $t = \Delta t$ according to (1.12) (1.13) with ω_h just computed. And so on ...

<u>Convergence results</u>
We recall that (4) has a unique smooth solution for smooth data in R^2 (this is not true in R^3).
In [1] the following result is shown :

44

<u>Theorem 1</u>
If Ψ_h is discretized by (1.14). If ω_h is piecewise constant in
x and t, if the solution of (4) is smooth (no shocks in particular) then for all ε there exists h^o such that

$$\max_{(x,t)} |\omega_h(x,t)-\omega(x,t)| \leq C(h^{1-\varepsilon}+\Delta t) \quad \forall h \leq h^o. \qquad (1.15)$$

and

$$\max_{t \in]0,T[} |u_h - u|_{L^2(\Omega)} \leq C(h^{1-\varepsilon}+\Delta t) \qquad (1.16)$$

<u>Theorem 2</u>
The algorithm is L^∞ stable in the sens that

$$\max_{x,t} |\omega_h(x,t)| \leq \max_{x,t} |\tilde\omega(x,t)|$$

<u>Remark</u>
Unfortunatly algorithm 1 is not conservative i.e. when u.n= 0

$$\int_\Omega \omega_h(x,t)dx \neq \int_\Omega \omega^o(x)dx . \qquad (1.17)$$

2. CONSERVATIVE SCHEMES

The algorithm just described has some mathematical properties.
Its main drawback is its lack of conservativity. It is a natural feature of Euler's equation to have

$$\int_\Omega \omega^m dx \quad \text{independent of t,} \quad \forall m . \qquad (2.1)$$

In many transient flow simulations one wishes to keep this property at least for m=1. There are at least 3 ways to do so but
with all three the proof convergence is an open question.

<u>Method 1</u>
The first way is to work with Dirac functions as in the classical vortex method (see Baker [4], for example). As explained
above ω is now given by

$$\omega(x,t) = \sum_i \delta(x-x_i(t) \omega_i^o \qquad (2.2)$$

Hence Ψ_h is the solution of

$$\int_\Omega \nabla\Psi_h \nabla\varphi_h \, dx = \sum \omega_i^o \varphi_h(x_i(t)) \qquad \forall\varphi_h \in H_{oh} \qquad (2.3)$$

This method is very well suited to certain problems like the
roll up of vortex sheets but it is not always easy to approximate $\omega^o(x)$ by a sum of Dirac functions.

<u>Method 2</u>
As in method 1 we can integrate the characteristics forward in
time only but approximate ω^o by a sum of step functions.

Let ξ_i^o be a set of points, one per triangle T_i. Let ω^o be approximated by

$$\omega^o(x) = \Sigma \ I(x-\xi_i^o)\omega_i^o \qquad\qquad (2.4)$$

where $I(x-\xi_i^o)$ equals 1 if x belongs to the same triangle as ξ_i^o
and zero otherwise.

Then as above, at later times one has

$$\omega(x,t) = \Sigma \ I(x-\xi_i^o(t))\omega_i^o \ \frac{\sigma_i}{\sigma_{j(i)}} \qquad\qquad (2.5)$$

where $\xi_i(t)$ is the position at time t of the particle that
started from ξ_i^o ; where σ_i is the area of T_i and $j(i)$ is the
index of the triangle to which $\xi_i(t)$ belongs.

Unlike the algorithm presented in § 1, here $\omega(x,t)$ is zero on
all triangles which have no $\xi_i(t)$. Of course this procedure is
better on problems which have initially $\omega^o = 0$ in most triangles.
This will be the case of our vortex alleviation problem.

<u>Remark</u>
Both methods are difficult to implement if $\omega_\Gamma \neq 0$ (or $u.n \neq 0$).

<u>Method 3</u>
We modify our original algorithm in order to take conservativity
into account.

Conservativity is lost for the following reasons (see figure 3).
. 2 or more characteristics can end in the same triangles.
. all backward computed characteristics introduce additional
  vortices.
. the area of the starting triangles are different from the
  area of ending triangles.

This method can be formalized but it is easier to take a simple
example such as the one of figure 3.

At time $2\Delta t$ according to figure 3 one sets

$$\omega_4 = (\omega_1 \sigma_1 + \omega_2 \sigma_2)/\sigma_4 \qquad (2.6)$$

$$\omega_5 = \omega_3 \sigma_3/(\sigma_5 + \sigma_6) \qquad (2.7)$$

$$\omega_6 = \omega_3 \sigma_3/(\sigma_5 + \sigma_6) \qquad (2.8)$$

3. EXTENTION TO 3-D CALCULATIONS

In R^3, ω and Ψ are now vectors and Euler's system of equations is :

$$\frac{\partial \omega}{\partial t} + u.\nabla\omega + \omega.\nabla u = 0$$

$$\nabla \times \nabla \times \Psi = \omega \qquad (3.1)$$

$$u = \nabla \times \Psi$$

+ boundary conditions as in the 2-D case.

In principle the same technique can be extended namely if $X^{x,t}(\tau)$ denotes the characteristic solution of :

$$\frac{dX}{d\tau} = u \quad ; \quad X(t) = x \quad ; \qquad (3.2)$$

then the vector ω satisfies the differential system

$$\frac{d}{d\tau}(\omega(X^{x,t}(\tau),\tau))\big|_{\tau=t} + \omega.\nabla u = 0 \ .$$

$$\omega(X^{x,t}(0),0) = \omega^o(X^{x,t}(0)) \qquad (3.3)$$

Hence one has first to compute a characteristic, say by procedure 1 backward in time, then ω is computed by any Runge-Kutta method but forward in time starting from $X^{x,t}(0)$.

The computation of the vector Ψ poses some problem numerically because the solution is not unique. One possibility is to add a regularisation and solve for $\varepsilon \ll 1$:

$$-\varepsilon \nabla.(\nabla\Psi) + \nabla \times \nabla \times \Psi = \omega$$

$$\Psi\big|_\Gamma = \Psi_\Gamma \qquad (3.4)$$

A different method altogether is to transport u directly. Then the Euler equation is rewritten as follows :

$$\frac{d}{d\tau} u(X^{x,t}(\tau),\tau)\big|_{\tau=t} = - \nabla p$$

$$\nabla.u = 0 \qquad (3.5)$$

we use an implicit time-discretization of (3.5) :

$$\frac{1}{\Delta t} u^{n+1}(x) = - \nabla p^{n+1} + \frac{1}{\Delta t} u^n(X^{x,(n+1)\Delta t}(n\Delta t))$$

$$\nabla . u^{n+1} = 0 .$$

(3.6)

Now we approximate u^{n+1} by a piecewise quadratic continuous function (P^2-conforming element) and p^{n+1} by a piecewise linear discontinuous function ; then (3.6) is approximated by

$$\frac{1}{\Delta t}((u_h^{n+1},v_h))+((\nabla p_h^{n+1},v_h))= \frac{1}{\Delta t}((u^n(X^{x,(n+1)\Delta t}(n\Delta t)),v_h))$$

$$\forall q_h \in Q_h$$

(3.7)

$$- \varepsilon((p_h,q_h))+ ((\nabla . u_h^{n+1},q_h)) = 0$$

or equivalently

$$\frac{1}{\Delta t}((u_h^{n+1},v_h))+\varepsilon((\nabla . u_h^{n+1},\nabla . v_h)) = \frac{1}{\Delta t}(u^n(X^{x,(n+1)\Delta t}(n\Delta t),v_h))$$

$$\forall v_h \in V_{oh}$$

(3.8)

Here $((.,.))$ stands for the scalar product of $L^2(\Omega)$ and ε is a small regularization parameter.

In physical term (3.8) can be interpreted as
  . 1 step of transport of u^n
  . 1 step of projection on the subspace of V_{oh} of functions with nearly (ε) zero divergence.

Convergence results for this method are shown in [5] and numerical tests will be given in a later publication.

This last method is perhaps more simple ; it can be made conservative (depending upon the quadrature formula for the numerical computation of the right hand side of (3.8)) but it is more dissipative than the previous methods.

4. NUMERICAL EXAMPLES

Some simple applications of the previous methods can be found in [1]. Here we have taken one example from aeronautical engineering.

The problem is to study the behavior of an airfoil as it passes heavy turbulences created either by another airplane or by bad weather.

The Kutta-Joukowski condition

On non simply connected domains such as the one of figure 4 the stream function Ψ is defined up to a constant value. Our flow must satisfy the Kutta-Joukowski condition, that is there should be no flow across the wake of an obstacle. This means when we have a trailing edge like here, that the normal derivative of the velocity at the trailing edge must be null.

This implies $|\vec{u}^+.\vec{n}^+| = |\vec{u}^-.\vec{n}^-|$ where $\vec{n}^+$ and $\vec{n}^-$ are the normals to the upper and lower side of the trailing edge. Since by our approximation method we have constant velocities over the triangles this condition cannot be exactly satisfied. However it enables us to compute the circulation in the following manner :

At a given time step :

1) Take φ_1 solution of

$$- \Delta\varphi_1 = \omega \quad \text{on} \quad \Omega \quad ,$$
$$\varphi_1\big|_{\Gamma_i} = 0$$
$$\varphi_1\big|_{\Gamma_e} = y\, u_o - x\, v_o$$

2) Take φ_2 solution of

$$- \Delta\varphi_2 = 0$$
$$\varphi_2\big|_{\Gamma_i} = 0$$
$$\varphi_2\big|_{\Gamma_e} = 1$$

The stream function we are looking for is given by

$$\Psi = \varphi_1 + \lambda(t)\, \varphi_2$$

where $\lambda(t)$ is a constant such that

$$|\nabla(\varphi_1 + \lambda\, \varphi_2)^+|^2 = |(\varphi_1 + \lambda\, \varphi_2)^-|^2$$

those values being computed on the upper and lower triangle with one side at the trailing edge.

Since this a quadratic equation in λ there are two solutions the circulation is the one for which the normal velocities are in opposite directions.

<u>Computations</u>

We consider a simple Joukowski airfoil defined by the transformation :

$$x = 2\,d\,\cos\theta \quad ; \quad y = \varepsilon\,d(2\sin\theta - \sin 2\theta)$$

We choose $d = 2.5$ and $\varepsilon = .1$ for our specific examples. We want to simulate the flight of our airfoil passing near a vortex. It can be viewed as the very simplified simulation of a plane flying into or next to vortex created by another plane or by an atmospheric turbulence. Such transient problems cannot be treated by most classical methods. In such case one is most interested by the lift factor :

$$L(t) = \rho\,|\vec{u}_{\infty}|\,.\,\lambda(t)$$

where $\vec{u}_{\infty}$ is the velocity at infinity, ρ the air density and $\lambda(t)$ is the circulation as defined above. For the sake of simplicity we took $\vec{u} = (1,0)$ and $\rho=1$.

Figure 5 gives the mesh for one Finite Element Method and a zoom of the mesh near the airfoil.

We wanted first to illustrate the effect of using the conservative scheme of method 2 in § 2. A vortex, $\omega = 5$, was placed on two triangles slightly before the airfoil, hence passing over it and due to its orientation pushing the airfoil downward. The computation (with the same $\Delta t = 2.$) was performed twice, one with the non-conservative scheme and once with the conservative one.

Figure 6 gives the computed lift factor for both cases ; the upper curve corresponds to the conservative case.
We see on the lower curve that the non-conservative scheme gives a vortex with an intensity increasing with time, even after passing over the airfoil. This can be easily explained by the fact that its path reaches larger and larger elements (cf. figure 5). Figure 6 gives also the corresponding paths followed by the two pair of vortices. The lower pair follows a smooth path along streamlines, it corresponds of course to the conservative scheme.

Next in order to illustrate the convergence in Δt of our method we run the same problem with the conservative scheme for $\Delta t=2.$, as before, $\Delta t= 1.$ and $\Delta t = 0.5$.

Figure 7 gives the lift factor curves as the pair of vortices passes over the airfoil for the three time steps, over the time interval $[0, 20]$. The convergence as Δt is lowered is quite good. Note that the lift factor is minimum when the vortex is near the trailing, a well known aerodynamic fact.

Finally to visualize the effect on the stream lines we computed a strong vortex passing over the airfoil. Starting from the same triangles as before we took $\omega = 20$. Figure 8 illustrates the path of the two vortices. Note that they push strongly the airfoil downwards and have a far more erratic path than those computed before (figure 6). Figure 9 gives at different time the streamlines of the corresponding problem.

REFERENCES

[1] C. BARDOS, M. BERCOVIER, O. PIRONNEAU. The Vortex Method With Finite Elements, INRIA Report n° 15 (1980).

[2] J.P. CHRISTIANSEN. Numerical Solution of Hydrodynamics by the method of point vortices J. of Comp. Phys. 13, 863-879 (1973).

[3] A.J. CHORIN. Numerical study of slightly viscous flow, J. of Fluid Mech. 57, 784-796 (1973).

[4] G. BAKER. The clouds and cells technique applied to the roll up of vortex sheets, J. of Comp. Phys. 75-95 (1979).

[5] O. PIRONNEAU. On the transport-diffusion algorithms and its applications to the Navier-Stokes Equations (to appear).

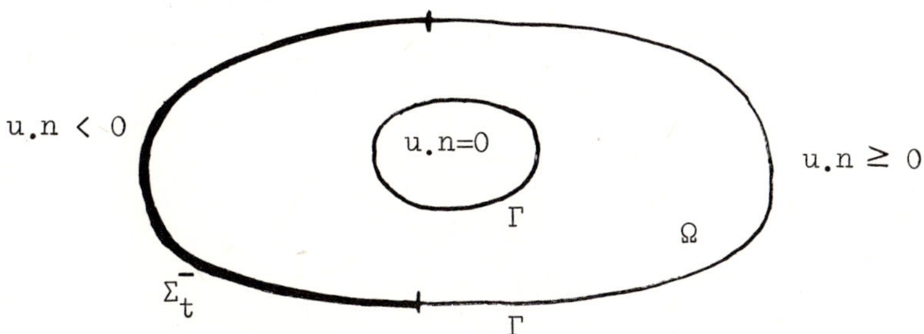

Figure 1 : Consistant boundary conditions for
Euler's equation.

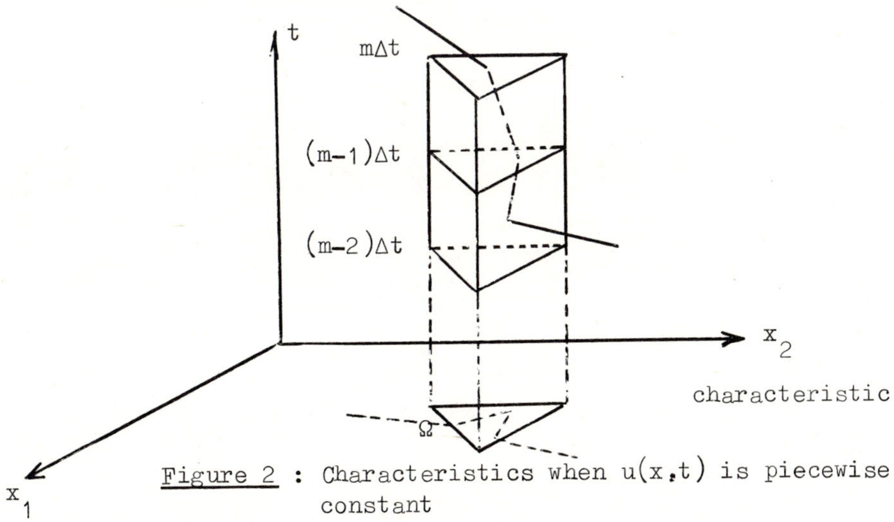

Figure 2 : Characteristics when $u(x,t)$ is piecewise
constant

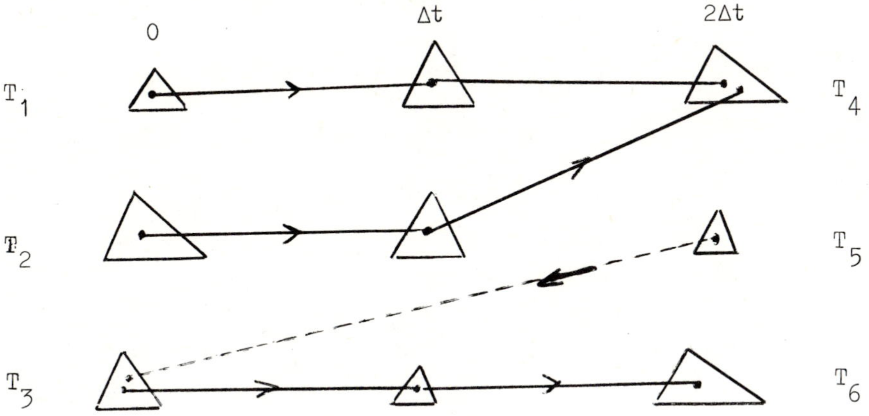

Figure 3 : A possible configuration of forward and
backward characteristics from t=0 to 2Δt.

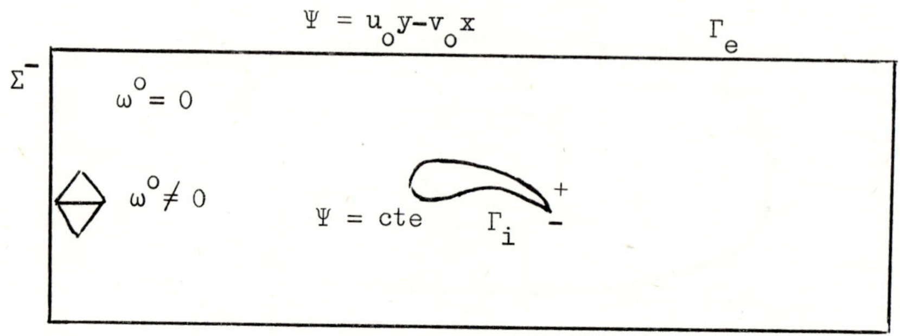

<u>Figure 4</u> : Boundary conditions for the vortex
alleviation problem.

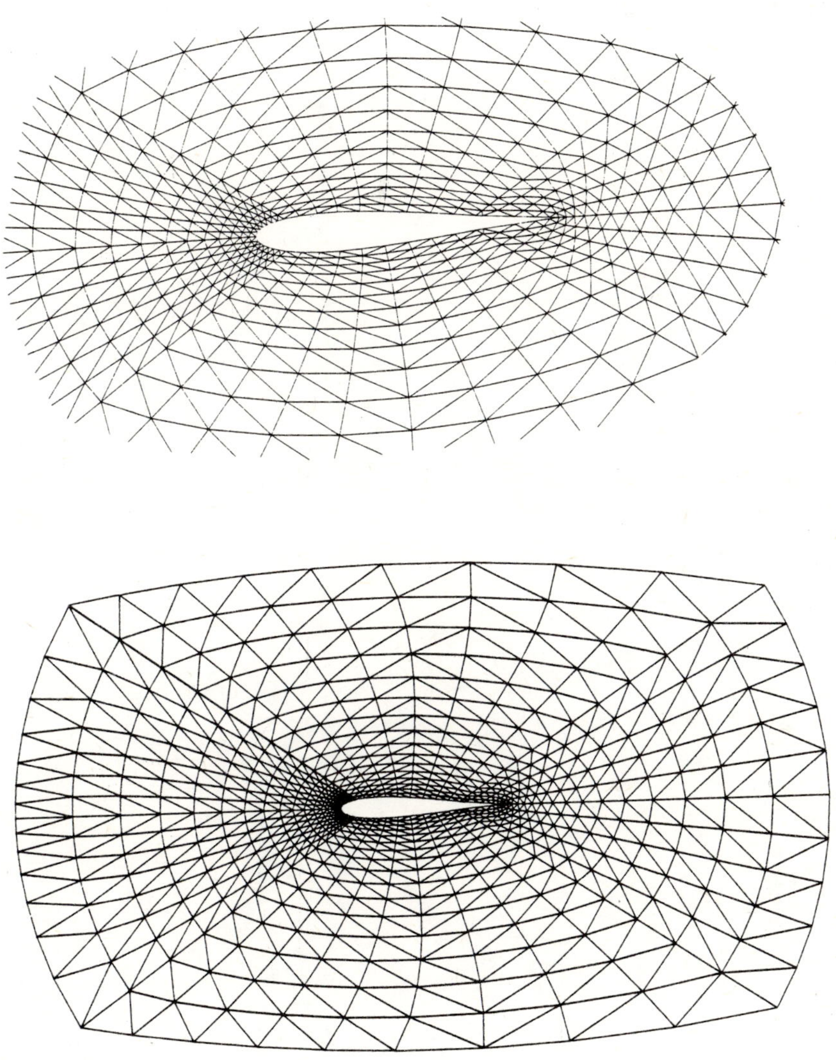

Figure 5

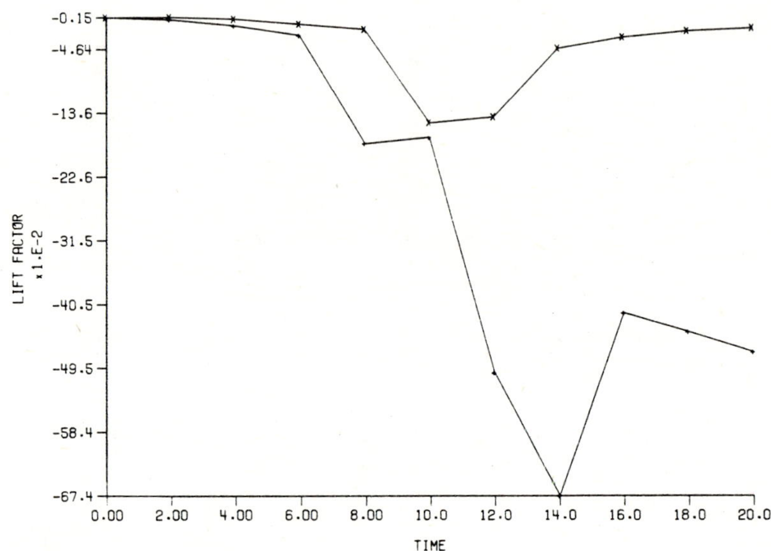

<u>Figure 6</u>

-0.11
-1.26
-3.54
-5.83
-8.12
-10.4
-12.7
-15.0
-17.3
LIFT FACTOR
(1.E-2
0.00
2.00
4.00
6.00
8.00
10.0
12.0
14.0
16.0
18.0
20.0
TIME

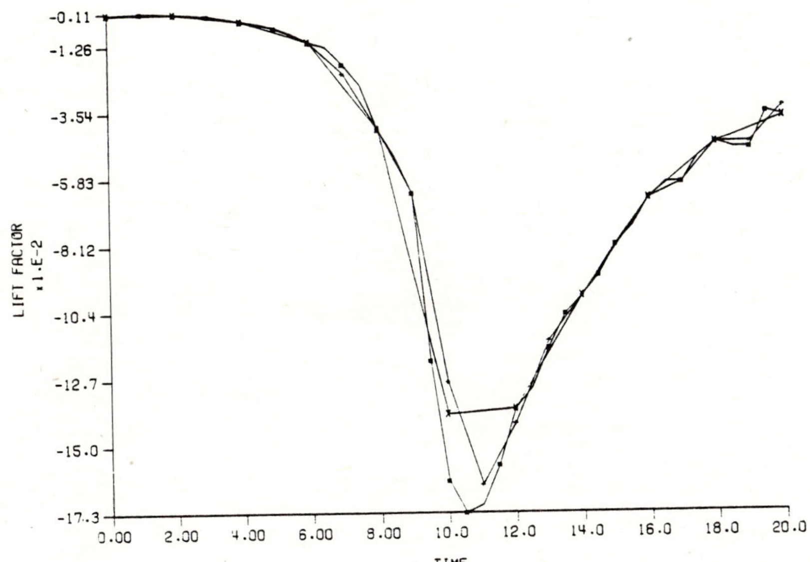

Figure 7

C

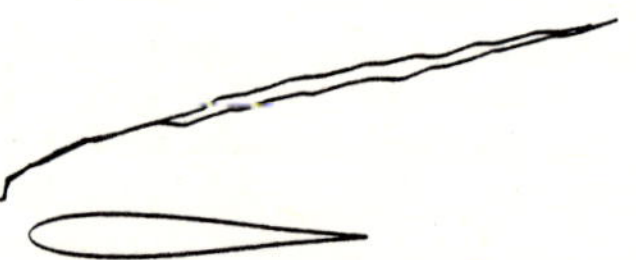

Figure 8

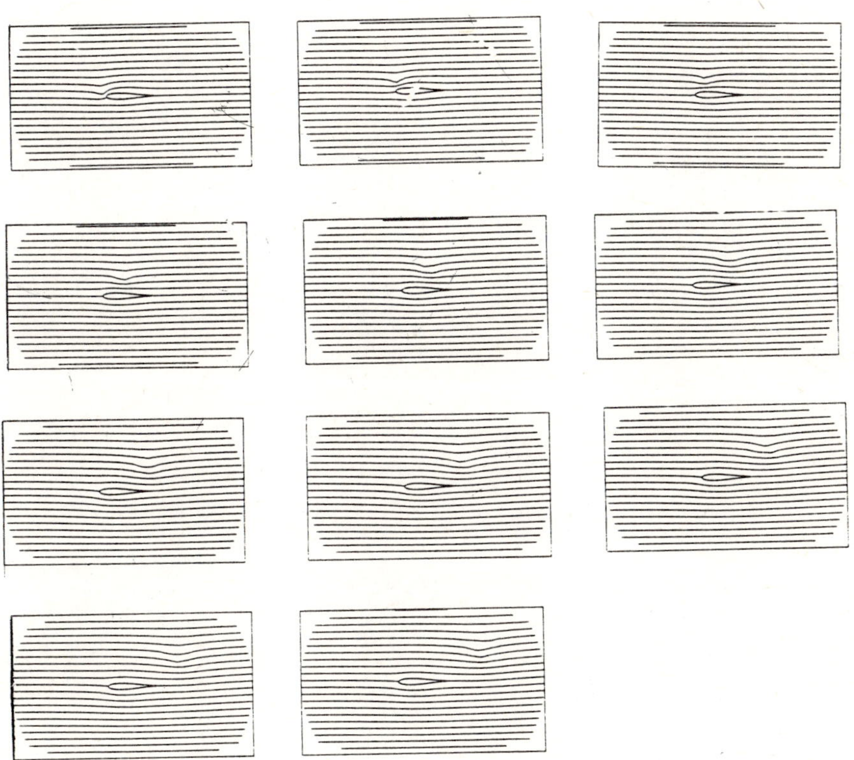

Figure 9

CHAPTER 3

DIFFERENTIAL ANALYSERS OF STRONG DISCONTINUITIES IN
ONE-DIMENSIONAL GAS FLOW
N.N.Yanenko and E.V.Vorozhtsov

Institute of Theoretical and Applied Mechanics
U.S.S.R. Academy of Sciences, Novosibirsk 630090,
U.S.S.R.

INTRODUCTION

At present finite-difference shock-capturing methods are one of the most efficient means of mathematical modeling of gas flows with discontinuities.
One of characteristic features of these methods
arising in the course of their realization on an
electronic computer consists in the fact that the
amount of the numerical information obtained as a
result of the solution of a problem exceeds by several orders of magnitude what is of real interest
for a research worker (Neumann (1966), Rusanov
(1973)). Another feature of the shock-capturing techniques is their low accuracy in the vicinity of
strong discontinuities which are spread thereat
within several intervals of the computing mesh.
These circumstances give rise to two practical problems: the one on interpretation of the numerical
results obtained and the one on the increase in accuracy of the numerical solution in the vicinity of
strong discontinuities. In this connection the
questions of the development and the foundation of
the differential analysers of strong discontinuities, that is algorithms for the determination of
the locations of singularities within the computational domain by the shock-capturing computational
results, are of present interest.
It should be noted that at present there exist a
few algorithms for localization of singularities
when using homogeneous difference schemes. These
are, first of all, the algorithms for localization
of shock waves and contact discontinuities by the
coalescence of different isolines, see, e.g., the

book by Liubimov and Rusanov (1970), Proceedings
of the Fourth International Conference on Numerical Methods in Fluid Dynamics (1975), the book
edited by Godunov (1976). Some authors, for example
Kosarev (1971), Ivanov (1975), Vorozhtsov, Ermolin
and Fomin (1979) have determined the shock wave location by maximum gradients of the flow parameters.
Rozhdestvenskii and Yanenko (1968), Laval (1969),
Wilkins (1970,1971a,1971b) localized the shock
front by maximum of the artificial viscosity. Miranker and Pironneau (1975,1976) have shown at the
example of the Burgers equation how the problem of
localization of a shock wave can be reduced to some variational problem. Usakov (1979,1980) has applied the Miranker and Pironneau's approach to the
localization of discontinuities in one-dimensional
filtration problems. In some papers (for example,
Gridneva et al (1975), O'Keefe et al (1975)) the
evolution of a contact discontinuity was traced out
using marker particles.

In 1976 the paper by present authors was published in which the first theory was described apparently that justified some methods for locating
shock waves by the shock-smoothing computation results of the one-dimensional gas dynamic problems.
In the course of further investigations the authors
also succeeded in constructing a theory of differential analysers of contact discontinuities. Below a review of basic results obtained by the authors in the field during last five years is presented.

THEORY OF DIFFERENTIAL ANALYSERS OF SHOCK WAVES

Before stating the main notions of this theory,
let us consider a comparatively simple example. Let
$u_\mu(x,t)$ be the exact solution of the equation

$$\partial u_\mu / \partial t + u_\mu \, \partial u_\mu / \partial x = \mu \, \partial^2 u_\mu / \partial x^2, \ \mu = \text{const} > 0 \qquad (1)$$

with the initial condition

$$u_\mu(x,0) = \begin{cases} u_1, & x < x_0 \\ u_2, & x > x_0 \end{cases} \qquad (2)$$

where u_1, u_2, x_0 are constants and $u_1 > u_2$; x
is the spatial coordinate, t is the time. Let us
indicate the following properties of the solution
of the problem (1),(2): (A) for arbitrary fixed
$t > 0$ there exists such a point (x_c, t) at

which the solution graphs of the problem (1),(2)
obtained for the different values of the constant
$\mu > 0$ intersect, see Fig.1. (B) $\max\limits_{x} |\partial u_\mu / \partial x| =$
$= |\partial u_\mu / \partial x|_{x=x_c}$. (C) Whitham (1977) has shown that
at $t \to \infty$ the solution of
the problem (1),(2) converges in the domain $u_1 > x/t > u_2$
to the progressive wave type solution $U_\mu(\xi)$ of the
equation (1) where

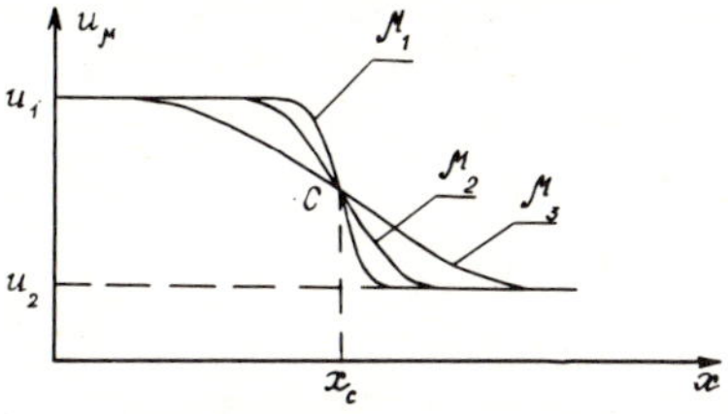

$$\xi = x - 0.5(u_1 + u_2)t,$$

$$U_\mu(\xi) = 0.5(u_1 + u_2) -$$

Figure 1

$$- 0.5(u_1 - u_2)\,\text{th}\left[(u_1 - u_2)\xi / (4\mu)\right].$$

Other results on the investigation of the convergence of the solution $u_\mu(x,t)$ to the progressive
wave type solution may be found in the papers by
Il'in and Oleinik (1960) and Peletier (1971). (D)
The locus of points (x_c, t) forms in the (x,t)
plane a line which is the discontinuity trajectory
in the solution $u(x,t)$ of the problem

$$\partial u / \partial t + u(\partial u / \partial x) = 0, \qquad (3)$$

$$u(x,0) = \begin{cases} u_1, x < x_0 \\ u_2, x > x_0, \end{cases} \quad u_1 > u_2 .$$

Thus, despite the fact that the discontinuity in
the solution of the problem (1),(2) is smoothed at
$t > 0$, there exists a point at which the quantity
$u_\mu(x_c, t_0)$, $t_0 > 0$, does not depend on μ in the
profiles of $u_\mu(x, t_0)$ obtained for different μ ,
and the abscissa of this point coincides with the
exact position of the discontinuity in the solution
of the problem (3). In our papers (cf.Yanenko, Vorozcov, Fomin (1976), Fomin, Vorozhtsov and Yanenko
(1976), Yanenko, Fomin, Vorozhtsov (1978), Yanenko,
Kovenya et al (1979)) this point was called the
centre of a smeared discontinuity.

All our further considerations are connected
with a problem on localization of strong discontinuities in the numerical solution of a system of
the Euler equations for an inviscid compressible
gas. Let us write this system as follows

$$\partial \vec{u} / \partial t + \partial \varphi(\vec{u}) / \partial x = 0 \qquad (4)$$

where

$$\vec{u} = \begin{pmatrix} \rho \\ \rho w \\ \rho E \end{pmatrix}, \quad \varphi(\vec{u}) = \begin{pmatrix} \rho w \\ p + \rho w^2 \\ w p + \rho w E \end{pmatrix}, \quad E = \mathcal{E} + w^2/2 , \tag{5}$$

ρ is density, $\mathcal{E}$ is specific internal energy, p is pressure, w is velocity. To close the system (4), (5) the equation of state is used in the form

$$p = F(\rho, \mathcal{E}) \tag{6}$$

or in the form

$$\mathcal{E} = f(p, \rho) . \tag{7}$$

Let us consider a finite difference shock-smoothing scheme consistent with the equation system (4)-(6). Generally, the numerical solution obtained by such a scheme depends on the grid parameters h, τ where h is a step of the uniform spatial grid, τ is the time step. Employing the algorithm presented, e.g., in the book by Shokin (1979) let us calculate the first differential approximation which corresponds to the above difference scheme. We shall consider the solutions of the progressive wave type of the equations of the differential approximation, i.e. the solutions depending only upon the variable $x' = x - Dt$ where D is the stationary shock wave speed. As is known, the solution of the progressive wave type is determined with accuracy up to the translation along the x' axis. Let

$$\vec{w}(x', x'_0, h, \tau) = \{ w_1(x', x'_0, h, \tau), \dots, w_m(x', x'_0, h, \tau) \}$$

be a vector function of the solution of the progressive wave class of the equations of the differential approximation type having the form

$$F_i(\vec{w}, d\vec{w}/dx', h, \tau) = 0 , \quad i = 1, \dots, m, \tag{8}$$

at the boundary conditions

$$\vec{w}(x', x'_0, h, \tau) = \begin{cases} \vec{W_1} , & x' \longrightarrow -\infty \\ \vec{W_2} , & x' \longrightarrow +\infty \end{cases} \tag{9}$$

where x'_0 is an arbitrary constant having the same dimension as the variable x', and $\vec{W_1}$, $\vec{W_2}$ are constant vectors satisfying the Hugoniot conditions the subscripts "1" and "2" refer to the states behind and before the shock front, respectively. We call the solution $\vec{w}(x', x'_0, h, \tau)$ a smeared shock wave.

<u>Definition 1.</u> Let us call a smeared shock wave centre the point in the solution $\vec{w}(x', x'_0, h, \tau)$

of a progressive wave class of the equations (8) of the differential approximation type at the boundary conditions (9) at which the values of the components of the vector $\vec{w}$ do not depend on h and τ for each fixed value x_0' .

We say further that the smeared shock wave centre is unique if there exists the only value of vector $\vec{w}$ components which does not depend on the values of constant quantities h , τ , x_0' .

Unlike the smeared shock wave we understand the finite difference shock wave as a solution of the finite difference equations approximating the shock wave and calculated at the discrete grid points in the (x, t) plane. Analogous to the notion of a smeared shock wave center, let us introduce the notion of the finite difference shock wave centre as a point which (i) is located in the finite difference shock wave zone, (ii) belongs to the solutions of the finite difference equations at the same t obtained by using two different values of constant steps h_1 , τ_1 and h_2 , τ_2 where $|h_1 - h_2| + |\tau_1 - \tau_2| \neq 0$ and (iii) is determined with the help of linear interpolation at the points that are not nodes.

<u>Definition 2.</u> By the differential analyser of a shock wave we mean the algorithm which enables us to find the finite difference shock wave centre coordinates in a computational cell by shock-smoothing calculation.

It should be noted that at present there is no mathematical means which could be appropriate for studying the question on the existence and uniqueness of a finite difference shock wave centre at the level of a difference scheme itself, that is at the "discrete" level. At the same time, as it was shown by us in the papers cited above, it is possible to carry out an investigation of the existence and uniqueness of a smeared shock wave centre (see the above introduced Definition 1), that is an investigation "at the differential level". After that one needs only checking qualitative and quantitative conclusions of the theory by direct computations using various difference schemes. Numerous examples of such numerical tests contained in the Vorozhtsov's thesis (1976) and in the Vorozhtsov's paper (1977) confirm the validity of the theoretical conclusions obtained.

<u>Existence and uniqueness of the smeared shock wave centre in the solution of a gas dynamic system with the artificial viscosity introduced additively into the pressure</u>

The stationary gas dynamic equation system derived from the equation system (4),(5) by substituting the pressure p by the quantity $p+q$ where q is a pseudoviscosity which was proposed by Neumann and Richtmyer (1950) seems to be the most simple mathematical model describing the effects of "smearing" of a gas dynamic shock wave resulting from the introduction of an artificial viscosity. In this connection it seemed natural at first to carry out an investigation of a question on the existence of the smeared shock wave centre in the solution of the above mentioned system. Consider the problem of a steady shock wave. Following to the classical work by J.von Neumann and R.D.Richtmyer (1950), we shall study the solutions of the above equation system which depend only on the variable $x' = x - Dt$, that is the progressive wave type solutions. D is the speed of a stationary shock wave. Let us set $u' = u - D$ where u is the gas speed. Integrating the system once, we obtain:

$$\rho u = C_1 = m , (p+q) + mu = C_2 , \qquad (10)$$

$$m\left[p/(\rho(\gamma-1)) + 0.5\, u^2 \right] + (p+q)u = C_3 .$$

Here $\gamma = \mathrm{const} > 1$, C_1 , C_2 , C_3 are integration constants, and the primes at the quantities x , u are omitted for brevity of notations. For the definiteness the shock wave is assumed to move from the left to the right; then $D > 0$, $m < 0$. The values determining the state behind the shock front are denoted by the subscript "1", the subscript "2" denotes the state before the shock front; let $\vartheta = 1/\rho$.

Theorem 1. Let the artificial viscosity q entering Equations (10) be as follows:

$$q = \begin{cases} Q\left(h\dfrac{du}{dx}, h\dfrac{dp}{dx}, h\dfrac{d\rho}{dx}, p, \rho\right), & \dfrac{du}{dx} < 0 \\ 0 & , \dfrac{du}{dx} \geqslant 0 \end{cases} \qquad (11)$$

and the system of Equations (10)-(11) is uniquely resolvable with respect to $(d\vartheta/dx)$ so that

$$d\vartheta/dx = g(C_1, C_2, C_3, \vartheta, h)$$

where $g(C_i, \vartheta, h)$ is positive within the interval $(\vartheta_1, \vartheta_2)$ and the function

$$\Phi(C_i, \vartheta, h) = \int \frac{d\vartheta}{g(C_i, \vartheta, h)} \qquad (12)$$

is continuous with respect to ϑ and changes its sign within $(\vartheta_1, \vartheta_2)$, then there exists the unique smeared shock wave centre.

From this theorem proved in papers by Yanenko, Vorozcov and Fomin (1976) and Fomin, Vorozhtsov and Yanenko (1976) some conclusions follow which are important for justification of an algorithm of an approximate shock front localization by maximum of the artificial viscosity. In particular, it is shown that if the inequality $|\Phi_1(\sqrt{\vartheta_1\vartheta_2})| \leqslant 1$ is satisfied where $\Phi_1(\vartheta)$ is a function related to the function $\Phi(C_i,\vartheta,h)$, defined by the formula (12), by the relationship $\Phi(C_i,\vartheta,h) = h\Phi_1(\vartheta)$, then the distance along the axis of the point of maximum of the artificial viscosity from the smeared shock wave centre does not exceed h. If the conditions of Theorem 1 are satisfied and, besides, the equality $\Phi_1''(\vartheta(x_0)) = 0$ holds where ϑ_0 is the value of the function $\vartheta(x)$ at the wave centre, then the smeared shock wave centre coincides with the point of extremum of functions

$$|du/dx|, \quad d\vartheta/dx, \quad |d(p+q)/dx|.$$

By the use of the results obtained some artificial viscosities of the form (11) were analyzed, see Table 1.

Table 1

q	$ah^2\rho\left(\dfrac{du}{dx}\right)^2$	$-ahc_0\rho\dfrac{du}{dx}$	$-ahc\rho\dfrac{du}{dx}$	$-ah\rho\dfrac{du}{dx}\left(\sigma_1 c + \sigma_2 h\left|\dfrac{du}{dx}\right|\right)$	$-ah\rho\left(1+bcp\times\dfrac{du}{dx}\Big/\dfrac{dp}{dx}\right)\dfrac{du}{dx}$
Does the smeared shock wave centre exist?	Yes	Yes	Yes	Yes at (13), (14), (15)	Yes at (16)
Practical criterion for finding the smeared shock wave centre	by $\max\left\|\dfrac{du}{dx}\right\|$ at $\dfrac{du}{dx}<0$	by $\max\left\|\dfrac{du}{dx}\right\|$ at $\dfrac{du}{dx}<0$	by $\max q$ at $du/dx<0$ $\|\Phi_1(\sqrt{\vartheta_1\vartheta_2})\|<1$	by $\max q$ at $du/dx<0$ $\|\Phi_1(\sqrt{\vartheta_1\vartheta_2})\|<1$	by $\max q$ at $du/dx<0$ $\|\Phi_1(\sqrt{\vartheta_1\vartheta_2})\|<1$

$$(1+\eta)^2(\gamma+1)\left[4\sigma_2 - a\sigma_1^2(\gamma-1)\right]^2 - 4\eta\left[a\sigma_1^2\gamma(\gamma-1) - 4\sigma_2(\gamma+1)\right](a\sigma_1^2\gamma - 4\sigma_2) \leqslant 0; \qquad (13)$$

$$a>0; \quad 4a\sigma_2(\gamma+1) < a^2\sigma_1^2\gamma(\gamma-1); \quad \sigma_2\neq 0; \qquad (14)$$

$$a^2 \sigma_1^2 \gamma(\gamma-1) < B(\zeta,\gamma,a,\sigma_1,\sigma_2); \qquad (15)$$

$$\mathcal{B} > \left[\zeta - \frac{\gamma+1}{\gamma-1}\right]\left\{(\gamma-1)\Big/\left[2\gamma\left(\frac{\gamma+1}{\gamma-1}\zeta+\eta^2-\frac{\gamma+1}{\gamma}\zeta(1+\zeta)\right)\right]\right\}^{1/2}. \quad (16)$$

In the inequalities (13)-(16) $\eta = \vartheta_2/\vartheta_1$; the form
of the function $B(\cdot)$ is not presented here in view
of a bulky form of this function (see Vorozhtsov,
Fomin, Yanenko (1976)). Figure 2 shows density and
velocity plots in the vicinity of the shock wave
obtained by the two-step Lax-Wendroff scheme with
the quadratic artificial viscosity q introduced
additively into the pressure. Here $h_1 = 1/40$, $h_2 = 2h_1$,
$h_3 = 0.5\,h_1$, the Courant number
$K \cong 0.25$ and the shock wave has
already propagated the distance $\approx 23\,h_1$ (Fomin, Vorozhtsov and Yanenko (1976)).
It is clear that the finite-
-difference shock wave centre
determines the shock front
position with some error;
however, the results of numerous computations show that
the accuracy of the shock wave localization on the basis

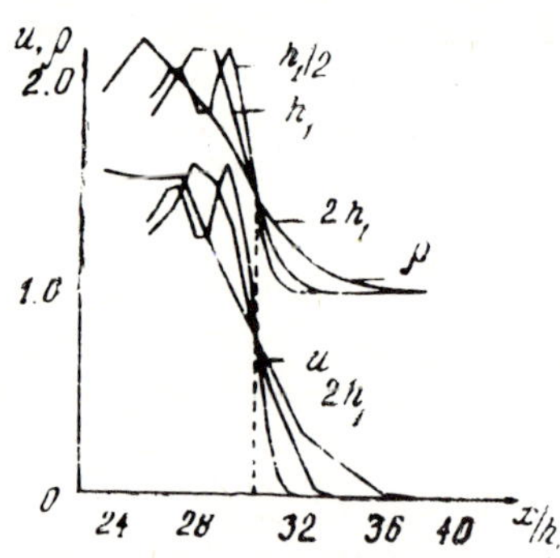

Figure 2.

of the centre notion is higher
than that of a scheme. For example, in the computation presented in the Figure 2 the above mentioned
error in determining the discontinuity position did
not exceed $0.1\,h_1$. Note that it proves to be possible to establish the relative disposition of the
point of $\max q$ and the smeared shock wave centre
by using the above stated Theorem 1. Thus, for the
quadratic viscosity $q = a h^2 \rho\,[\min\,(\partial u/\partial x,0)]^2$
the point of maximum of the artificial viscosity
lies to the left of the smeared shock wave centre,
as this was shown by Vorozhtsov (1977). Results of
numerical computations presented by Vorozhtsov
(1977) agree well with this theoretical conclusion.

<u>Approximation viscosity and the smeared shock wave
centre existence</u>
In the presence of the so-called approximation
or scheme viscosity in the finite-difference shock-
-capturing scheme the question on the existence of
the smeared shock wave centre was investigated by
using differential approximations of a difference
scheme. Since at present conservative difference
schemes are mainly used in gas dynamic computations,
a multiparametric family of conservative difference
schemes approximating the hyperbolic system (4) was

introduced (Fomin, Vorozhtsov and Yanenko (1976)).

Let $w(x,t)$ be the vector-function coinciding with the solution of finite difference equations approximating system (4) in the mesh nodes and being determined from the requirement of existence of derivatives

$$\partial^{j+i} w(x,t)/\partial x^i \partial t^j, \quad 1 \leqslant i+j \leqslant 3, \quad i \geqslant 0, j \geqslant 0$$

for the rest of points (x,t). Let $T_{\pm h}$, $T_{\pm \tau}$ be the translation operators along the x and t axis, such that

$$T_{\pm h} w(x,t) = w(x \pm h, t); \quad T_{\pm \tau} w(x,t) = w(x, t \pm \tau),$$

τ is the time step, I identity operator.

<u>Theorem 2.</u> If the finite-difference scheme in entire steps approximating the system (4) is representable in the form

$$\frac{1}{\tau} C \left\{ \beta_1 \left[T_\tau - (\alpha_3 T_h + \alpha_4 I + \alpha_5 T_{-h}) \right] + \beta_2 \left[\alpha_6 T_h + \alpha_7 I + \right. \right.$$
$$\left. + \alpha_8 T_{-h}) - T_{-\tau} \right] \right\} w(x,t) + \frac{1}{h} \left[\alpha_1 (T_h - I) + \alpha_2 (I - T_{-h}) \right] \times \tag{17}$$
$$\times \left[\beta_3 T_\tau + \beta_4 I + \beta_5 T_{-\tau} \right] \varphi(w(x,t)) = \frac{1}{h} \left(T_{h/2} - T_{-h/2} \right) \times$$
$$\times \left\{ \frac{1}{h} \Omega(w(x,t))(T_{h/2} - T_{-h/2}) \left[\beta_6 T_\tau + \beta_7 I + \beta_8 T_{-\tau} \right] w(x,t) \right\}$$

and the following conditions are satisfied:
(a) C is a three point operator such that its second differential approximation $\tilde{C}$ is of the form

$$\tilde{C} = I + \sum_{i=1}^{2} h^i \frac{\partial^i}{\partial x^i} a_i(w) I$$

where $a_i(w), \Omega(w)$ are matrices of 3×3 dimension which may depend not only on w but also on the derivatives of functions $w(x,t)$;
(b) elements of the matrix Ω have magnitude of order $O(\tau) + O(h)$;
(c) $\alpha_i, \beta_i, i = 1, \ldots, 8$ are constants such that

$$\alpha_1 + \alpha_2 = 1, \alpha_3 + \alpha_4 + \alpha_5 = 1, \alpha_6 + \alpha_7 + \alpha_8 = 1,$$
$$\beta_1 + \beta_2 = 1, \beta_3 + \beta_4 + \beta_5 = 1, \beta_6 + \beta_7 + \beta_8 = 1,$$
$$0 \leqslant \alpha_j \leqslant 1, \quad 0 \leqslant \beta_j \leqslant 1, j = 1, 2, \ldots, 8,$$

then the first and second differential approximations of the difference scheme (17) can be presented in divergence form.

Thus, by the "equations of the differential approximation type" either the system (10) or the

68

system of the equations of a difference scheme differential approximation is meant.

In the previous section we have considered the progressive wave type solutions of the system which differs from the equation system (4)-(6) by the presence of terms with artificial viscosity q , see the equation system (10). A **natural** question arises on the applicability of the progressive wave type solutions of the first differential approximation (f.d.a.) equations for the description of the numerical solution properties in the smeared shock wave. This question is closely related to the one on establishing an analogue of the property (C) of the solution to a problem (1)-(2) for systems of equations of the first differential approximation of the difference schemes consistent with the system (4)-(6). Let the Γ -form of the first differential approximation of a conservative difference scheme from the family (17) approximating the system (4) have the form

$$\frac{\partial w}{\partial t} + \frac{\partial \varphi(w)}{\partial x} = -\frac{\tau}{2}\frac{\partial^2 w}{\partial t^2} + \frac{\partial}{\partial x}\left[C(w,h,\tau)\frac{\partial w}{\partial x}\right] \quad (18)$$

where C is a matrix. Then the f.d.a. Π -form may be written in the form

$$\frac{\partial w}{\partial t} + \frac{\partial \varphi(w)}{\partial x} = \frac{\partial}{\partial x}\left[B(w,h,\tau)\frac{\partial w}{\partial x}\right] \quad (19)$$

where

$$B(w,h,\tau) = C(w,h,\tau) - (\tau/2)A^2(w), A = \frac{\partial \varphi}{\partial w}. \quad (20)$$

As in the previous section, let us consider solutions of systems (18),(19) of the form $w = w(y)$, $y = x - Dt$ where $D = \text{const}$. Then the systems (18), (19) are transformed into a system of ordinary differential equations

$$B_1(w,h,\tau)dw/dy = \varphi(w) - D\cdot w + G \quad (21)$$

where G is a constant vector (the integration constant). In the case of the use of the f.d.a. Π --form $B_1 \equiv B$ where the matrix B is determined according to (20), in the case of the f.d.a. Γ --form

$$B_1 = C(w,h,\tau) - (\tau/2)D^2 I , \quad (22)$$

I is the unit matrix. In order that the solution of the system (21) tends to constant values at $y \to \pm\infty$ according to (9) it is necessary (Rozhdestvenskii and Yanenko (1968)) for points W_1 , W_2 be stationary points of the system (21), that turns to relations

$$G = -\left[\varphi(\vec{W}_1) - D\cdot\vec{W}_1\right] = -\left[\varphi(\vec{W}_2) - D\cdot\vec{W}_2\right]$$

from which the Hugoniot conditions immediately follow. We have carried out the numerical integration of the system (21) at the boundary conditions (9) when using two first-order schemes: the Lax scheme and S.K.Godunov's scheme. The numerical results obtained proved to be very close to the finite-difference solution obtained in the (x, t)-computations by the above schemes in the case of shock waves of finite intensity. It turned out that the matrix B_1 defined by the equation (22) is applicable in (21), as well as the matrix B_1 defined according to (20). These results give also experimental evidence of the presence of the property (C) analogue of the solution to a problem (1)-(2) in the grid solutions obtained by using the above considered difference schemes.

Note that the problem of theoretical investigation of the proximity degree of the solutions of finite-difference equations obtained in the (x,t) -computations and of the solutions of the boundary value problem (21),(9) is very difficult because of the presence of a nonlinearity in difference equations (17), in the system (21), and because of the presence of singular points in the system (21) which can be situated also within the shock transition zone (Vorozhtsov and Yanenko (1980)).

In the case of shock waves of weak intensity the numerical solution by the Lax scheme and Godunov's scheme proved to be unsteady. This manifested itself in the increase of the smoothed shock transition width with the increase in t . Thus, the progressive wave type solutions of the f.d.a. equations prove to be inapplicable in this case.

The following theorem proved by Fomin, Vorozhtsov and Yanenko (1976) identifies a class of schemes from the family (17) for which the smeared shock wave centre exists.

<u>Theorem 3.</u> If the constant coefficients α_i, β_i in the scheme (17) and the matrices $a_1(w)$, $\Omega(w)$ satisfy the conditions

(a) $\beta_1(\alpha_3 - \alpha_5) - \beta_2(\alpha_6 - \alpha_8) = 0$;

(b) $hDa_1(w) + \tau(\beta_3 - \beta_5)DA(w) - \frac{h}{2}(\alpha_1 - \alpha_2)A(w) + \Omega(w) =$
$= \beta_9(D)I$,

(c) $\varepsilon = (h^2/2\tau)\left[\beta_1(\alpha_3 + \alpha_5) - \beta_2(\alpha_6 + \alpha_8)\right] - \frac{\tau}{2}(\beta_1 - \beta_2)D^2 + \beta_9 > 0$,

then the unique smeared shock wave centre exists in the progressive wave type solution of the first differential approximation equations of the scheme (17).

With the aid of the above theorems the question is investigated on the existence of the smeared

shock wave centre when using some well-known finite-difference schemes approximating the system of
one-dimensional gas dynamic equations in Eulerian
variables. These results are summarized in Table 2,
see Vorozhtsov, Fomin, Yanenko (1976).

$$\tau/h = \mathrm{const}\ ;\ \tau \leqslant h/(\max_{x}(|u|+c)).\qquad(23)$$

In (23) c is the local sound speed. Note that the
condition $\tau/h = \mathrm{const}$ is only a necessary condition for the existence of the smeared shock wave
centre when employing the Harlow's method, the FLIC
method, the "coarse particles" method, the Lax-Wendroff method, the MacCormack method; the search for
other necessary as well as sufficient conditions
for the above methods proves to be extremely difficult in virtue of the extraordinary complexity of
the approximation viscosity structure for these
schemes, cf., e.g., the papers by Fomin, Vorozhtsov
Yanenko (1979), Belotserkovskii and Davydov (1971),
Lerat and Peyret (1975). At the same time our numerous computational experiments with the above
schemes indicate the existence of the finite-difference shock wave centre under conditions (23). Moreover, the shock front position is determined with
a high accuracy ($\leqslant 0.2h$), cf. Vorozhtsov (1976),
Vorozhtsov and Yanenko (1980). All these experimental facts need theoretical background. As illustra-
tions to the Table 2 in the Figures 3-6 the computational examples are presented for problems on the
motion of steady (Figs 3-5) and unsteady (Fig.6)
shock waves. When the condition $\tau/h = \mathrm{const}$ was satisfied the abscissa of the point of intersection
of the solution graphs at different τ and h coin-
cided with the exact position of a shock front
within an error $\leqslant 0.2h$.

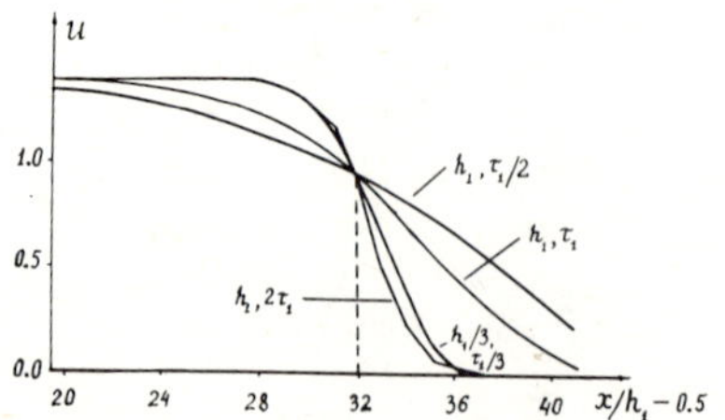

Figure 3. Lax scheme.

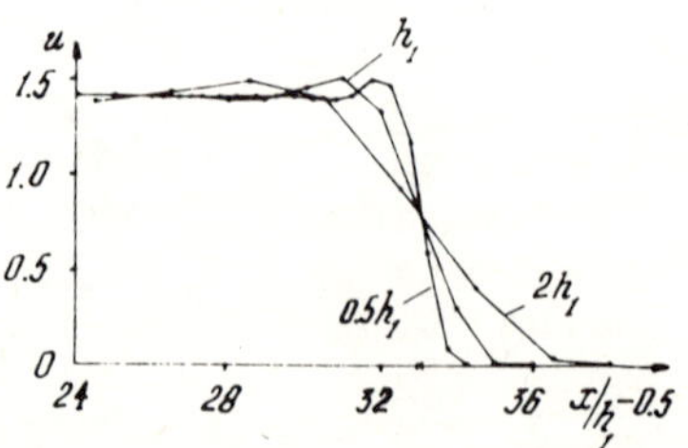

Figure 4. Harlow's
method, $\tau/h = \mathrm{const}$.

Table 2

Method	Order of approxi-mation	Does the sme-ared shock wave centre exist?	Method	Order of approxima-tion	Does the sme-ared shock wave centre exist?
Lax (1954)	$O(\tau)+O\left(\frac{h^2}{2\tau}\right)$	Yes, at $\tau/h < 1/D$	Lax and Wendroff (1960)	$O(\tau^2)+O(h^2)$	Yes, at (23)
Rusanov (1961)	$O(\tau)+O(h)$	Yes, at (23)	Burstein (1964)	$O(\tau^2)+O(h^2)$	Yes
Harlow (1964)	$O(\tau)+O(h)$	Yes, at (23)	Two-step Lax-Wendroff, Rich-tmyer and Mor-ton (1967)	$O(\tau^2)+O(h^2)$	Yes, at (23)
FLIC Gentry et al (1966)	$O(\tau)+O(h)$	Yes, at (23)	MacCormack (1969)	$O(\tau^2)+O(h^2)$	Yes, at (23)
"coarse particles" method, Belotserkov-skii and Davydov (1971)	$O(\tau)+O(h)$	Yes, at (23)	MacCormack with a scalar quadratic viscosity	$O(\tau^2)+O(h^2)$	Yes

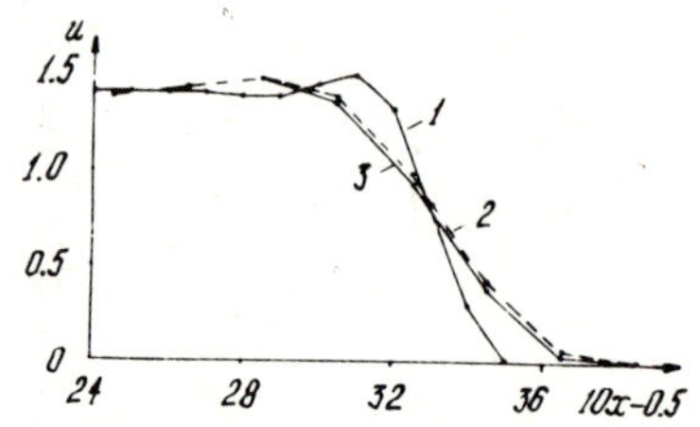

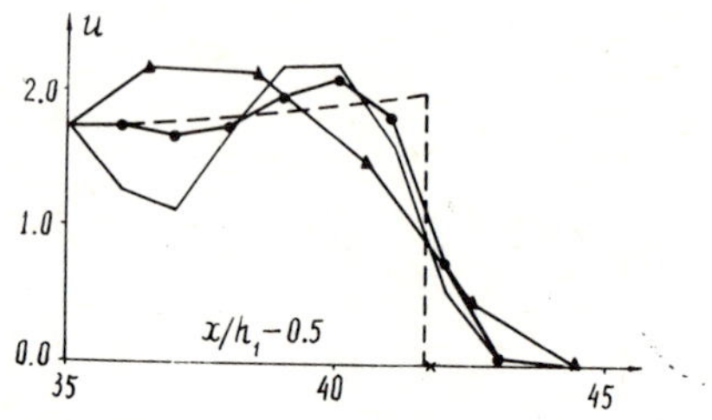

Figure 5   Figure 6

Figure 5. Harlow's method, $\tau/h \neq const$.
Fig.6. MacCormack scheme. The problem on the self-
-similar motion in an inhomogeneous atmosphere caused by an impulsive plane impact (Zel'dovich and
Raiser (1966)). (− − −) exact solution, $t = 100\tau_1$;
(———) numerical solution, $\tau = \tau_1$, $h = h_1$;
(−•−•−) numerical solution, $\tau = 2\tau_1$, $h = h_1$;
(−▲−▲−) numerical solution, $\tau = 2\tau_1$, $h = 2h_1$. The
cross on the x axis shows the abscissa of the finite-difference shock wave centre obtained at $\tau/h = const$.

The next theorem (Vorozhtsov and Yanenko (1980))
generalizes the technique of an approximate shock
front localization by maximum of a scalar artificial
viscosity   described in the previous section for
the case of the scheme viscosity.
   Theorem 4. If the matrix $A(w) = \partial\varphi/\partial w$ has m real
eigenvalues $\lambda_1(w) < ... < \lambda_m(w)$ where m is the number
of the equations in the system (4) and
(a) $\lambda_1(w) < \lambda_2(w) < ... < \lambda_m(w)$ within the smeared
shock wave zone;
(b) the functions $\lambda_j(w), j = 1,...,m$ are continuous and
monotone in the smeared shock wave zone;
(c) there exists such a number k , $1 \leqslant k \leqslant m$ ·, that
$$\lambda_k(\vec{W}_1) > D > \lambda_k(\vec{W}_2),$$
then at the maximum point of the quantities
$$|\partial Q_i/\partial x|$$
where
$$Q = B(w,h,\tau)\partial w/\partial x \ , \ Q = (Q_1,...,Q_m) \ , \ i = 1,...,m \ ,$$
and the matrix $B(w,h,\tau)$ enters the f.d.a. (19)
the relation
$$\lambda_k(w) = D$$
takes place.
By using Rusanov's results (1973) and Zemplén's
theorem (Rozhdestvenskii and Yanenko (1968)) the

following property

$$\max_x \| B\, \partial w/\partial x \| = \| B\partial w/\partial x \|_{x=x_f}$$

is easily deduced where x_f is the exact shock
front abscissa, see also Yanenko, Kovenya et al
(1979). Note that for the first time a possibility
of direct application of the first differential approximation for purposes of shock front localization was shown by numerical experiments by Vorozhtsov (1972). In particular, algorithms for the shock
wave localization were realized based on the determination of the shock front point as a point of ma-
ximum of the quantity $\| B\partial w/\partial x \|$ where the matrix
B was computed from the f.d.a. p-form of some
first-order difference schemes.

The generalization of the above theory of differential analysers of one-dimensional shock waves
for the multidimensional case is associated with
necessity of allowing for some specific factors inherent in multidimensional computations on the Eulerian grid. As it was pointed out by Roache (1976)
in the case of a shock located at an angle to the
computing mesh lines the Rankine-Hugoniot conditions are not necessarily exactly satisfied, since
in virtue of the shock smearing the gradient of the
normal momentum flux can be spread in the direction
tangential to the shock. Therefore one cannot automatically apply the one-dimensional theory described above to the considerations of the difference
solution properties along the beams orthogonal to
the shock front, even in the case when, according
to the exact solution, the tangential velocity component behind and before the front of the shock is
equal to zero at chosen points. Until now there are
no investigations of the structure of smeared two-
dimensional curvilinear shock waves within the framework of the first differential approximation method (Shokin (1979)). Despite this fact, a series
of algorithms for two-dimensional shock wave localization was developed for practical computations
(Vorozhtsov (1972), Yanenko, Kroshko et al (1976),
Yanenko, Fomin, Vorozhtsov (1978), Fomin, Ermolin
et al (1979)). These algorithms have taken into account the recommendations obtained earlier from the
one-dimensional theory and were successfully applied by the authors of the papers cited to the numerical computations of some two-dimensional high-
-velocity impact problems.

A comparatively more simple problem from the
point of view of theoretical considerations proved
to be the problem on the determination of the properties of curvilinear shock waves smearing in
shock-capturing computations depending upon the

shock wave orientation with respect to the Eulerian
rectangular computing mesh lines. Information of
such a kind is important for the numerical investigations of gas dynamic problems involving interactions of different discontinuities either with one
other, or with rigid walls. In the paper by Fomin,
Vorozhtsov and Yanenko (1979) it is shown by the
use of the first differential approximation method
that in the particle-in-cell method computations
(Harlow (1964)) the width of a smeared cylindrical
shock wave zone reaches its minimum at some non-zero inclination angle of the radius-vector of a
shock front point with respect to the x -axis. Note that the results of the theoretical investigation carried out are applicable also to the FLIC
method (Gentry, Martin, Daly (1966)), since the difference scheme of the FLIC method was employed by
Fomin, Vorozhtsov and Yanenko (1979) for the theoretical analysis.

THEORY OF DIFFERENTIAL ANALYSERS OF CONTACT DISCONTINUITIES

It was already observed earlier (cf., e.g., Rusanov (1969), Harten (1977)) that the contact discontinuity is approximated by some transition zone
in which the solution is smoothed or "smeared" like
the shock wave when using difference schemes in Eulerian variables. However, it turned out that the
size of this transition zone, unlike the smeared
shock wave zone, increases proportionally to $t^{1/(z+1)}$
as t increases, where z is the order of approximation of the finite-difference scheme. In this connection the questions of development of algorithms
for contact discontinuities localization in the computational domain by the shock-capturing computation results of gas dynamics problems are also of
present interest for contact discontinuities.
Following the paper by Yanenko and Vorozcov
(1979) let us present the method of investigation
of the numerical solution properties in the neighbourhood of a contact discontinuity. With the aid
of this method it then will not be difficult to derive also algorithms of differential analysers of
contact discontinuities. For the system of Equations (4)-(5),(7) let us consider the following
Riemann problem:

$$t=0 , \ w(x,t)=u_0, \ p(x,t)=p_0, \ \varrho(x,t)=\begin{cases} \varrho_1, & x<x_0 \\ \varrho_2, & x>x_0 \end{cases} \quad (24)$$

where ω_0 , p_0 , ϱ_1 , ϱ_2 , x_0 are constants and $\varrho_1 \neq \varrho_2$. The exact solution of this problem has the form

$$t \geq 0, \ \omega(x,t)=\omega_0, \ p(x,t)=p_0, \quad \varrho(x,t)=\begin{cases} \varrho_1, & x < x_0+\omega_0 t \\ \varrho_2, & x > x_0+\omega_0 t \end{cases} \quad (25)$$

and thus contains only the contact discontinuity. For the definiteness the case $\omega_0 > 0$ is considered below, i.e. the contact discontinuity moves from the left to the right. The function $f(p,\varrho)$ entering the equation of state (7) is supposed to be three times continuously differentiable at $p \geq 0$, $\varrho \geq 0$. As in the previous section we use the means of differential approximations of a difference scheme. Let us consider a class of difference schemes approximating the system (4)-(5) with an order of accuracy $O(h^2)+O(\tau^2)$, $1 \leq z \leq 2$, whose f. d.a. Γ-form is as follows:

$$\Gamma w \equiv \frac{\partial w}{\partial t} + \frac{\partial \varphi(w)}{\partial x} - \qquad (26)$$

$$- \sum_{\substack{i,j \\ i+j=z}} h^i \tau^j F_{ij}\left(w, \frac{\partial w}{\partial x}, \frac{\partial w}{\partial t}, \ldots, \frac{\partial^{z+1} w}{\partial x^k \partial t^{z+1-k}}\right) = 0$$

where

$$w = \left\{ w_1, w_2, w_3 \right\}, \ w_1 \equiv \varrho, \ w_j = f_j(\varrho, \omega, p, \mathcal{E}(p,\varrho)), j=2,3,$$

$$F_{ij} = \left\{ F_{ij1}, F_{ij2}, F_{ij3} \right\}, \ 0 \leq K \leq z+1,$$

h is the step of uniform computing mesh on the x-axis, τ is time step. Let us express the derivatives $\partial^m w/\partial x^k \partial t^{m-k}$, $m-K > 0$, which enter in F_{ij}, in terms of the derivatives with respect to x making use of the algorithm described in the book by Shokin (1979), for example.

$$\partial^m w/\partial x^k \partial t^{m-k} = f_{m,K}(w, \partial w/\partial x, \ldots, \partial^m w/\partial x^m).$$

As a result we obtain from (26) the f.d.a. Π -form

$$\Pi w \equiv \partial w/\partial t + \partial \varphi(w)/\partial x - \qquad (27)$$

$$- \sum_{\substack{i,j \\ i+j=z}} h^i \tau^j \mathcal{F}_{ij}(w, \partial w/\partial x, \ldots, \partial^{z+1} w/\partial x^{z+1}) = 0$$

where

$$\mathcal{F}_{ij}\left(w, \frac{\partial w}{\partial x}, \ldots, \frac{\partial^{z+1} w}{\partial x^{z+1}}\right) = F_{ij}\left(w, \frac{\partial w}{\partial x}, f_{1,0}\left(w, \frac{\partial w}{\partial x}\right), \ldots,\right.$$

$$f_{z+1,K}\left(w, \frac{\partial w}{\partial x}, \ldots, \frac{\partial^{z+1} w}{\partial x^{z+1}}\right), \quad \mathcal{F}_{ij} = \{\mathcal{F}_{ij1}, \mathcal{F}_{ij2}, \mathcal{F}_{ij3}\}.$$

__Definition 3.__ Let us call the quasilinearization operation L of the f.d.a. (26) or (27) the operation

$$\Gamma L w = 0, \quad \Pi L w = 0$$

where

$$L w = \{w_1, L w_2, L w_3\}, \quad L w_j = f_j(\varrho, u_0, p_0, \varepsilon(p_0, \varrho)),$$
$$j = 2, 3$$

and u_0, p_0 are constant velocity and pressure values entering in (24). Let

$$U = \left\{1, u_0, 0.5\, u_0^2 + f(p_0, \varrho) + \varrho \frac{\partial f}{\partial \varrho}(p_0, \varrho)\right\}. \qquad (28)$$

__Definition 4.__ The first differential approximation (26) will be called K-consistent if

$$\Gamma L w = U \cdot \left[\frac{\partial \varrho}{\partial t} + u_0 \frac{\partial \varrho}{\partial x} - \right.$$
$$\left. - \sum_{\substack{i,j \\ i+j=z}} h^i \tau^j F_{ij1}\left(L w, \frac{\partial L w}{\partial x}, \frac{\partial L w}{\partial t}, \ldots, \frac{\partial^{z+1} L w}{\partial x^K \partial t^{z+1-K}}\right)\right] = 0,$$

that is the left hand sides of the equations of a system $\Gamma L w = 0$ differ from each other only by a scalar multiplier. K-consistence of the f.d.a. Π-form (27) is defined analogously. Note that the initial equation system (4),(5) possesses the K-consistence property. It can be easily proved that from the K-consistence of the f.d.a. Γ-form (26) it follows the K-consistence of the f.d.a. Π-form (27). From the definition 4 it follows that the investigation of the K-consistent first differential approximation (27) subject to initial conditions (24) reduces to the investigation of a single equation

$$\frac{\partial \varrho}{\partial t} + u_0 \frac{\partial \varrho}{\partial x} = \sum_{\substack{i,j \\ i+j=z}} h^i \tau^j F_{ij1}\left(L w, \frac{\partial L w}{\partial x}, \ldots, \frac{\partial^{z+1} L w}{\partial x^{z+1}}\right). \qquad (29)$$

The equation (29) is studied below at the initial condition

$$\varrho(x,0) = \begin{cases} \varrho_1, & x < x_0 \\ \varrho_2, & x > x_0 \end{cases} \qquad (30)$$

and boundary conditions

$$\lim_{x \to +\infty} \frac{\partial^{m-1} \varrho}{\partial x^{m-1}} = \lim_{x \to -\infty} \frac{\partial^{m-1} \varrho}{\partial x^{m-1}} = 0, \quad m = 2, \ldots, z \qquad (31)$$

which are imposed for $z \geqslant 2$.

__Definition 5.__ Let us call the contact strip G_{KD} an open simply connected domain in the (x, t)

plane which contains a contact discontinuity and throughout which the following conditions are satisfied for each fixed $t > 0$:

   a) $\operatorname{sign} \partial \varrho / \partial x = \operatorname{sign}(\varrho_2 - \varrho_1)$;

   b) $\varrho_1 (1-\delta) > \varrho > \varrho_2 (1+\delta)$ if $\varrho_1 > \varrho_2$;

       $\varrho_1 (1+\delta) < \varrho < \varrho_2 (1-\delta)$, if $\varrho_1 < \varrho_2$

where δ is a small positive number and $\varrho(x,t)$ is a solution of the problem (29)-(31).

<u>Definition 6.</u> By a central line of the contact strip G_{KD} we mean a line $L_{KD} \subset G_{KD}$ on which the value of the function $\varrho(x,t)$, where ϱ is a solution of (29)-(31), is independent of the mesh parameters h and τ .

Let us denote by $w_h(x,t,h,\tau)$ a vector-valued function which coincides at the mesh nodes of the (x,t)-plane with the solution of the finite-difference equations approximating the problem (4)-(5) (7),(24) and which is defined at the remaining (x,t) points by interpolation.

<u>Definition 7.</u> By the contact strip G_K we mean an open simply connected domain in the (x,t) plane which contains a contact discontinuity and throughout which

$$\operatorname{sign}\left[\varrho_h (x+h,t,h,\tau) - \varrho_h (x-h,t,h,\tau)\right] =$$
$$= \operatorname{sign}(\varrho_2 - \varrho_1) .$$

<u>Definition 8.</u> By a central line of the contact strip G_K we mean a line $L_K \subset G_K$ which, for each fixed $t > 0$, is the geometric locus of points of intersection of the profiles $\varrho_h(x,t,h_1,\tau_1)$ and $\varrho_h(x,t,h_2,\tau_2)$, where $|h_1 - h_2| + |\tau_1 - \tau_2| \neq 0$.

<u>Definition 9.</u> By a differential analyser of a contact boundary we mean an algorithm which makes it possible to find the coordinates of the points of the line L_K in the computational domain in accordance with the results of a straight-through calculation of gas dynamic problems.

<u>Definition 10.</u> Let us call the quantity

$$X(t,h,\tau,u_o,p_o) = |\varrho_1 - \varrho_2| \Big/ \max_{x,x \in G_{KD}} |\partial \varrho / \partial x| \tag{32}$$

the width after Prandtl of the contact strip where $\varrho(x,t)$ is the solution of the problem (29)-(31).

<u>Analysis of the schemes in the presence of the f.d.a. K-consistence</u>

<u>Theorem 5</u> (Yanenko and Vorožcov (1979)). If
i) the first differential approximation of a difference scheme of rth order of accuracy, $1 \leqslant z \leqslant 2$, is K-consistent;
ii) Equation (29) has the form

$$\partial \varrho / \partial t + u_0 \partial \varrho / \partial x = (-1)^{z+1} \mu (h, \tau, u_0, p_0) \partial^{z+1} \varrho / \partial x^{z+1} ; \qquad (33)$$

iii) $\mu (h, \tau, u_0, p_0) > 0$ for $u_0 > 0$, $p_0 > 0$ and for the values of h, τ satisfying the difference scheme stability condition;

iv) $\tau = 0(h)$,

then there exists a unique central line L_{KD} of the contact strip, and this line coincides with the trajectory of the contact discontinuity.

A proof of this theorem consists in obtaining an exact solution of the problem (33),(30),(31) and a subsequent study of its properties; moreover, in the case $z = 1$

$$\varrho (x, t) = 0.5 (\varrho_1 + \varrho_2) + 0.5 (\varrho_2 - \varrho_1) \Phi (\xi (x, t)),$$

for $z = 2$

$$\varrho (x, t) = (2 \varrho_2 + \varrho_1)/3 + (\varrho_2 - \varrho_1) \int_0^{\xi} Ai (\lambda) d\lambda \qquad (34)$$

where

$$\xi = (x - x_0 - u_0 t)/[(5 - z) \mu t]^{1/(z+1)} ,$$

$\Phi (\xi)$ is the probability integral (Tihonov and Samarskii (1977)), and $Ai (\lambda)$ is the Airy function (Smirnov (1969)). For practical calculations, using schemes of first and second orders of accuracy satisfying the conditions of the Theorem 5, we can derive the following three algorithms for a contact discontinuity differential analyzer.

Algorithm I. Localization of the line L_K directly on the basis of Definition 8.

Algorithm II. Determination for each $t > 0$ of the point of the line L_K as the point of $max |\partial^2 \varrho / \partial x^2|$ for $x \in G_K$.

Algorithm III. On the line L_{KD} we have the property $\varrho = 1/2 (\varrho_1 + \varrho_2)$ for $z = 1$ and $\varrho = (2\varrho_2 + \varrho_1)/3$ for $z = 2$. Therefore, for known values of ϱ_1 and ϱ_2 we can find an approximate value of the abscissa of a point of the line L_K through inverting the interpolation function $\varrho_h (x, t, h, \tau)$.

In Table 3 we summarize the results of our studies for a series of difference schemes of first and second orders of accuracy. We have shown that satisfaction of the equation

$$(d^2 / d\varrho^2) [\varrho \cdot f(p_0, \varrho)] = 0, \qquad (35)$$

where $f(p, \varrho)$ occurs in the equation of state (7), is a necessary condition for K-consistency of first differential approximations of difference schemes belonging to a broad class, including the difference schemes considered in Table 3. We remark that there is a deep relationship existing between the dispersion properties and the Gibbs phenomenon,

established by Chin (1975) for a semidiscrete approximation of the wave equation, and the properties
of the solution (34).

The results of numerous calculations of the problem (4)-(5),(7),(24), using the schemes listed in
Table 3, agree qualitatively and quantitatively
with the theory presented above; in particular,
they confirm the practical applicability of the Algorithms I-III for localization of a contact discontinuity. Moreover, the position of the contact
discontinuity is determined with an error $< h$, in
spite of the fact that the width of the contact
strip increases in proportion to $t^{1/(2+1)}$. On the
central line L_K of the contact strip the numerical
solution for the density ϱ_h depends "most weakly"
on h and τ . An example of computation of the problem (4)-(5),(7),(24) is presented in the Fig.7, in
(7) $f(p,\varrho) = p/(\varrho(\gamma-1))$, $\gamma = 2$. Solid lines in
the Figure 7
show the numeri-
cal solution ob-
tained by the
one-step Lax-We-
ndroff scheme.
The point of in-
tersection of
the density pro-
files obtained
by using the ab-
ove mentioned
difference sche-
me at various h

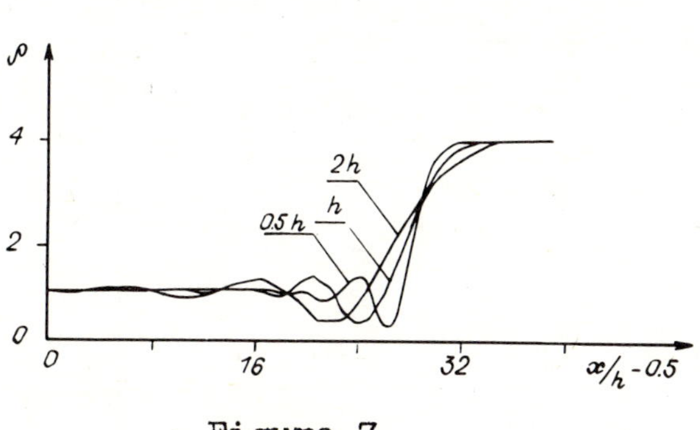

Figure 7

lies at a distance from the exact position of a
contact discontinuity (cf. solution (25)) not ex-
ceeding h .

Earlier Rusanov (1969) has pointed out the form
of the dependence of the thickness of the contact
strip on the time t . The theory presented above
gives a constructive way to determine the dependence of the contact strip thickness on the difference scheme parameters h and τ and on the time t for
each specific difference scheme (see Table 3). From
the study presented here an important conclusion
can be drawn concerning the fact that the formulation of a difference scheme preserving a constant
contact strip width in a gas is connected, first of
all, with the formulation of such a scheme for the
scalar linear equation

$$\partial\varrho/\partial t + u_o\,\partial\varrho/\partial x = 0.$$

Table 3

Method	Order of approxima-tion	cf. Equation (33)	Width X (32)	Practical criterion for determination of the line L_K
Lax (1954)	$O(\tau)+O\left(\frac{h^2}{2\tau}\right)$	$\frac{h^2}{2\tau}-\frac{\tau}{2}u_0^2$	$2(\pi \eta t)^{1/2}$	Use $\max\left|\partial\varrho/\partial x\right|$ for $\mathrm{sign}(\partial\varrho/\partial x)=\mathrm{sign}(\varrho_2-\varrho_1)$
"breakdown of dis-continuity", Godu-nov et al (1961)	$O(\tau)+O(h)$	$\left|u_0\right|\left(\frac{h}{2}-\frac{\tau}{2}\left|u_0\right|\right)$	$2(\pi \eta t)^{1/2}$	
FLIC, Gentry et al (1966)	$O(\tau)+O(h)$	$\left|u_0\right|\left(\frac{h}{2}-\frac{\tau}{2}\left|u_0\right|\right)$	$2(\pi \eta t)^{1/2}$	
"coarse particles" Belotserkovskii et al	$O(\tau)+O(h)$	$\left|u_0\right|\left(\frac{h}{2}-\frac{\tau}{2}\left|u_0\right|\right)$	$2(\pi \eta t)^{1/2}$	
Lax-Wendroff one-step, Lax and Wendroff (1960)	$O(\tau^2)+O(h^2)$	$\frac{\left|u_0\right|}{6}(h^2-\tau^2 u_0^2)$	$\dfrac{(3\eta t)^{1/3}}{Ai(\xi_0)}$	Use $\max\left|\partial^2\varrho/\partial x^2\right|$ for $\mathrm{sign}(\partial\varrho/\partial x)=\mathrm{sign}(\varrho_2-\varrho_1)$
Lax-Wendroff two-step, Richtmyer and Morton (1967)	$O(\tau^2)+O(h^2)$	$\frac{\left|u_0\right|}{6}(h^2-\tau^2 u_0^2)$	$\xi_0\cong-1{,}01879$ $Ai(\xi_0)\cong$	
MacCormack (1969)	$O(\tau^2)+O(h^2)$	$\frac{\left|u_0\right|}{6}(h^2-\tau^2 u_0^2)$	$0{.}535657$	

K-consistence property of the difference schemes of gas dynamics

The K-consistence property was defined above within the framework of the method of first differential approximation of a difference scheme. In the course of further investigations it became clear that it is possible to introduce an analogue of the K-consistence property directly for finite-difference schemes which approximate the system (4),(5). Suppose that an explicit difference scheme approximating the system (4),(5) is representable in the form

$$\varrho^{n+1} = H_1(u^n, p^n, \varrho^n, \mathcal{E}^n, T_1, h, \tau);\tag{36}$$

$$(\varrho u)^{n+1} = H_2(u^n, p^n, \varrho^n, \mathcal{E}^n, T_1, h, \tau);\tag{37}$$

$$(\varrho E)^{n+1} = H_3(u^n, p^n, \varrho^n, \mathcal{E}^n, T_1, h, \tau)\tag{38}$$

where $u^n = u(x, t^n)$ etc., $t^n = n \cdot \tau$, T_1 is the translation operator along the x-axis, that is $T_1 u(x,t) = u(x+h, t)$, $T_1^m u(x,t) = u(x+mh, t)$.

Definition 11. Let us call the quasilinearization operation L of the difference expression
$$H(u^n, p^n, \varrho^n, \mathcal{E}^n, T_1, h, \tau)$$
the following operation:
$$LH \equiv H(u_0, p_0, \varrho^n, f(p_0, \varrho^n), T_1, h_1, \tau)\tag{39}$$

where the function $f(p, \varrho)$ enters the equation of state (7) and u_0, p_0 are constant values of the velocity and of the pressure entering in (24).

Definition 12. We shall say that the difference momentum equation (37) is K-consistent with the difference continuity equation (36), if
$$LH_2 = u_0 LH_1.\tag{40}$$

Definition 13. We call the difference scheme (36)-(38) approximating the system of equations (4)-(6) K-consistent if:
a) the momentum equation (37) is K-consistent with the difference continuity equation;
b) $\quad F(LH_1, (LH_3 - 0.5(LH_2)^2/LH_1)/LH_1) \equiv p_0 \tag{41}$
where the function $F(\varrho, \mathcal{E})$ enters the equation of state (6).

Example 1. Applying the quasilinearization operation L to the difference equations of the FLIC method (Gentry et al (1966)) we obtain in the case $u_0 > 0$:

$$\varrho^{n+1} = \varrho^n - \varkappa(I - T_{-1})\varrho^n;$$

$$(\varrho u)^{n+1} = \varrho^n u_0 - u_0 \varkappa(I - T_{-1})\varrho^n;\tag{42}$$

$$(\rho E)^{n+1} = \rho^n \cdot f(p_0, \rho^n) + \rho^n \frac{u_0^2}{2} -$$

$$- \varkappa (I - T_{-1})\left((\rho\varepsilon)^n + \frac{u_0^2}{2}\rho^n\right)$$

where $\varkappa = u_0\tau/h$, I is the identity operator.
Let, for the definiteness, the function $F(\rho,\varepsilon)$ in
(6) have the form

$$F(\rho,\varepsilon) = \alpha\rho\varepsilon + B_1 \cdot (\rho - \rho_0) \qquad (43)$$

where α, B_1, ρ_0 are some positive constants. Then
the function $f(p,\rho)$ entering in (7) has obviously
the form

$$f(p,\rho) = \left[p - B_1(\rho - \rho_0)\right]/\alpha\rho ,$$

and it is easy to be convinced that the substitution of the formulas (42) into (43) yields the
identity (41). It can be seen from this simple example that the presence of the K-consistence property of the difference scheme (36)-(38) depends
on the form of the equation of state (6) or (7).

Theorem 6. A necessary condition for K-consis-
tence of the difference momentum equation in the
one-step Lax-Wendroff scheme is satisfaction of the
equation $L R_{i+1/2}^n = 0$ where

$$R_{i+1/2}^n = \frac{\partial F(\rho_{i+1/2}^n, \varepsilon_{i+1/2}^n)}{\partial \rho}(T_1 - I)\rho_i^n +$$

$$+ \frac{\partial F(\rho_{i+1/2}^n, \varepsilon_{i+1/2}^n)}{\partial \varepsilon}(T_1 - I)\varepsilon_i^n , \qquad (44)$$

$$\rho_{i+1/2}^n = (I + T_1)\rho_i^n/2 , \qquad \varepsilon_{i+1/2}^n = (I + T_1)\varepsilon_i^n/2 ,$$

and the subscript i denotes the node number on the
axis.

It is easy to check that if the function $F(\rho,\varepsilon)$
entering the equation of state (6) has the form
(43), then $L R_{i+1/2}^n = 0$. However, if the equation of
state has the form

$$p = \alpha\rho\varepsilon + \sum_{K=1}^{m} B_K(\rho - \rho_0)^K \qquad (45)$$

where α, B_K, ρ_0 are constants, then at $B_3 \neq 0$, $m=3$
the relationship $L R_{i+1/2}^n$ does already not take
place. For purposes of comparison let us investigate the K-consistence of the first differential
approximation of the one-step Lax-Wendroff scheme.
The f.d.a. Γ-form of this scheme has the form
(Fomin, Vorozhtsov and Yanenko (1976))

$$\frac{\partial w}{\partial t} + \frac{\partial \varphi(w)}{\partial x} = -\frac{\tau^2}{6}\frac{\partial^3 w}{\partial t^3} - \frac{h^2}{6}\frac{\partial^3 \varphi}{\partial x^3} . \qquad (46)$$

Applying the quasilinearization operation L to both sides of the equation (46) we obtain the system

$$U\left(\frac{\partial \varrho}{\partial t} + w_0 \frac{\partial \varrho}{\partial x}\right) = -\frac{\tau^2}{6}\frac{\partial^2}{\partial t^2}\left(U\frac{\partial \varrho}{\partial t}\right) - \frac{h^2}{6}w_0\frac{\partial^2}{\partial x^2}\left(U\frac{\partial \varrho}{\partial x}\right) \qquad (47)$$

where the vector U is defined by (28). It is easy to see from (47) that the K-consistence of the f.d.a. (46) takes place if the equation (35) is satisfied. In this case it is easy to get the f.d. a. Π-form from the f.d.a. (47):

$$U(\partial\varrho/\partial t + w_0\, \partial\varrho/\partial x) = -\mu U \partial^3\varrho/\partial x^3 \qquad (48)$$

where

$$\mu(h,\tau,u_0) = (h^2/6)u_0 - (\tau^2/6)u_0^3 . \qquad (49)$$

From (48) and from the definition (28) of the vector U it follows that in the equation system (46) the f.d.a. of the momentum equation is K-consistent with the f.d.a. of the continuity equation independently of the form of the equation of state. This contradicts the result formulated above in the form of the Theorem 6. However, there is in fact no contradiction, because, by the definition (Shokin (1979)), the first differential approximation takes into account only the terms with the least powers of the grid parameters h, τ contained in the expansions of difference operators entering in the finite-difference scheme. Consider a somewhat more general equation of state than (45), namely,

$$p = \alpha\varrho\varepsilon + b(\varrho) \qquad (50)$$

where $b(\varrho)$ is supposed to be four times continuously differentiable function. Let us write also the next group of terms with respect to the order of smallness alongside with the terms of the first differential approximation of the momentum equation in the case of the one-step Lax-Wendroff scheme. Applying the quasilinearization operation L to the equation obtained, we get the following equation

$$w_0\frac{\partial\varrho}{\partial t} + w_0^2\frac{\partial\varrho}{\partial x} = -\frac{\tau^2}{6}w_0\frac{\partial^3\varrho}{\partial t^3} - \frac{h^2}{6}u_0^2\frac{\partial^3\varrho}{\partial x^3} -$$

$$\qquad (51)$$

$$-\frac{\tau^3}{24}u_0\frac{\partial^4\varrho}{\partial t^4} + \frac{\tau h^2}{24}u_0^3\frac{\partial^4\varrho}{\partial x^4} +$$

$$+\frac{\tau h^2 u_0}{2}\left[-\frac{1}{4}\frac{d^3 b}{d\varrho^3}\cdot\left(\frac{\partial\varrho}{\partial x}\right)^2\frac{\partial^2\varrho}{\partial x^2} - \frac{1}{24}\frac{d^4 b}{d\varrho^4}\cdot\left(\frac{\partial\varrho}{\partial x}\right)^4\right] .$$

The result of application of the quasilinearization operation L to the corresponding differential approximation of the continuity equation is as follows:

$$\frac{\partial \varrho}{\partial t} + u_0 \frac{\partial \varrho}{\partial x} = -\frac{\tau^2}{6}\frac{\partial^3 \varrho}{\partial t^3} - \frac{h^2}{6} u_0 \frac{\partial^3 \varrho}{\partial x^3} - \frac{\tau^3}{24}\frac{\partial^4 \varrho}{\partial t^4} + \frac{\tau h^2}{24} u_0^2 \frac{\partial^4 \varrho}{\partial x^4} . \tag{52}$$

It follows from (51) that if $\ b'''(\varrho) \neq 0$, then the differential approximation (51) is not κ-consistent with the equation (52). Thus, the equation (51) "signals" the violation of the κ-consistence of the difference momentum equation.

Methods for suppression of oscillations caused by the κ-inconsistence

The equation (35) was already pointed out above as one of the necessary conditions for κ-consistence. Let us now consider at the example of the Lax scheme the case when the equation (35) is not satisfied. As a result of the quasilinearization of the f.d.a. of the Lax scheme we obtain the following equation system:

$$\frac{\partial \varrho}{\partial t} + u_0 \frac{\partial \varrho}{\partial x} = \wp \ \frac{\partial^2 \varrho}{\partial x^2} ; \tag{53}$$

$$u_0 \left(\frac{\partial \varrho}{\partial t} + u_0 \frac{\partial \varrho}{\partial x} \right) = \wp \, u_0 \frac{\partial^2 \varrho}{\partial x^2} ; \tag{54}$$

$$\left[0.5\, u_0^2 + f(p_0 \varrho) + \varrho \frac{\partial f}{\partial \varrho}(p_0, \varrho) \right] \left(\frac{\partial \varrho}{\partial t} + u_0 \frac{\partial \varrho}{\partial x} \right) =$$

$$= \wp \frac{\partial}{\partial x} \left\{ \left[0.5 u_0^2 + f(p_0, \varrho) + \varrho \frac{\partial f}{\partial \varrho}(p_0, \varrho) \right] \frac{\partial \varrho}{\partial x} \right\} . \tag{55}$$

Solving the equation (53) at the initial condition (30), we find a function $\varrho(x,t)$ according to Theorem 5, and $\ \partial\varrho/\partial x \neq 0, \ \partial\varrho/\partial t \neq 0$. Let us now multiply both sides of the equation (53) by the quantity $-[0.5\, u_0^2 + f(p_0, \varrho) + \varrho\frac{\partial f}{\partial \varrho}(p_0, \varrho)]$ and add the result to the both sides of the equation (55). We obtain the formula $(\partial\varrho/\partial x)^2 = 0$ which contradicts the above inequality $\partial\varrho/\partial x \neq 0$. It is obvious that this contradiction is a consequence of our assumption made in the process of the quasilinearization operation L that in the contact strip $p(x,t) = const,$ $u(x,t) = const$. Thus, to avoid the contradiction indicated in the case considered it is necessary to suppose that the pressure and the velocity will not be constant in the contact strip in the absence of κ-consistence but they will be some functions of x , t . This means the nonmonotonicity of the numerical solution in the contact strip in the absence of κ-consistence. This nonmonotonicity will be displayed even in the case when the difference scheme employed is known as the "monotone" one.

Let

$$B(p_0, \varrho) \equiv (d^2/d\varrho^2)\left[\varrho \cdot f(p_0, \varrho) \right] \tag{56}$$

in accordance with (35). The equation of state is often taken not in the form (7), but in the form (6). Let us find the expression for $B(p_0, \varrho)$ in the case of application of the equation of state (6). Let us differentiate the both sides of the equation (6) with respect to ϱ assuming that $p = \mathrm{const}$. We obtain

$$\partial f / \partial \varrho = -(\partial F / \partial \varrho)/(\partial F / \partial \varepsilon). \tag{57}$$

With regard for (57) we can rewrite (56) in the form

$$B(p_0, \varrho) = -\frac{\partial F}{\partial \varrho}\Big/\frac{\partial F}{\partial \varepsilon} - \frac{\partial}{\partial \varrho}\left(\varrho\,\frac{\partial F}{\partial \varrho}\Big/\frac{\partial F}{\partial \varepsilon}\right). \tag{58}$$

Assuming that the argument ε in (6) depends in its turn on $\varrho: \varepsilon = f(p_0, \varrho)$, let us calculate the second derivatives entering (58):

$$\frac{d}{d\varrho}\left(\frac{\partial F}{\partial \varrho}\right) = \frac{\partial^2 F}{\partial \varrho^2} - \frac{\partial^2 F}{\partial \varrho \partial \varepsilon}\cdot\left(\frac{\partial F}{\partial \varrho}\Big/\frac{\partial F}{\partial \varepsilon}\right), \tag{59}$$

$$\frac{d}{d\varrho}\left(\frac{\partial F}{\partial \varepsilon}\right) = \frac{\partial^2 F}{\partial \varepsilon \partial \varrho} - \frac{\partial^2 F}{\partial \varepsilon^2}\cdot\left(\frac{\partial F}{\partial \varrho}\Big/\frac{\partial F}{\partial \varepsilon}\right).$$

Substituting formulas (59) into (58) we obtain finally:

$$B(p_0, \varrho) = -2 F_\varrho / F_\varepsilon - \varrho\cdot\left(F_\varepsilon F_{\varrho\varrho} - F_{\varrho\varepsilon}F_\varrho - F_\varrho F_{\varepsilon\varrho} + F_\varrho^2 F_{\varepsilon\varepsilon}/F_\varepsilon\right)\Big/\left(F_\varepsilon\right)^2 \tag{60}$$

where

$$F_\varepsilon \equiv \frac{\partial F(\varrho, \varepsilon)}{\partial \varepsilon}, \quad F_{\varrho\varrho} \equiv \frac{\partial^2 F(\varrho, \varepsilon)}{\partial \varrho^2} \qquad \text{etc.}$$

Employing formulas analogous to (59) for the third derivatives of the function $F(\varrho, \varepsilon)$, we find an expression for the derivative $dB(p_0, \varrho)/d\varrho$:

$$\frac{dB(p_0, \varrho)}{d\varrho} = 3\left(B + 2 F_\varrho/F_\varepsilon\right)\left(\frac{1}{\varrho} - F_{\varepsilon\varrho}/F_\varepsilon + F_{\varepsilon\varepsilon}F_\varrho/F_\varepsilon^2\right) +$$
$$+ \varrho\left(F_\varepsilon\right)^{-2}\cdot\left[\left(F_\varrho^3/F_\varepsilon^2\right)F_{\varepsilon\varepsilon\varepsilon} + 3 F_\varrho\left(F_{\varepsilon\varrho\varrho} -\right.\right.$$
$$\left.\left. - F_{\varepsilon\varepsilon\varrho}F_\varrho/F_\varepsilon\right) - F_\varepsilon F_{\varrho\varrho\varrho}\right]. \tag{61}$$

In the process of derivation of the formula (61) we have assumed the equality of mixed derivatives of the function $F(\varrho, \varepsilon)$.

Below we formulate a series of theorems showing the ways of diminution of the order of smallness of a difference $p(x, t) - p_0$ in the contact strip in the presence of the f.d.a. K -inconsistence caused by the violation of the condition (35).

Theorem 7. If
a) the solution $w_h(x, t)$ of the problem (4)-(6), (24) obtained by the Lax scheme is such that at $t \geqslant \tau$

$$|\partial^j \varrho_h / \partial x^j| \leqslant K_{1j}(t), \quad |\partial^j \varrho_h \varepsilon_h / \partial x^j| \leqslant K_{2j}(t),$$
$$0 < K_{\ell j} < \infty, \quad j = 0, 1, 2, 3, \quad \ell = 1, 2, \quad |\varrho_h(x, t)| > 0$$

at $(x, t) \in G_K$:

86

b) the function $F(\varrho, \varepsilon)$ in the equation of state
(6) satisfies the requirements:
$$|\partial^3 F/\partial\varrho^\kappa \partial\varepsilon^{3-\kappa}| \leq K_3 , \quad (x,t) \in G_\kappa , \quad K_3 = const ,$$
$$0 < K_3 < \infty , \quad \kappa = 0,1,2,3 ,$$
$$\partial^m F/\partial\varrho^\kappa \partial\varepsilon^{m-\kappa} = \partial^m F/\partial\varepsilon^{m-\kappa} \partial\varrho^\kappa , \quad m = 2,3 ,$$
$$1 \leq \kappa \leq m-1, \quad |\partial F/\partial\varepsilon| > 0 , \quad (x,t) \in G_\kappa ;$$

c) the corrective term q is introduced into the difference energy equation (38) of an initial scheme
as follows:
$$(\varrho E)^{n+1} = H_3 (w^n, p^n, \varrho^n, \varepsilon^n, T_1, h, \tau) - \tau q^n \tag{62}$$
where
$$q^n = \mu^n B^n \left[\left((T_1 - T_{-1})/(2h) \right) \varrho^n \right]^2 , \tag{63}$$
$$\mu^n = h^2/(2\tau) - (\tau/2) (u^n)^2 , \tag{64}$$

B^n is computed by the formula (60) at $t = t^n$;

d) at $(x, t^n) \in G_\kappa$ the estimations
$$| F (\varrho_h (x, t^n), \varepsilon_h (x, t^n)) - p_0 | = 0 (h^3),$$
$$| w_h (x, t^n) - w_0 | = 0 (h^3).$$
take place,
then at $(x, t^{n+1}) \in G_\kappa$ the following estimate takes
place:
$$| F (\varrho_h (x, t^{n+1}), \varepsilon_h (x, t^{n+1})) - p_0 | = 0 (h^3). \tag{65}$$

<u>Corollary 1.</u> Instead of the corrective term
of the form
$$q = \mu B (\partial\varrho/\partial x)^2, \tag{65a}$$
see (63), the terms of the form
$$q = \mu \cdot \Big\langle \beta \Big\{ \frac{\varrho}{F_\varepsilon} \Big[\Big(F_\varrho - \frac{\varepsilon}{\varrho} F_\varepsilon \Big) \varrho_{xx} + \frac{1}{\varrho} F_\varepsilon (\varrho \varepsilon)_{xx} \Big] \Big\} +$$
$$+ (1-\beta) B (p, \varrho) \cdot \sum_{\kappa=0}^{2} \gamma_\kappa \Big[\Big(-\frac{F_\varepsilon}{F_\varrho} \Big) \varepsilon_x \Big]^{2-\kappa} (\varrho_x)^\kappa \Big\rangle \tag{66}$$
can be considered where β , γ_κ are constant weight
coefficients satisfying the requirements $0 \leq \beta \leq 1, \gamma_0 + \gamma_1 + \gamma_2 = 1$.

<u>Corollary 2.</u> If the corrective term q is not introduced in the equation (38) by the formulas (62)-
-(64),(66), and $B(p_0, \varrho) \neq 0$, then at $(x, t^{n+1}) \in G_\kappa$
the following estimate takes place:
$$| F (\varrho_h (x, t^{n+1}), \varepsilon_h (x, t^{n+1})) - p_0 | = 0 (h^2).$$

<u>Remark.</u> Theorem 7 is valid also for the Godunov's scheme (Godunov, Zabrodin, Prokopov (1961),
in subdomains of supersonic flow), for the FLIC method (Gentry et al (1966)), for some versions of
the computational scheme of the "coarse particles"
method (Belotserkovskii and Davydov (1971)). One
should only set the corresponding name of the scheme in the condition a) and calculate μ^n according
to the Table 3.

The following theorem shows a possibility of
further increase in the order of smallness of a difference standing in the left hand side of the relationship (65) by the proper choice of a difference approximation of the corrective term q in (62)

and at some limitation on the function $F(\varrho, \varepsilon)$ entering the equation of state (6).

Theorem 8. If

a) the solution $W_h(x,t)$ of the problem (4)-(6),(24) obtained by the FLIC method is such that at $t \geqslant \widetilde{t}$
$$|\partial^j \varrho_h / \partial x^j| \leqslant K_{1j}(t), \quad |\partial^j \varrho_h \varepsilon_h / \partial x^j| \leqslant K_{2j}(t),$$
$$0 < K_{\ell j} < \infty, \quad j = 0,1,2,3,4, \quad \ell = 1,2, \quad |\varrho_h(x,t)| > 0$$
at $(x,t) \in G_K$;

b) the function $F(\varrho, \varepsilon)$ in the equation of state (6) has the form
$$F(\varrho, \varepsilon) = \alpha \varrho \varepsilon + \sum_{k=1}^{2} B_k (\varrho - \varrho_0)^K$$
where α, B_k $(K = 1,2)$ are constants and $\alpha \neq 0$;

c) the corrective term q is introduced into the difference energy equation (38) of an initial scheme by the formula (62) where
$$q^n = \eta^n B^n \left[((T_1 - T_{-1})/(2h)) \varrho^n \right] \left\{ \left[(T_1 - T_{-1})/2 + \right. \right.$$
$$\left. \left. \alpha (T_1 - 2I + T_{-1}) \right] \varrho^n / h \right\}^2, \tag{67}$$
$$\eta^n = 0.5 |u^n| (h - \widetilde{t} |u^n|), \quad \alpha = -\operatorname{sgn}(u^n), \tag{68}$$
B^n is computed by the formula (60) at $t = t^n$;

d) at $(x, t^n) \in G_K$ the estimations
$$|F(\varrho_h(x,t^n), \varepsilon_h(x,t^n)) - p_0| = 0(h^4),$$
$$|u_h(x,t^n) - u_0| = 0(h^4)$$
take place,

then at $(x, t^{n+1}) \in G_K$ the following estimate takes place: $\quad |F(\varrho_h(x,t^{n+1}), \varepsilon_h(x, t^{n+1})) - p_0| = 0(h^4).$

Remark. Theorem 8 is valid also for the Godunov's scheme (in subdomains of supersonic flow) and for some versions of the computational scheme of the "coarse particles" method (Belotserkovskii and Davydov (1971)).

The following theorem shows that for estimating the difference $|P_h(x, t^{n+1}) - p_0|$ one can successfully apply the results of the theory of spline functions (Ahlberg et al (1972), Stechkin and Subbotin (1976)) At first let us introduce some notations. Consider the numerical solution of the problem (4)-(6),(24) obtained by the FLIC method at $t = t^n > 0, \ t^n = n\widetilde{t}$ Suppose for the definiteness that $u_0 > 0$ in (24). Let us consider on the x axis an interval belonging to the contact strip and such that within it the density $\varrho_h(x, t^n)$ changes monotonically. Let us arrange the corresponding density grid values in the mesh nodes on the x axis in an increasing sequence:
$$\varrho_0^n < \varrho_1^n < \ldots < \varrho_N^n. \tag{69}$$
Note that the numbering of ϱ^n values in (69) generally does not coincide with the node numbering in computing mesh on the x axis. Let us approximate the function $b(\varrho)$ entering the equation of state (50) by a parabolic spline $S_2(\varrho)$ in the interval

$[\varrho_0^n, \varrho_N^n]_n$. We require that: (1) $S_2''(\varrho) = S_{2K}''$ at
$(\varrho_{K-1}^n + \varrho_K^n)/2 \leq \varrho \leq (\varrho_K^n + \varrho_{K+1}^n)/2$, $K = 1, \ldots, N-1$,
S_{2K}'' are constants; (2) $S_2(\varrho)$ should have a continuous first derivative; (3) $S_2(\varrho_K^n) = b(\varrho_K^n)$, $K = 0, \ldots, N$;
(4) and, finally, we require for the spline $S_2(\varrho)$ to have a continuous second derivative at the points $(\varrho_0^n + \varrho_1^n)/2$ and $(\varrho_{N-1}^n + \varrho_N^n)/2$. Stechkin and Subbotin (1976) have established the conditions for existence and uniqueness of the constructed spline $S_2(\varrho)$ With regard for these conditions we have proved the following theorem.

Theorem 9. If
a) the solution $W_h(x,t)$ of the problem (4)-(6), (24) obtained by the FLIC method is such that at $t \geq \tau$

$$|\partial \varrho_h / \partial x| \leq K_1 , \quad K_1 = const, \quad 0 < K_1 < \infty ;$$

b) $\varrho_* = \varrho_0^n > 0$, $u_0 > 0$ where u_0 is the velocity value entering the initial conditions (24);
c) the function $F(\varrho, \varepsilon)$ in the equation of state (6) has the form

$$F(\varrho, \varepsilon) = \alpha \varrho \varepsilon + b(\varrho)$$

where $\alpha = const > 0$, $b(\varrho)$ is a three times continuously differentiable function for $\varrho > 0$;
d) $|F(\varrho_h(x,t^n), \varepsilon_h(x,t^n)) - p_0| \leq K_2 h^3$, $K_2 = const$;
$|u_h(x,t^n) - u_0| \leq K_3 h^3$, $K_3 = const$, $x \in G_K$,
$$0 < K_j < \infty, \quad j = 2,3;$$

e) the time step τ satisfies the inequality

$$\tau \leq \min \left\{ \frac{h}{2u_0}, \frac{u_0 K_1 - 2 K_3 h^2 \varrho_N^h}{2 h K_2} \right\}$$

f) $(\max \Delta \varrho_i)/\Delta \varrho_i = O(1)$, $\Delta \varrho_i = \varrho_{i+1}^n - \varrho_i^n$, $i = 0, \ldots, N-1$;
g) a corrective term q is introduced into the difference energy equation (38) of an initial scheme as follows:

$$(\varrho E)^{n+1} = H_3(u^n, p^h, \varrho^h, \varepsilon^h, T_1, h) - \tau q^n,$$
$$q^n = \mu^n B^n \left[((T_1 - T_{-1})/(2h)) \varrho^n \right]^2,$$
$$\mu^n = 0.5 |u^n|(h - \tau |u^n|),$$

then for $\varrho_h(x,t^{n+1}) \in [\varrho_0^h, \varrho_N^n]$ the following estimate is valid:

$$|F(\varrho_h(x,t^{n+1}), \varepsilon_h(x,t^{n+1})) - p_0| = O(h^3).$$

Theorem 10. If
a) the solution $W_h(x,t)$ of the problem (4)-(6),(24) obtained by the one-step Lax-Wendroff scheme is such that at $t \geq \tau$

$$|\partial^j \varrho_h / \partial x^j| \leq K_{1j}(t), \quad |\partial^j \varrho_h \varepsilon_h / \partial x^j| \leq K_{2j}(t),$$
$$0 < K_{\ell j} < \infty, \quad j = 0, \ldots, 4, \quad \ell = 1,2, \quad |\varrho_h(x,t)| > 0$$

at $(x,t) \in G_K$;
b) the function $F(\varrho, \varepsilon)$ in the equation of state (6) satisfies the requirements:

$$|\partial^4 F / \partial \varrho^k \partial \varepsilon^{4-k}| \leq K_3, \quad (x,t) \in G_K, \quad K_3 = const,$$

$$0 < K_3 < \infty, \quad \kappa = 0, \ldots, 4,$$
$$\partial^m F/\partial \varrho^\kappa \partial \varepsilon^{m-\kappa} = \partial^m F/\partial \varepsilon^{m-\kappa} \partial \varrho^\kappa, \quad m = 2,3,4,$$
$$1 \leqslant \kappa \leqslant m-1, \quad |\partial F/\partial \varepsilon| > 0, \quad (x,t) \in G_\kappa;$$

c) the corrective term q is introduced into the difference energy equation (38) of an initial scheme
as follows:

$$(\varrho E)^{n+1} = H_3 \left(u^n, p^n, \varrho^n, \varepsilon^n, T_1, h, \tau\right) + \tau q^n \tag{70}$$

where

$$q^n = \mu^n \left(\frac{dB}{d\varrho}\right)^n \cdot \left[(T_1 - T_{-1})\varrho^n/(2h)\right]^3 + \tag{71}$$
$$+ 3 \mu^n B^n \cdot \left[(T_1 - T_{-1})\varrho^n/(2h)\right] \cdot \left[(T_1 - 2I + T_{-1})\varrho^n/(h^2)\right],$$

$B^n, (dB/d\varrho)^n$ are computed by the formulas (60),(61)
respectively, at $t = t^n$;
d) at $(x,t^n) \in G_\kappa$ the estimations

$$\left| F\left(\varrho_h(x,t^n), \varepsilon_h(x,t^n)\right) - p_0 \right| = 0(h^4),$$
$$\left| u_h(x,t^n) - u_0 \right| = 0(h^4)$$

take place,
then at $(x, t^{n+1}) \in G_\kappa$ the following estimate takes
place:

$$\left| F\left(\varrho_h(x,t^{n+1}), \varepsilon_h(x,t^{n+1})\right) - p_0 \right| = 0(h^4).$$

<u>Corollary 1.</u> Instead of the corrective term
of the form

$$q = \mu \left[\frac{dB}{d\varrho}\left(\frac{\partial \varrho}{\partial x}\right)^3 + 3B \frac{\partial \varrho}{\partial x} \frac{\partial^2 \varrho}{\partial x^2}\right], \tag{73}$$

which is approximated in (71), the terms of the
form

$$q = \mu \cdot \frac{dB}{d\varrho} \cdot \sum_{\kappa=0}^{3} \alpha_\kappa \left(-\frac{F_\varepsilon}{F_\varrho} \varepsilon_x\right)^{3-\kappa} (\varrho_x)^\kappa + 3\mu B \{\varrho_x\} \{\varrho_{xx}\},$$

or

$$q = \mu \cdot \left[(\varrho \varepsilon)_{xxx} + (\varrho F_\varrho/F_\varepsilon - \varepsilon)\varrho_{xxx}\right] \tag{74}$$

can be considered where the following notations are
employed:

$$\{\varrho_x\} = \sum_{m=0}^{1} \beta_m \left(-\frac{F_\varepsilon}{F_\varrho} \varepsilon_x\right)^{1-m} (\varrho_x)^m,$$

$$\{\varrho_{xx}\} = \sum_{z=0}^{1} \gamma_z \left[-\frac{1}{F_\varrho}\left(F_{\varrho\varrho}\varrho_x^2 + 2F_{\varrho\varepsilon}\varrho_x\varepsilon_x + F_{\varepsilon\varepsilon}\varepsilon_x^2 + F_\varepsilon\varepsilon_{xx}\right)\right]^{1-z} (\varrho_{xx})^z,$$

$\alpha_\kappa, \beta_m, \gamma_M$ are constant weight coefficients, such
that

$$\alpha_0 + \alpha_1 + \alpha_2 + \alpha_3 = 1, \quad \beta_0 + \beta_1 = 1, \quad \gamma_0 + \gamma_1 = 1.$$

<u>Corollary 2.</u> If one sets $q = 0$ in (70), then for
$B(p_0, \varrho) \neq 0$ the following estimate

$$\left| F\left(\varrho_h(x, t^{n+1}), \varepsilon_h(x,t^{n+1})\right) - p_0 \right| = 0(h^3)$$

is valid in the contact strip.

$\underline{\text{Remark to the Theorems } 7,8,10.}$ Employing Theorem 5, it is easy to unite the formulas for (65a) and (73) into a more general formula

$$q = (-1)^{z+1} \mu \cdot \sum_{k=0}^{z-1} C_z^k \cdot \left\{ \frac{\partial^{z-k-1}}{\partial x^{z-k-1}} \left[B(p_0,\varrho) \frac{\partial \varrho}{\partial x} \right] \right\} \left(\frac{\partial^{k+1} \varrho}{\partial x^{k+1}} \right) \qquad (75)$$

where z is the order of approximation of a difference scheme,

$$\partial B(p_0,\varrho)/\partial x = \left[dB(p_0,\varrho)/d\varrho \right] \partial \varrho /\partial x ,$$
$$\partial^0 q(x,t)/\partial x^0 \equiv q(x,t), \quad 0! = 1, \quad C_z^k = \frac{z!}{k!(z-k)!} .$$

Then instead of the initial energy equation, see Equations (4),(5), one should approximate in the contact strip an equation

$$\partial \varrho E/\partial t + \partial(pu + \varrho u E)/\partial x = -q$$

where q is determined by (75).

For purposes of checking theoretical conclusions contained in this section we have carried out numerous computations of the problem (4)-(6),(24) employing difference schemes listed in Table 3, and different equations of state which do not satisfy the condition (35). In some cases the difference solution for the pressure differed from the exact pressure value in the contact strip by more than 50%. The difference solution for the velocity proved to be less sensible to the presence of the K --inconsistence. Here we present only a few numerical results which we have obtained when using the equation of state (Sedgwick et al (1978))

$$p = \left(a_1 + \frac{a_2}{\frac{\varepsilon}{I_0 \eta^2} + 1} \right) \varepsilon \varrho + A_1 \varsigma + A_2 \varsigma^2 \qquad (76)$$

where

$\eta = \varrho/\varrho_0 = \varsigma + 1$, a_1, a_2, A_1, A_2, I_0 are constants for the particular material. The equation (76) is supposed to be valid (Sedgwick et al (1978)) for a material being in the condensed state $(\varrho/\varrho_0 > 1)$ or in any cold state $(\varepsilon < I_s, I_s$ is a constant). As a material we have taken aluminium for computations and the following values of constants in (24):

$$u_0 = 1 cm/\mu s , \quad p_0 = 3,5 MB , \quad \varrho_1 = 7,5 \, g/cm^3 ,$$
$$\varrho_2 = 3 g/cm^3 .$$

The results presented in the Figures 8-10 are obtained for $h = 4\tau = 0.02$ which corresponds to the Courant number $K \cong 0.63$. The moment of time $t = 40\tau$ for all three figures, solid lines represent the numerical solution obtained by using an initial difference scheme, the black triangles show the difference solution obtained by the use of the algorithms for suppression of the K-inconsistence and which were formulated in the form of Theorems 7

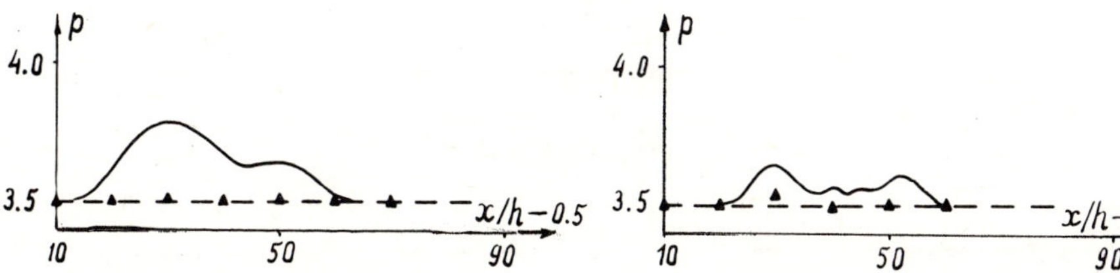

Figure 8. Lax scheme

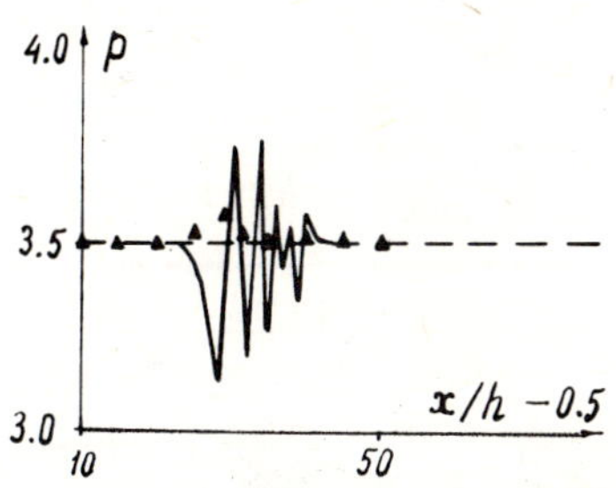

Figure 10. One-step
Lax-Wendroff scheme

Figure 9. FLIC method.
and 10. When using the one-step Lax-Wendroff scheme we have considered the difference momentum equation in the form

$$(\varrho w)_i^{n+1} =$$
$$= H_2(w_i^n, p^n, \varrho^n, \varepsilon_i^n, T_1, h, \tau) - \quad (77)$$
$$- \frac{1}{2}\left(\frac{\tau}{h}\right)^2\left(w_{i+1/2}^n R_{i+1/2}^n - \right.$$
$$\left. - w_{i-1/2}^n R_{i-1/2}^n \right)$$

where $w_{i\pm1/2}^n = 0.5\,(w_i^n + w_{i\pm1}^n)$ and the quantity $R_{i+1/2}^n$ was computed by the formula (44). Fig.10 shows the result of joint application of the formula (77) and the formula for q (see Equation (70)) of the form (74) where the derivatives were approximated by centred differences. If in the equation of state (76) the coefficient α_2 is nullified, then some "mathematical" gas is obtained to which the Theorem 8 is applicable. It follows from Table 4 that the use of an algorithm for eliminating the K-inconsistence, which was formulated in the form of Theorem 8, leads to a substantial diminution of the relative error in grid pressure values in the contact strip in comparison with the algorithm of Theorems 7,9. This error is given in Table 4 in per cents for both difference schemes considered.

Table 4 illustrates also the influence of the joint application of the formula (77) for R and the formula (74) for q , the equation of state (76) with $\alpha_2=0$. Since the term q determined by the formula (75) is not in divergence form, its implementation in computations leads to the violation of the conservation of the total gas energy in the computational domain. However, it turned out that the size of the corresponding relative error in energy is insignificant: in our computations it did not exceed 1% for first-order schemes and 0.2% for second-order schemes. Note that the use of the term

Table 4

Difference scheme	and its modification	t/τ		
		20	40	60
FLIC	$q \equiv 0$	4.38	3.77	3.47
	$q \neq 0, \alpha = -1$ in (67)	0.93	0.76	0.66
	$q \neq 0, \alpha = 0$ in (67)	1.54	1.17	0.99
Lax-Wendroff, one-step	$q \equiv 0, R \equiv 0$	16.11	8.63	7.57
	$q \neq 0, R \neq 0$	4.63	4.16	4.12
	$q \neq 0, R = 0$	11.53	6.44	5.35

q (75) has not lead in practice to the deterioration of the computational stability when using difference schemes from Table 3 and the equations of state which we have considered.

An analysis of the computation results has shown that an implementation in computations of algorithms for suppression of the K-inconsistence of difference schemes in the contact strip enables one to apply successfully algorithms I-III for localization of contact discontinuities which were derived above from the Theorem 5 under the assumption of the K-consistence presence. In our opinion, the overcoming of the K-inconsistence should become a necessary stage in constructing new finite-difference methods for the numerical solution of gas dynamics problems involving contact discontinuities.

References
Ahlberg, J.H., Nilson,E.N.(1972) The Theory of Splines and Their Applications. "Mir", Moscow.
Belotserkovskii,O.M., and Davydov,Yu.M.(1971) A nonstationary "coarse particle" method for gas-dynamical computations. USSR Comp.Math.and Math.Phys., 11, 1:241-274.
Burstein,S.Z.(1964) Numerical methods in multidimensional shocked flows. AIAA J.2,12:2111-2117.
Chin,R.C.Y.(1975) Dispersion and Gibbs Phenomenon Associated with Difference Approximations to Initial Boundary-Value Problems for Hyperbolic Equations. J.Comp.Phys., 18, 3:233-247.
Fomin,V.M., Ermolin,N.E., Kroshko,E.A., Fedorov,A.V., Chubarova,E.V., Chernysheva,R.T.(1979) Differential analyser and its application to two dimensional gas dynamic computations. In: "Čislennye me-

tody mehaniki splošnoĭ sredy", Novosibirsk, 10, 3:138-141.
Fomin,V.M., Vorozhtsov,E.V., and Yanenko,N.N.(1976) Differential Analysers of Shock Waves: Theory Computers & Fluids, 4, 3/4:171-183.
Fomin,V.M., Vorozhtsov,E.V., and Yanenko,N.N.(1979) On the Properties of Curvilinear Shock Waves "Smearing" in Calculations by the Particle-in-Cell Method. Computers and Fluids 7, 2:109-121.
Gentry,R.A., Martin,R.E., Daly,B.J.(1966) An Eulerian Differencing Method for Unsteady Compressible Flow Problems. J.Comput.Phys., 1, 1:87-118.
Godunov,S.K.(Editor), Zabrodin,A.V., Ivanov,M.Ya., Kraiko,A.N., Prokopov,G.P.(1976) Numerical Solution of Multidimensional Gas Dynamics Problems. "Nauka", Moscow.
Godunov,S.K., Zabrodin,A.V., Prokopov,G.P.(1961) The Difference Scheme for Two-Dimensional Unsteady Problems in Gas Dynamics and the Calculation of Flows with a Detached Shock Wave. Zhurn.Vych.Mat. i Mat.Fiz., 1, 6:1020-1050.
Gridneva,V.A., Korneyev,A.I., Trushkov,V.G.(1975) On the Numerical Method for Computation of the Flow of Elastic-Plastic Media with Large Deformations. In:Proc.5th Sci.Conf.Math.and Mech., Tomsk Univ., Vol.2, Tomsk, Tomsk University, p.115.
Harlow,F.H.(1964) The particle-in-cell computing method for fluid dynamics. Methods in Comp.Phys., Academic Press, New York, 3, pp.319-343.
Harten,A.(1977) The Artificial Compression Method for Computation of Shocks and Contact Discontinuities. I. Single Conservation Laws. Comm.Pure and Appl.Math., 30, 5:611-638.
Il'in,A.M., Oleinik,O.A.(1960) Asymptotic behaviour of solutions of the Cauchy problem for some quasilinear equations for large values of time. Mat.Sbornik 51, pp.191-216.
Ivanov,M.Ya.(1975) On the solution of two- and three-dimensional problems on the transonic flow around bodies. Zhurn.Vych.Mat.i Mat.Fiz., 15, 5: 1222-1240.
Kosarev,V.I.(1971) On the computation of supersonic steady gas flows with imbedded shocks. Zhurn. Vych.Mat.i Mat.Fiz., 11,5:1262-1271.
Laval,P.(1969) Méthode de pseudo-viscosité et problème de l'amorçage d'une tuyère. Rech.Aérospat., No.131, pp.3-16.
Lax,P.D.(1954) Weak Solutions of Nonlinear Hyperbolic Equations and Their Numerical Computation. Comm.Pure and Appl.Math.7, 1:159-193.
Lax,P.D., and Wendroff,B.(1960) Systems of Conservation Laws III. Comm.Pure and Appl.Math., 13,

94
2:217-237.
Lerat,A., Peyret,R.(1975) Propriétés dispersives
et dissipatives d'une classe de schémas aux diffé-
rences pour les systèmes hyperboliques non linéai-
res. Rech.Aérospat.No.2,pp.61-79.
Liubimov,A.N., Rusanov,V.V.(1970) Gas Flows Arou-
nd Blunt Bodies. Part I. Computing Method and Ana-
lysis of the Flows. "Nauka", Moscow.
MacCormack,R.W.(1969) The Effect of Viscosity in
Hypervelocity Impact Cratering. AIAA Paper 69-354.
Miranker,W.L., Pironneau,O.(1975) A Global Shock
Fitting Method. IRIA, Rapport de Recherche n°123.
Miranker,W.L., Pironneau,O.(1976) An Example of a
Global Shock Fitting Method. An Internat.J.Compu-
ters and Mathematics with Applications, 2,1:63-71.
Neumann,J.von(1966) Theory of Self-Reproducing
Automates. Univ.Illinois Press.
Neumann,J., Richtmyer,R.(1950) A method for the
numerical calculation of hydrodynamic shicks. J.
Appl.Phys.21,3::232-237.
O'Keefe J.D., Ahrens,Th.J.(1975) Shock effects
from a large impact on the Moon. Proc.6th Lunar
Sci.Conf.Houston, Tex., Vol.3. New York e.a.,
pp.2831-2844.
Peletier,L.A.(1971) Asymptotic Stability of Tra-
velling Waves. IUTAM Symp.on Instability of Conti-
nuous Systems. Springer-Verlag, Berlin, pp.418-422.
Proceedings of the Fourth International Conference
on Numerical Methods in Fluid Dynamics (1975).
Lecture Notes in Physics, 35. Springer-Verlag.
Richtmyer,R.D., and Morton,K.W.(1967) Difference
Methods for Initial Value Problems, 2nd Edn.Inter-
science-Wiley, New York.
Roache,P.J.(1976) Computational Fluid Dynamics.
Hermosa Publishers, Albuquerque.
Rozhdestvenskii,B.L., and Yanenko,N.N.(1968) Sys-
tems of Quasilinear Equations and Their Applicat-
ions toGas Dynamics. "Nauka", Moscow.
Rusanov,V.V.(1961) Calculation of Interaction of
Non-Steady Shock Waves with Obstacles, National
Research Council of Canada Technical Translation
1027, Zhurn.Vych.Mat.i MatFiz., 1, 2:267-279.
Rusanov,V.V.(1969) Difference Schemes of Third-Order Accuracy for "Across"-Computation of Discontinuous Solutions. Fluid Dynam.Transact., Vol.4. Wa-
rszawa, pp.285-294.
Rusanov,V.V.(1973) Processing and Analysis of Computation Results for Multidimensional Problems of
Aerohydrodynamics. Lect.Notes Phys. 18:154-162.
Sedgwick,R.T.,Hageman,L.J., Herrmann,R.G., and Wa-
ddell,J.L.(1978) Numerical Investigations in Penetration Mechnics. Int.J.Engng.Sci., 16, 11:859-869.

Shokin,Yu.I.(1979) Differential Approximation Method. "Nauka", Novosibirsk.
Smirnov,V.I.(1969) A Course on Higher Mathematics, Vol.III, Part 2. "Nauka", Moscow.
Stechkin,S.B., and Subbotin,Yu.N.(1976) Splines in Numerical Mathematics. "Nauka", Moscow.
Tihonov,A.N., Samarskii,A.A.(1977) Equations of Mathematical Physics. "Nauka", Moscow.
Usakov,Z.(1979) On the localization of a discontinuity in the numerical solution of the Buckley-Leverett problem. In:"Cislennye metody mehaniki splosnoi sredy", Novosibirsk, 10, 6:141-149.
Usakov,Z.(1980) On a numerical experiment on discontinuity localization in the solution of the filtration problem for a three-phase incompressible fluid. In:"Cislennye Metody reseniia zadac fil'trazii mnogofaznoi neszimaiemoi zidkosti. Trudy IV Vsesoiuznogo Seminara. Novosibirsk, pp.241-245.
Vorozhtsov,E.V.(1972) On differential analysers of shock fronts. In: "Komplexy programm matematiceskoi fiziki". Novosibirsk, pp.98-115.
Vorozhtsov,E.V.(1976) Differential Analysers of Shock Waves in Shock-Capturing Schemes for Gas Dynamics Problems. Thesis, Novosibirsk.
Vorozhtsov,E.V.(1977) Numerical tests of differential analysers of shock waves. - In: "Cislennye metody mehaniki splosnoi sredy", Novosibirsk, 8, 2:12-27.
Vorozhtsov,E.V., Ermolin,N.E., Fomin,V.M.(1979) Computation of two-dimensional chemically non-equilibrium gas mixtures in nozzles and jets. In: "Cislennye metody mehaniki splosnoi sredy", Novosibirsk 10, 2:30-39.
Vorozhtsov,E.V., Fomin,V.M., Yanenko,N.N.(1976) Differential analysers of shock waves. Applications of the theory. In: "Cislennye metody mehaniki splosnoi sredy", Novosibirsk, 7, 6:8-23.
Vorozhtsov,E.V., and Yanenko,N.N.(1980) On Some Algorithms for Shock Wave Recognition by Shock-Capturing Computational Results. Computers & Fluids 8, pp.313-326.
Whitham,G.B.(1974) Linear and Nonlinear Waves. Wiley-Interscience Publ.
Wilkins,M.L.(1970) Finite Difference Scheme for Calculating Problems in Two Space Dimensions and Time. J.Comput.Phys.5, pp.406-414.
Wilkins,M.L.(1971a) Calculation of Elastic-Plastic Flow. "Trudy Sekzii po cislennym metodam v gazovoi dinamike Vtorogo Mezdunarodnogo collokviuma po gazodinamike vzryva i reagiruiuscih sistem (Novosibirsk, August 19-23 (1969)". Vol1. Izd.VZ AN SSSR, Moscow, pp.408-517.

Wilkins,M.L.(1971b) Calculation of a Slender Body
Moving Through Air at Sypersonic and Subsonic Velo-
cities. AIAA J., 9, 10:2073-2075.
Yanenko,N.N., Fomin,V.M., Vorozhtsov,E.V.(1978)
Differential Analysers of Shock Waves in Shock-Cap-
turing Schemes for Gas Dynamics Problems."Institute
of Theoretical and Appl.Mechanics of the U.S.S.R.
Academy of Sciences Siberian Branch. Preprint",No.7
Yanenko,N.N., Kovenya,V.M., Lisejkin,V.D., Fomin,
V.M., and Vorozhtsov,E.V.(1979) On Some Methods
for the Numerical Simulation of Flows with Complex
Structure. Computer Methods in Appl.Mechanics and
Engineering, 17/18, Part III, pp.659-671.
Yanenko,N.N., Kroshko,E.A., Lisejkin,V.V., Fomin,
V.M., Shapeev,V.P. and Shitov,Yu.A.(1976) Methods
for the construction of moving grids for problems
of fluid dynamics with big deformations. "Lect.No-
tes Phys." No.59:454-459.
Yanenko,N.N., Vorozhtsov,E.V.(1979) On the theory
of differential analysers of contact discontinui-
ties. Dokl.AN SSSR, 247, 1:48-52.
Yanenko,N.N., Vorozcov,E.V. and Fomin,V.M.(1976)
Differential analysers of shock waves. Soviet Math.
Dokl. 17, 2:358-362.
Zel'dovich,Ya.B., and Raiser,Yu.P.(1966) Physics
of Shock Waves and High-Temperature Hydrodynamic
Phenomena. "Nauka", Moscow.

CHAPTER 4

SOLUTION OF THE UNSTEADY NAVIER-STOKES EQUATIONS BY A
FINITE-ELEMENT PROJECTION METHOD

J. Donea, S. Giuliani, H. Laval and L. Quartapelle

1 INTRODUCTION

In recent years, a great deal of effort has been directed toward the application of the finite element method (FEM) to the numerical solution of the incompressible Navier-Stokes equations. Among the most recent contributions we can mention methods based on the penalty function formulation (Heinrich, Marshall and Zienkiewicz, 1978; Hughes, Liu and Brooks, 1979; Bercovier and Engelman, 1979); methods approximating the pressure by piecewise polynomials which are either discontinuous (Olson and Tuann, 1978), or continuous (Olson and Tuann, 1978; Gresho, Lee, Stullich and Sani, 1978; Taylor and Ijam, 1979) across inter-element boundaries; a mixed-type formulation for the stream function-vorticity equations adapted to high Reynolds number calculations (Fortin and Thomasset, 1979); and finally, the least-squares formulation of the Navier-Stokes equations (Bristeau, Pironneau, Glowinski, Périaux and Perrier, 1979).

Despite this flourishing of new finite-element formulations, it seems that no attempt has been made to develop a finite element version of the finite difference method originally proposed by Chorin for the solution of the time-dependent Navier-Stokes equations in the velocity-pressure formulation (Chorin, 1968 and 1969). This is surprising in two respects: Chorin's method was the first method enabling cost-effective calculations of fully three-dimensional problems; and the method has been continuously referred to in the computational literature (see, e.g., Hirt and Cook, 1972; Takami and Kawahara, 1974; and, more recently, Goda, 1979). A possible reason why FEM researchers have disregarded Chorin's method can perhaps been traced back to the extensive use in Chorin's papers of splitting methods for the treatment of the spatial derivatives in different directions, methods which are not convenient in a FEM context. On the other hand, the fundamental characteristic of Chorin's formulation is independent of any particular spatial discretization technique and lies in the fact that the momentum equation and the incompressibility condition are treated separately according to a procedure consisting of two distinct steps (hereafter Chorin's method will be referred to as 'fractional-step method'). It is therefore possible, and it seemed worthwhile attempting, to investigate Chorin's method in a FEM perspective. The results of such an investigation will be reported in the present chapter.

As will be shown later, the use of a finite-element semidiscretization in the formulation of Chorin's method presents several advantages with respect to the finite difference approach: firstly, the Galerkin equation for pressure which arises from the finite-element semidiscretization is such that no pressure boundary conditions are explicitly required along the boundary segments on which the velocity is specified; secondly, the FEM weak form of the incompressibility condition leads to the explicit construction of the projection operator which plays a fundamental role in Chorin's method; finally, the stress boundary conditions appear as natural boundary conditions in the FEM formulation.

The content of this chapter is organized as follows. In Section 2, we first recall the governing equations expressed in the natural variables: velocity and pressure. Then, we introduce some useful Hilbert space notations and present the weak variational form of the governing equations. This is the starting point for the presentation made in Section 3 of the fractional-step method adapted to a finite-element semidiscretization. In the next section, Section 4, two different spatially discrete models based on 4-node and 9-node Lagrange isoparametric elements are introduced and the fully-discretized finite-element equations are derived in matrix form. The discrete equations of the incompressibility step are discussed in the necessary detail in Section 5. An explicit representation of the discrete counterpart of the projection operator is obtained by taking full advantage of the concept of generalized inverse. The effects of discretization in space and time are discussed in Sections 6 and 7 with the advection-diffusion equation in one dimension taken as a model equation. The consequences of combining the possible mass representa-

tions (consistent or diagonal) with two explicit time integration schemes (Euler and Adams-Bashforth) are briefly discussed
and stability conditions are derived for meshes consisting of
linear and parabolic elements and the diagonal mass representation. In the last section, Section 8, results of numerical
experiments are presented and compared to solutions obtained
by other numerical methods.

2. VARIATIONAL FORMULATION OF THE BASIC EQUATIONS

Let Ω be a bounded domain of R^d (d = 2 or 3) with a suitably
smooth boundary Γ. The mathematical description of the motion
of an incompressible viscous Newtonian fluid is given by the
Navier-Stokes equations

$$\frac{\partial \underline{u}}{\partial t} + (\underline{u} \cdot \underline{\nabla}) \underline{u} = \underline{\nabla} \cdot \underline{\sigma} \tag{2.1}$$

$$\underline{\nabla} \cdot \underline{u} = 0, \tag{2.2}$$

where

$$\sigma_{ij} = - p\delta_{ij} + 2\nu D_{ij}(\underline{u}), \tag{2.3}$$

$$D_{ij}(\underline{u}) = \frac{1}{2} \left(\frac{\partial u_i}{\partial x_j} + \frac{\partial u_j}{\partial x_i} \right) . \tag{2.4}$$

Here, $\underline{u}$ is the velocity, $\underline{\sigma}$ the stress tensor per unit density
of the fluid, p the pressure per unit density, and ν the coef-
ficient of kinematic viscosity. Equations (2.1) - (2.4) are
supplemented by the boundary conditions

$$\underline{u}\big|_{\Gamma_1} = \underline{b} , \tag{2.5}$$

$$\underline{n} \cdot \underline{\sigma}\big|_{\Gamma_2} = \underline{c} , \tag{2.6}$$

where $\underline{n}$ is the outward unit vector normal to the boundary, $\underline{b}$
represents the velocity vector specified on Γ_1 and $\underline{c}$ is the
traction vector specified on Γ_2. Γ_1 and Γ_2 are subsets of
boundary Γ satisfying $\overline{\Gamma_1 \cup \Gamma_2} = \Gamma$ and $\Gamma_1 \cap \Gamma_2 = \phi$. The initial
condition for problem (2.1) - (2.4) is

$$\underline{u}(\underline{x},0) = \underline{u}_o(\underline{x}), \tag{2.7}$$

with the initial velocity field $\underline{u}_o(\underline{x})$ satisfying the incompressibility condition, i.e. $\underline{\nabla} \cdot \underline{u}_o = 0$.

When the velocity is prescribed on the entire boundary, i.e., when $\Gamma_1 \equiv \Gamma$ and $\Gamma_2 \equiv \phi$, the velocity boundary data $\underline{b}$ must satisfy condition

$$\int d\Gamma \ \underline{n} \cdot \underline{b} = 0 \tag{2.8}$$

which follows from (2.2) by application of Green's formula, and the pressure is defined up to an arbitrary additive constant.

In view of the different roles which will be played by the normal and tangential components of the velocity boundary conditions in the present approach to the evolution equations, let (2.5) be written as

$$\underline{n} \cdot \underline{u} \Big|_{\Gamma_1} = \underline{n} \cdot \underline{b}, \tag{2.9}$$

$$\underline{n} \times \underline{u} \Big|_{\Gamma_1} = \underline{n} \times \underline{b} \ . \tag{2.10}$$

We introduce some Hilbert spaces necessary for a variational formulation of problem (2.1) - (2.7) (see, e.g., Temam, 1977). Let $H^1(\Omega)$ be the standard Sobolev space of scalar functions defined on Ω

$$H^1(\Omega) = \{u | u \in L^2(\Omega), \ u' \in L^2(\Omega)\}, \tag{2.11}$$

where the prime denotes the first derivative with respect to the spatial variables. Similarly, let $H^1(\Omega)^d$ be the Sobolev space of vector functions

$$H^1(\Omega)^d = \{\underline{u} | \underline{u} \in L^2(\Omega)^d, \ \underline{u}' \in L^2(\Omega)^d\} \ . \tag{2.12}$$

In the spaces $L^2(\Omega)$ and $L^2(\Omega)^d$ of scalar and vector square-integrable functions, the scalar product is defined, in the normal way, by

$$(u,v) = \int d\Omega \ uv, \tag{2.13}$$

$$(\underline{u},\underline{v}) = \int d\Omega \ \underline{u} \cdot \underline{v} = \sum_{i=1}^{d} (u_i, v_i) \ . \tag{2.14}$$

A variational formulation of problem (2.1) - (2.7) is obtained by multiplying (2.1) by a test function $\underline{v} \in H^1(\Omega)^d$, $\underline{v}\big|_{\Gamma_1} = 0$, and (2.2) by a test function $q \in L^2(\Omega)$ (see, e.g., Raviart, 1979). After integration by parts and noting that D_{ij} is symmetric and that (2.6) holds, the stress term gives

$$(\underline{\nabla} \cdot \underline{\sigma}, \ \underline{v}) = (p, \underline{\nabla} \cdot \underline{v}) - 2\nu((\underline{u},\underline{v})) + \int d\Gamma_2 \ \underline{c} \cdot \underline{v}, \tag{2.15}$$

where we have defined

$$((\underline{u},\underline{v})) = \sum_{i,j=1}^{d} (D_{ij}(\underline{u}), D_{ij}(\underline{v})). \qquad (2.16)$$

Thus, problem (2.1) - (2.7) can be restated as follows: given $\underline{b}$, $\underline{c}$ and $\underline{u}_0$, find $\underline{u}(\underline{x}, t) \in H^1(\Omega)^d$ and $p(\underline{x}, t) \in L^2(\Omega)$ such that

$$(\frac{\partial \underline{u}}{\partial t},\underline{v})+((\underline{u}\cdot\underline{\nabla})\underline{u},\underline{v}) = (p,\underline{\nabla}\cdot\underline{v}) - 2\nu((\underline{u},\underline{v})) + \int d\Gamma_2\underline{c}\cdot\underline{v},$$
$$\forall \underline{v} \in H^1(\Omega)^d, \ \underline{v}\big|_{\Gamma_1} = 0, \qquad (2.17)$$

$$(\underline{\nabla}\cdot\underline{u},q) = 0, \qquad \forall q \in L^2(\Omega), \qquad (2.18)$$

$$\underline{u}\big|_{\Gamma_1} = \underline{b}, \qquad (2.19)$$

$$\underline{u}(\underline{x},0) = \underline{u}_0(\underline{x}). \qquad (2.20)$$

Under suitable smoothness conditions for the boundary data $\underline{b}$ and $\underline{c}$, the Navier-Stokes problem (2.17) - (2.20) admits at least one solution (see, f.i. Temam, 1977).

3. TIME DISCRETIZATION : FRACTIONAL-STEP METHOD

In this section we describe a discretization in time of problem (2.17) - (2.20) by means of a finite difference method. We will consider the method proposed by Chorin (1968) which is based on an orthogonal decomposition theorem due to Ladyzhenskaya (1969). This theorem states that any vector field which is the gradient of a scalar function is orthogonal to any solenoidal vector field with zero normal component on the boundary. By virtue of this theorem, the momentum equation and the incompressibility condition in the Navier-Stokes problem can be treated according to a procedure consisting of two distinct phases or fractional steps. Let us denote by $\underline{u}^n$ and p^n the velocity and pressure solution of the discretized equations (to be defined) at time t^n, where $t^n = t^{n-1} + \Delta t^n$ and $t^0 = 0$. From $\underline{u}^n$ and the boundary conditions the fields $\underline{u}^{n+1}$ and p^{n+1} at time t^{n+1} are calculated as follows. First, an intermediate velocity field $\underline{u}^{n+\frac{1}{2}}$, not satisfying the condition of incompressibility, is calculated as the solution of a discretized version of the momentum equation without the pressure term. Then, this intermediate velocity field is decomposed into a divergenceless field and a gradient vector field. The former is the final divergenceless velocity field $\underline{u}^{n+1}$, whereas the latter is related to the gradient of the pressure field p^{n+1}. As originally pointed out by Chorin (1969, p. 351-352), in this fractional-step method the velocity boundary conditions cannot be satisfied in a pointwise manner. In fact, it turns out that the final solenoidal velocity field $\underline{u}^{n+1}$ resulting from the second step is such that the boundary conditions for the normal component of the velocity are satisfied exactly, whereas the conditions for the tangential components are satisfied in a weak sense only (see, also, Temam, 1977, p.397-398).

More specifically, $\underline{u}^{n+1}$ and p^{n+1}, $n = 0,1,2,\ldots$, are defined by the following algorithm. The calculation starts from $\underline{u}^0 = \underline{u}_0$, with $\underline{\nabla}\cdot\underline{u}_0 = 0$. Then from $\underline{u}^n$ and boundary data $\underline{b}^{n+1} = \underline{b}(\underline{x},t^{n+1})$ and $\underline{c}^n = \underline{c}(\underline{x},t^n)$, find the intermediate velocity field $\underline{u}^{n+\frac{1}{2}} \in H^1(\Omega)^d$ such that

$$\left(\frac{\underline{u}^{n+\frac{1}{2}} - \underline{u}^n}{\Delta t^{n+1}}, \underline{v}\right) = -((\underline{u}^n\cdot\underline{\nabla})\underline{u}^n, \underline{v}) - 2\nu((\underline{u}^n,\underline{v})) +$$
$$+ \int d\Gamma\, \underline{c}^n\cdot\underline{v}\,, \qquad (3.1)$$
$$\forall \underline{v} \in H^1(\Omega)^d, \quad \underline{v}|_{\Gamma_1} = 0,$$

$$\underline{u}^{n+\frac{1}{2}}\Big|_{\Gamma_1} = \underline{b}^{n+1}. \qquad (3.2)$$

The final velocity $\underline{u}^{n+1}$ and the pressure p^{n+1} are obtained by adding to $\underline{u}^{n+\frac{1}{2}}$ the dynamical effect of the pressure, determined so that the incompressibility condition remains satisfied, and by imposing the boundary condition (2.9) for the normal component of the velocity: find $\underline{u}^{n+1} \in H^1(\Omega)^d$ and $p^{n+1} \in L^2(\Omega)$ such that

$$\left(\frac{\underline{u}^{n+1} - \underline{u}^{n+\frac{1}{2}}}{\Delta t^{n+1}}, \underline{v}\right) - (p^{n+1},\underline{\nabla}\cdot\underline{v}) = 0\,, \qquad (3.3)$$
$$\forall \underline{v} \in H^1(\Omega)^d, \quad \underline{n}\cdot\underline{v}|_{\Gamma_1} = 0,$$

$$(\underline{\nabla}\cdot\underline{u}^{n+1},q) = 0, \quad \forall q \in L^2(\Omega), \qquad (3.4)$$

$$\underline{n}\cdot\underline{u}^{n+1}\Big|_{\Gamma_1} = \underline{n}\cdot\underline{b}^{n+1}. \qquad (3.5)$$

It is important to notice how the boundary conditions on the velocity are imposed in the present fractional-step method. In the first step, all the components of the boundary velocity $\underline{b}^{n+1}$ are imposed in determining $\underline{u}^{n+\frac{1}{2}}$, whereas, in the second step, only the normal component $\underset{\sim}{n} \cdot \underset{\sim}{b}^{n+1}$ is imposed on the final velocity field $\underset{\sim}{u}^{n+1}$. As a result, the boundary conditions (2.10) for the tangential components of the velocity will be satisfied by $\underset{\sim}{u}^{n+1}$ in a weak sense only. We can say that, in the present formulation, the boundary conditions for the normal component are of an essential type, whereas those for the tangential components are of a natural-like character.

It is worth poiting out that other time integration schemes different from the simple two-level, fully-explicit one employed in (3.1) can also be considered without introducing any modification in the formulation of the equations of the second step (3.3) – (3.5).

4. SPATIAL DISCRETIZATION BY FINITE ELEMENTS

A fully discretized form of problem (3.1) to (3.5) is obtained by replacing spaces $H^1(\Omega)^d$ and $L^2(\Omega)$ by their finite dimensional counterparts. We shall consider two different spatially discrete models of the equations governing two-dimensional plane and axisymmetric problems. Both models are based on Lagrangian quadrilateral finite elements and make use of interpolations of different order for the velocity and the pressure. Since $p \in L^2(\Omega)$, continuity of the pressure on inter-element boundaries is not requested. On the other hand, $\underset{\sim}{u} \in H^1(\Omega)^d$ implies that the interpolation of the velocity must have a C^o-continuity.

The first spatially discrete model is based on 4-node isoparametric elements with bilinear local velocity field and elementwise constant pressure. The second model uses 9-node isoparametric elements with biquadratic local velocity field and bilinear discontinuous pressure interpolation. The shape functions for $\underset{\sim}{u}$ and p will be denoted by $\underset{\sim}{L}_j$ and N_k, respectively, and their expressions can be found in Zienkiewicz (1977). Let $\underset{\sim}{U}$ and $\underset{\sim}{P}$ denote the vectors of nodal values of the finite dimensional approximation of $\underset{\sim}{u}$ and p, respectively. Equations (3.1) and (3.2) become, in matrix notation,

$$\underset{\sim}{M}(\underset{\sim}{U}^{n+\frac{1}{2}} - \underset{\sim}{U}^n)/\Delta t^{n+1} = -\left[\underset{\sim}{A}(\underset{\sim}{U}^n) + \nu\underset{\sim}{D}\right]\underset{\sim}{U}^n + \underset{\sim}{F}^n, \qquad (4.1)$$

$$\underset{\sim}{U}^{n+\frac{1}{2}}\Big|_{\Gamma_1} = \underset{\sim}{B}^{n+1}, \qquad (4.2)$$

where $\underset{\sim}{M}$ is the consistent mass matrix, $\underset{\sim}{A}$ the advection matrix, $\underset{\sim}{D}$ the diffusion matrix resulting from the scalar product $(D_{ij}(\underset{\sim}{u}), D_{ij}(\underset{\sim}{v}))$, and $\underset{\sim}{F}^n$ is the vector of nodal loads due to the stress boundary conditions. Furthermore, $\underset{\sim}{U}|_{\Gamma_1}$ denotes the vector of velocity values at boundary nodes on Γ_1 and $\underset{\sim}{B}$ is the corresponding vector of nodal values of the velocity boundary data $\underset{\sim}{b}$. Similarly, the spatially discrete form of problem (3.3) - (3.5) is

$$\underset{\sim}{M}(\underset{\sim}{U}^{n+1} - \underset{\sim}{U}^{n+\frac{1}{2}})/\Delta t^{n+1} - \underset{\sim}{C}\,\underset{\sim}{P}^{n+1} = 0, \tag{4.3}$$

$$\underset{\sim}{C}'^{T}\,\underset{\sim}{U}^{n+1} = 0, \tag{4.4}$$

$$\underset{\sim}{n}\cdot\underset{\sim}{U}^{n+1}\Big|_{\Gamma_1} = \underset{\sim}{n}\cdot\underset{\sim}{B}^{n+1}, \tag{4.5}$$

where $\underset{\sim}{C}$ and $\underset{\sim}{C}'$ are rectangular matrices whose elements are defined by integrals of the type $\int d\Omega \underset{\sim}{\nabla}\cdot\underset{\sim}{L}_j N_k$. The matrices $\underset{\sim}{C}$ and $\underset{\sim}{C}'$ differ in that the degrees of freedom of the velocity corresponding to the component normal to the boundary on Γ_1 are excluded in the construction of $\underset{\sim}{C}$, whereas they are retained in $\underset{\sim}{C}'$. The essential boundary condition (4.5) can be embodied into equation (4.4) so that system (4.3) - (4.5) can be rewritten in symmetric form as

$$\underset{\sim}{M}(\underset{\sim}{U}^{n+1} - \underset{\sim}{U}^{n+\frac{1}{2}})/\Delta t^{n+1} - \underset{\sim}{C}\,\underset{\sim}{P}^{n+1} = 0, \tag{4.6}$$

$$\underset{\sim}{C}^{T}\,\underset{\sim}{U}^{n+1} = \underset{\sim}{E}^{n+1}, \tag{4.7}$$

where

$$\underset{\sim}{E}^{n+1} = -(\underset{\sim}{C}'^{T} - \underset{\sim}{C}^{T})\,\underset{\sim}{B}^{n+1}.$$

Equations (4.6) - (4.7) represent a linear system of coupled equations for the unknowns $\underset{\sim}{U}^{n+1}$ and $\underset{\sim}{P}^{n+1}$. The dimension of the system is equal to the number of the degrees of freedom of the velocity and pressure altogether.

Notice that $\underset{\sim}{U}$ in (4.6) - (4.7) does not involve the normal component on Γ_1 and therefore the meaning of $\underset{\sim}{U}$ here is slightly different from that used in (4.1) - (4.2). However, the two quantities have been denoted by the same symbol for notational simplicity.

5. PROJECTION OPERATOR

In this section we present an equivalent formulation of the equations (3.3) - (3.5) of the second step that will be useful for the analysis of the peculiar structure of the linear system of algebraic equations (4.6) - (4.7). The particular case of homogeneous velocity boundary conditions on the entire boundary, i.e., $\underset{\sim}{u}|_{\Gamma} = 0$, will be addressed in order to simplify the analysis, while retaining all the essential features of the fundamental incompressibility problem.

The orthogonal decomposition theorem, which is fundamental for Chorin's fractional-step method, can be stated as follows (see Temam, 1977, p. 15-16):

$$L^2(\Omega)^d = H \oplus H^\perp , \qquad (5.1)$$

where H is the space

$$H = \{u \mid u \in L^2(\Omega)^d, \quad \nabla \cdot u = 0, \quad n \cdot u \big|_\Gamma = 0\}, \qquad (5.2)$$

and $H^\perp$ is the orthogonal complement of H (in $L^2(\Omega)^d$) and can be characterized as follows

$$H^\perp = \{u \mid u \in L^2(\Omega)^d, \quad u = \nabla p, \quad p \in H^1(\Omega)\}. \qquad (5.3)$$

In (5.1), the symbol $\oplus$ denotes an orthogonal sum, which means that vectors u_1 and u_2 belonging to different subspaces are mutually orthogonal, i.e., $\int d\Omega u_1 \cdot u_2 = 0$. Temam (1977, p. 397-398) has shown that the solution u^{n+1} of problem (3.3) - (3.5) with homogeneous velocity boundary conditions can be characterized equivalently as the solution of the following problem: find $u^{n+1} \in H$ such that

$$(u^{n+1}, w) = (u^{n+\frac{1}{2}}, w), \quad \forall w \in H. \qquad (5.4)$$

Relation (5.4) defines u^{n+1} as the orthogonal projection of $u^{n+\frac{1}{2}}$ on the space H in $L^2(\Omega)^d$, and therefore it can be rewritten as

$$u^{n+1} = \mathscr{P}_H u^{n+\frac{1}{2}} , \qquad (5.5)$$

where $\mathscr{P}_H$ denotes the orthogonal projection operator in $L^2(\Omega)^d$ on the space H. As now will be shown, the matrix form of equations (4.6) - (4.7) resulting from the finite-element spatial discretization of the decomposition problem (second step of the fractional-step method) provides an explicit representation of the spatially discrete counterpart of the projection operator $\mathscr{P}_H$. The construction of this operator also suggests a convenient elimination scheme for solving the linear system (4.6) - (4.7) without storing its exceedingly large matrix of coefficients. Equation (4.6), solved with respect to U^{n+1}, gives

$$U^{n+1} = U^{n+\frac{1}{2}} + \Delta t^{n+1} M^{-1} C P^{n+1} . \qquad (5.6)$$

Now, substituting U^{n+1} from (5.6) into (4.7), and assuming homogeneous velocity boundary conditions, we obtain the following linear system which governs the pressure field:

$$(C^T M^{-1} C) P^{n+1} = - (\Delta t^{n+1})^{-1} C^T U^{n+\frac{1}{2}} . \qquad (5.7)$$

The matrix $(C^T M^{-1} C)$ exhibits a rank deficiency which is in relation with the well-known fact that, in the case of purely Dirichlet boundary conditions, the pressure is defined up to an additive constant. System (5.7) can be solved by resorting to the mathematical tools provided by the theory of the generalized inverses (Rao and Mitra, 1971). The solution P^{n+1} of (5.7) can be expressed as

$$P^{n+1} = - (\Delta t^{n+1})^{-1} (C^T M^{-1} C)^- C^T U^{n+\frac{1}{2}} , \qquad (5.8)$$

where $(C^T M^{-1} C)^-$ denotes the generalized inverse of the square matrix $(C^T M^{-1} C)$. By back-substitution of P^{n+1} into (5.6) we find that

$$U^{n+1} = \left[I - M^{-1} C (C^T M^{-1} C)^- C^T \right] U^{n+\frac{1}{2}} \qquad (5.9)$$

or, equivalently,

$$U^{n+1} = \left[I - C^{T-} C^T \right] U^{n+\frac{1}{2}} . \qquad (5.10)$$

Here $C^{T-} = M^{-1} C (C^T M^{-1} C)^-$ is a generalized inverse of the rectangular matrix C^T and I is the usual identity matrix. By defining

$$\mathcal{L} = C^{T-} C^T = M^{-1} C (C^T M^{-1} C)^- C^T , \qquad (5.11)$$

and

$$\mathcal{P} = I - \mathcal{L} = I - M^{-1} C (C^T M^{-1} C)^- C^T , \qquad (5.12)$$

(5.10) can be rewritten in the form

$$U^{n+1} = \left[I - \mathcal{L} \right] U^{n+\frac{1}{2}} = \mathcal{P} U^{n+\frac{1}{2}} . \qquad (5.13)$$

It is readily shown that $\mathcal{L}^2 = \mathcal{L}$ and henceforth also $\mathcal{P}^2 = \mathcal{P}$, so that $\mathcal{L}$ and $\mathcal{P}$ are projection operators in the finite dimensional spaces of the considered approximation. Furthermore, by definition of $\mathcal{P}$, one has that $C^T \mathcal{P} U = 0$ for any U. Therefore, $\mathcal{P}$ is the discretized counterpart of the operator projecting on the space H (5.2) of divergenceless velocity fields with zero normal component on the boundary. It must be noticed that, through equation (5.10) or (5.13) the formal elimination of the pressure variable from the equations is attained, the discrete form of the incompressibility condition being satisfied due to the 'filtering' action of the operator $\mathcal{P}$ in (5.13). However, owing to the very large dimension of the matrix operator $\mathcal{P}$, it is more convenient, for reasons of computational efficiency, to retain the pressure in the numerical scheme and to simulate the projection (5.13) by the following procedure.

Firstly, evaluate the right hand side of (5.7) by multiplying the known vector $\underline{U}^{n+\frac{1}{2}}$ by matrix $\underline{C}^T$. Then, calculate the solution $\underline{P}^{n+1}$ of the linear system (5.7) with a number of equations equal to the number of the degrees of freedom of the pressure (minus one). Finally, determine the vector $\underline{U}^{n+1}$ by means of equation (5.6). This scheme can be regarded as a block Gaussian elimination scheme which exploits the special structure of the original linear system (4.6) - (4.7). Matrices $\underline{C}$ and $\underline{C}^T$ need not be stored since the usual element-by-element multiplication technique, typical of the finite element method, can be employed to compute $\underline{C}^T\underline{U}$ and $\underline{C}\underline{P}$, and also to generate matrix $(\underline{C}^T\underline{M}^{-1}\underline{C})$ once and for all at the beginning of the computation.

6. DIAGONAL MASS REPRESENTATION

To arrive at a computationally simple and efficient method for solution of equations (4.1) - (4.2) and (4.6) - (4.7), it is highly attractive to use an explicit time integration procedure. In these conditions, the only system to be solved is the linear system (5.7) governing the discrete pressure field and this reflects the implicit character of pressure in an incompressible flow.

To make explicit schemes viable, it is necessary to replace the consistent mass matrix by some approximated mass matrix which appears in diagonal form. In this section we discuss the effect of replacing the consistent mass matrix $\underline{M}$ by its diagonal approximation $\underline{M}_d$ which is obtained by adding all terms of each row of the consistent mass matrix and placing the results on the diagonal. The study of this approximation problem in the fractional-step method would require, at least in principle, consideration of the complete procedure comprising the two steps in order to include possible interaction effects. However, in accordance with the split character of the fractional-step method, we will consider separately the effect of the diagonal mass representation on the momentum equation and on the pressure equation which involves the projection operator. The first effect will be discussed in connection with a linearized model of the momentum equation: the advection-diffusion equation in one dimension. The analysis of this scalar one-dimensional linear equation can be generalized without substantial difficulties to the case of a vector equation in two and three dimensions. In addition, the analysis provides results applicable locally in fully non-linear situations. The effect of the diagonal mass approximation on the projection operation will be only briefly touched upon, the main justification for its adoption relying almost entirely upon the analysis of the case of a uniform mesh and upon the satisfactory results of the numerical experiments.

6.1 Advection-diffusion equation

Let us consider the linearized advection-diffusion equation in one dimension

$$\frac{\partial u}{\partial t} + v\,\frac{\partial u}{\partial x} - \nu\frac{\partial^2 u}{\partial x^2} = 0, \tag{6.1}$$

where v and ν are constant coefficients. Assuming an initial condition in the form of a spatial Fourier component, namely, $u(x,o) = \hat{u}\,e^{ikx}$, the exact solution of the partial differential equation (6.1) is

$$u(x,t) = \hat{u}\,e^{ikx-\lambda t}, \tag{6.2}$$

where

$$\lambda = \nu k^2 + ivk = \delta + i\omega. \tag{6.3}$$

By definition, $\delta = \nu k^2$ is the exact damping parameter and $\omega = vk$ is the exact frequency.

For a uniform mesh of piecewise linear or quadratic finite elements, the Galerkin formulation provides spatially discrete equations for the nodal values $u_j(t)$, $j = \ldots, 1,2,\ldots$ The equations for the linear element are

$$(1+r'D_1)\dot{u}_j + \frac{v}{2h}\,A_1 u_j - \frac{\nu}{h^2}\,D_1 u_j = 0, \tag{6.4}$$

whereas the equations for the parabolic approximation are of two kinds according to whether a corner (j even) or a mid-side (j odd) node is considered, namely

$$(1+r''(2D_1 - D_2))\dot{u}_j + \frac{v}{2\ell}(2A_1 - \frac{1}{2}A_2)u_j$$

$$- \frac{v}{\ell^2}(2D_1 - \frac{1}{4}D_2)u_j = 0, \qquad (6.5_1)$$

$$(1+r''D_1)\dot{u}_j + \frac{v}{2\ell}A_1 u_j - \frac{v}{\ell^2}D_1 u_j = 0, \qquad (6.5_2)$$

where h and 2ℓ = h are the dimensions of the linear and quadratic elements, respectively, and the following difference operators have been introduced

$$A_i u_j = u_{j+i} - u_{j-i} , \qquad (6.6)$$

$$D_i u_j = u_{j+i} - 2u_j + u_{j-i} . \qquad (6.7)$$

In (6.4) and (6.5), r'=1/6 and r''=1/10 for the consistent mass representation, whereas r' = r'' = 0 for the diagonal mass representation.

The exact solution of the semidiscrete equations (6.4) of the linear approximation is sought in the factorized form

$$u_j(t) = \bar{u}(t)e^{ikjh} , \qquad (6.8)$$

where $\bar{u}(t)$ is an unknown function of time. Substituting (6.8) into (6.4), we obtain an ordinary differential equation for $\bar{u}(t)$ in the form

$$(1-2r'(1-\cos p))\dot{\bar{u}} + \left[\frac{i\omega\sin p}{p} + \frac{\delta 2(1-\cos p)}{p^2} \right]\bar{u} = 0 \qquad (6.9)$$

where p = kh is the dimensionless wave number. The solution of (6.9) is therefore

$$\bar{u}(t) = \bar{u}_o e^{-\bar{\lambda}t} \qquad (6.10)$$

where $\bar{u}_o$ is fixed by the 'initial' condition, i.e. $\bar{u}_o = \bar{u}(0) = \hat{u}$, and the semidiscrete response complex variable $\bar{\lambda} = \bar{\delta} + i\bar{\omega}$ is given by

$$\frac{\bar{\delta}}{\delta} = \frac{2(1-\cos p)/p^2}{1-2r'(1-\cos p)} , \qquad (6.11)$$

$$\frac{\bar{\omega}}{\omega} = \frac{\sin p /p}{1-2r'(1-\cos p)} . \qquad (6.12)$$

The dependence of the damping ratio $\bar{\delta}/\delta$ on the wave number p

110

is shown in Fig. 6.1 for the two approximations of the mass matrix. The consistent mass representation enhances the damping whereas the diagonal mass representation attenuates the damping with respect to its exact value. Both effects are greater at higher spatial frequencies. As far as the frequency ratio $\bar{\omega}/\omega$ is concerned, Fig. 6.2 shows that the frequency in the region of short wavelengths is depressed in both cases, the consistent mass representation providing a frequency response better than that of the diagonal mass representation.

The case of the parabolic finite element is treated in a slightly different manner owing to the presence of two kinds of nodes. Let us introduce the two-component vector $\underset{\sim}{v}_m = (u_{2m}, u_{2m-1})$ for any integer m, i.e. m = ...1,2,... . The solution of (6.5) is sought in the form

$$\underset{\sim}{v}_m(t) = \underset{\sim}{\bar{v}}(t)e^{ikmh}, \tag{6.13}$$

which leads to a system of two ordinary differential equations

$$M\underset{\sim}{\dot{\bar{v}}} + (i\omega A + \delta D)\underset{\sim}{\bar{v}} = 0 \tag{6.14}$$

where the matrices M, A and D are defined by

$$M = \begin{bmatrix} 1-2r''(1+\cos p) & 2r''(1+e^{ip}) \\ r''(1+e^{-ip}) & (1-2r'') \end{bmatrix}$$

$$A = \frac{1}{p}\begin{bmatrix} -\sin p & -2(1-e^{ip})/i \\ (1-e^{-ip})/i & 0 \end{bmatrix}$$

$$D = \frac{1}{p^2}\begin{bmatrix} 2(7+\cos p) & -8(1+e^{ip}) \\ -4(1+e^{-ip}) & 8 \end{bmatrix}$$

with $p = kh = 2k\ell$. The solution of (6.14) is $\underset{\sim}{\bar{v}}(t) = \underset{\sim}{\bar{v}}_o e^{-\bar{\lambda}t}$, with $\underset{\sim}{\bar{v}}_o$ and $\bar{\lambda}$ solutions of the generalized eigenvalue problem

$$(-\bar{\lambda}M + i\omega A + \delta D)\underset{\sim}{\bar{v}}_o = 0 \tag{6.15}$$

The vanishing of the determinant of (6.15) gives, in the case of pure diffusion,

$$\frac{\bar{\delta}}{\delta} = \frac{11+C-6r''\alpha \mp \left[(11+C-6r''\alpha)^2 - 48\alpha(1-2r''\beta)\right]^{\frac{1}{2}}}{p^2(1-2r''\beta)}, \tag{6.16}$$

and, in the case of pure advection,

$$\frac{\bar{\omega}}{\omega} = \frac{-(1+6r'')S \pm \left[(1+6r'')^2 S^2 + 16\alpha(1-2r''\beta)\right]^{\frac{1}{2}}}{2p\,(1-2r''\beta)} \qquad (6.17)$$

where, for notational simplicity, we have introduced $S = \sin p$, $C = \cos p$, $\alpha = 1-\cos p$ and $\beta = 2+\cos p$. The first sign in (6.16) and (6.17) corresponds to the "physical" solution, whereas the second corresponds to a spurious mode of the parabolic interpolation. The latter has, in the limit $p \to 0$, an infinite damping and a phase velocity of opposite sign and of magnitude twice that of the exact one (Laval, 1980). The damping ratio $\bar{\delta}/\delta$ and the frequency ratio $\bar{\omega}/\omega$ of the fundamental mode of the quadratic element are plotted in Figs. 6.1 and 6.2, respectively. Comparison with the linear element shows that the damping response of the higher order element is better irrespective of the mass representation. On the other hand, the frequency response of the quadratic element with a diagonal mass approximation is inferior to the frequency response of the linear element in the consistent mass representation. In either case, the application of a time integration operator leads to a further discretization of the semidiscrete equations and one should thus consider the effect of time discretization on the damping and frequency responses before any final judgement can be made. This will be done in Section 7.

6.2 Diagonal mass matrix and projection operator

By virtue of the analysis of Section 5, the effect of the mass matrix representation on the equations of the second step can be discussed in connection with the equations providing the matrix representation of the projection operator. For a uniform mesh of rectangular elements, it can be shown that, if the consistent mass matrix $\underline{M}$ is replaced in (5.12) by its diagonalized counterpart $\underline{M}_d$, the same projection operator matrix $\underline{\mathscr{P}}$ is obtained, i.e., we have

$$\underline{\mathscr{P}} = \underline{I} - \underline{M}_d^{-1}\,\underline{C}\,(\underline{C}^T \underline{M}_d^{-1}\,\underline{C})^- \underline{C}^T, \qquad (6.18)$$

where $\underline{\mathscr{P}}$ is defined by (5.12). This result is valid both for bilinear and biquadratic finite elements. On the other hand, it is found that the replacement of $\underline{M}$ by $\underline{M}_d$ modifies the matrix of the linear system governing the pressure. Therefore, if a diagonal mass matrix is employed in the computational

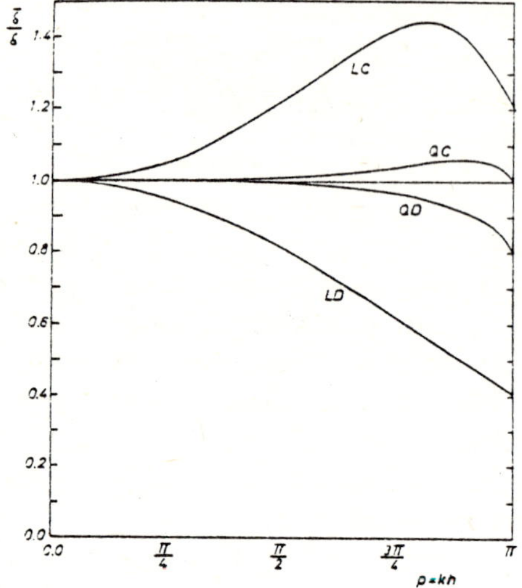

Figure 6.1 - Ratio of the semidiscrete damping parameter $\bar{\delta}$ to the exact damping parameter δ versus the dimensionless wave number. Comparison of Linear and Quadratic finite elements with Consistent and Diagonal mass representations

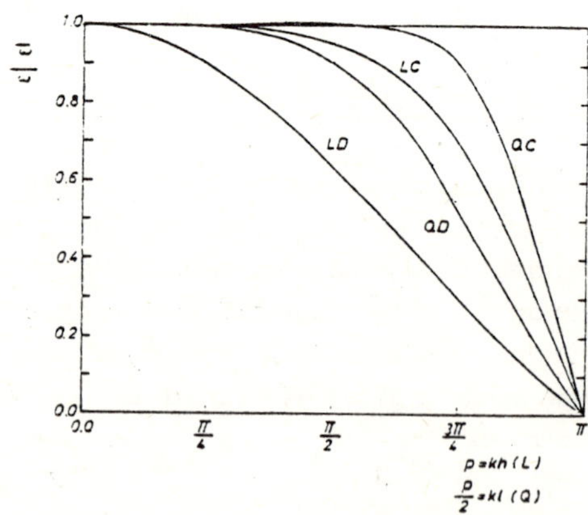

Fig. 6.2 - Ratio of the semidiscrete frequency $\bar{\omega}$ to the exact frequency ω versus the dimensionless wave number. Comparison of Linear and Quadratic finite elements with Consistent and Diagonal mass representations

scheme described at the end of Section 5, one actually calculates an approximate pressure $\underline{P}_d^{n+1}$ (a 'diagonalized pressure' we would like to say) which is related to the exact pressure $\underline{P}^{n+1}$ by the obvious relationship

$$(\underline{C}^T \underline{M}_d^{-1} \underline{C})\underline{P}_d^{n+1} = (\underline{C}^T \underline{M}^{-1} \underline{C})\underline{P}^{n+1}. \qquad (6.19)$$

But, if the diagonal mass approximation is used throughout the entire scheme, one finally recovers from $\underline{P}_d^{n+1}$ the same velocity field $\underline{U}^{n+1}$ as that which would have been obtained by using the consistent mass matrix at each stage of the computation.

For an arbitrary mesh, it is not clear whether (6.1) is still valid or whether approximations of $\underline{M}$ different from the diagonal one here considered can be devised such that the construction of the projection operator is invariant with respect to them. For practical purposes, in all our numerical experiments we have employed the diagonal mass approximation on nonuniform meshes. This has two implications: i) the condition of incompressibility is taken into account in a way that can, in principle, be different from that resulting from the consistent mass formulation; ii) the calculated pressure is in any case $\underline{P}_d$, i.e., the pressure affected by the diagonal mass approximation. Nevertheless, as will be shown later by the numerical results, we found no violation of the incompressibility condition and obtained regular pressure fields which are in excellent agreement with those obtained in the literature by means of other numerical methods.

7. TIME INTEGRATION SCHEMES AND STABILITY

In this section we introduce, always in the context of the one-dimensional advection-diffusion equation, two time-discretization schemes that we have used for explicit time integration of the Navier-Stokes equations: the two-level Euler scheme and the three-level Adams-Bashforth scheme. Since the application of a time integration operator introduces a further discretization of the (spatially) semidiscrete equations, we investigate the combined effect of the two approximations on the damping and frequency responses of the fully discretized equations. Furthermore, we provide the conditions for numerical stability of the equations resulting from the linear and quadratic spatial discretization combined with both time integration schemes.

7.1 <u>Euler explicit scheme</u>

From the discussion in Section 6.1, the solution of the semi-discrete equations (6.4) or (6.5) can be interpreted as solution of the simple ordinary differential equation

$$\dot{u} + \bar{\lambda}u = 0, \qquad (7.1)$$

where the semidiscrete response variable $\bar{\lambda} = \bar{\delta} + i\bar{\omega}$ depends on the considered spatial approximation: linear or quadratic interpolation, and consistent or diagonal mass matrix representation.

Let us discretize (7.1) by finite differences using the two-step, first-order accurate, Euler explicit scheme, viz.,

$$u^{n+1} = u^n - \bar{\lambda}\Delta t u^n. \qquad (7.2)$$

The stability condition for this scheme is obtained from $|G| \leqslant 1$ where $G = 1 - \bar{\lambda}\Delta t$ is the amplification factor. One obtains

$$\Delta t \leqslant \frac{2\bar{\delta}}{\bar{\delta}^2 + \bar{\omega}^2} . \qquad (7.3)$$

For pure advection ($\bar{\delta} = 0$) the scheme is unconditionally unstable, whereas for pure diffusion ($\bar{\omega} = 0$) the stability condition becomes

$$\Delta t \leqslant 2/\bar{\delta} . \qquad (7.4)$$

Substituting the expressions for $\bar{\delta}$ given by (6.9) and (6.13), it is easy to obtain the values of the stability limits of the diffusion equations with linear and quadratic elements in the diagonalized mass matrix approximation (see Table 7.1). In a similar manner, the stability conditions for the mixed advection-diffusion case can be calculated from (7.3) and they are reported in Table 7.1. To analyze the effect of the Euler explicit scheme (7.2) on the damping and frequency responses, we assume a solution of the discretized equations, in the form

$$u^n = \tilde{u} e^{-\tilde{\lambda}n\Delta t} , \qquad (7.5)$$

where $\tilde{u}$ is fixed by the initial condition, i.e., $\tilde{u} = u(o)$, and $\tilde{\lambda} = \tilde{\delta} + i\tilde{\omega}$, $\tilde{\delta}$ and $\tilde{\omega}$ being the fully discrete damping parameter and frequency, respectively. It follows immediately that

$$\frac{\tilde{\lambda}}{\bar{\lambda}} = - \frac{1}{\bar{\lambda}\Delta t} \ln (1 - \bar{\lambda}\Delta t), \qquad (7.6)$$

or, equivalently, separating real and imaginary parts,

$$\frac{\tilde{\delta}}{\bar{\delta}} = - \frac{1}{\bar{\delta}\Delta t} \ln\left[(1-\bar{\delta}\Delta t)^2 + (\bar{\omega}\Delta t)^2\right]^{\frac{1}{2}} , \qquad (7.7)$$

$$\frac{\tilde{\omega}}{\bar{\omega}} = \frac{1}{\bar{\omega}\Delta t} \tan^{-1}\left[\bar{\omega}\Delta t/(1-\bar{\delta}\Delta t)\right]. \qquad (7.8)$$

The ratio $\tilde{\delta}/\bar{\delta}$ is plotted in Fig. 7.1a as a function of $\bar{\delta}\Delta t$ for $\bar{\omega}\Delta t = 0.0$; 0.25 and 0.50. In a pure diffusion case, the effect of the explicit time integration is to increase the damping for all values of Δt. Since the effect of spatial discretization is to reduce the damping in the case of a diagonal mass representation (see Fig. 6.1), compensatory effects are obtained when combining a diagonal mass representation with an explicit time integration. In the case of advection-diffusion, explicit integration tends to depress the semidiscrete damping parameter, especially for small values of $\bar{\delta}\Delta t$. However, small values of $\bar{\delta}\Delta t$ correspond to small values of $\bar{\omega}\Delta t$, and the figure shows that the distortion of the damping parameter corresponding to modes with long wavelengths will be rather limited.

The ratio $\tilde{\omega}/\bar{\omega}$ is plotted in Fig. 7.1b as a function of $\bar{\omega}\Delta t$ for $\bar{\delta}\Delta t = 0.25$; 0.50; 0.75 and 1.0. The explicit time integration scheme has the effect of enhancing all the phase speeds. The effect is particularly pronounced for small values of $\bar{\omega}\Delta t$. If $\bar{\delta}\Delta t$ is small, the semidiscrete frequencies are not greatly perturbed. For higher values of $\bar{\delta}\Delta t$, the perturbation becomes more and more important. On the other hand, spatial discretization strongly depresses the frequencies of modes with short wavelengths (see Fig. 6.2). Therefore, advantageous compensatory effects are again obtained, especially in this range, when combining a diagonal mass representation and an explicit time marching scheme.

7.2 Adams-Bashforth scheme

As a possible alternative to the simple two-level Euler scheme, we consider also the three-level, second-order accurate, Adams-Bashforth scheme (Lambert, 1973). When applied to

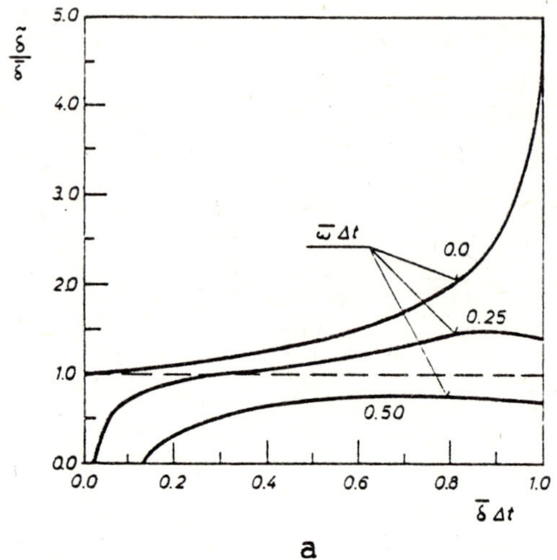

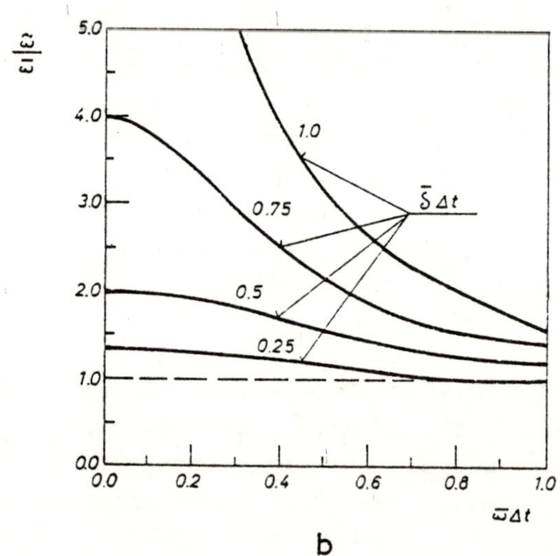

Figure 7.1 – Damping ratio $\tilde{\delta}/\bar{\delta}$ and frequency ratio $\tilde{\omega}/\bar{\omega}$ introduced by Euler's explicit integration scheme

eq. (7.1), this scheme gives

$$u^{n+1} = u^n - \bar{\lambda}\Delta t \left(\frac{3}{2} u^n - \frac{1}{2} u^{n-1}\right). \tag{7.9}$$

To study this multi-level scheme, let us introduce the two-component vector $\underset{\sim}{w} = (w_1, w_2)$ and define $w_1^n = u^n$, $w_2^n = u^{n-1}$ (Richtmyer and Morton, 1967). Then eq. (7.9) can be rewritten

$$w_1^{n+1} = w_1^n - \bar{\lambda}\Delta t \left(\frac{3}{2} w_1^n - \frac{1}{2} w_2^n\right), \tag{7.10$_1$}$$

$$w_2^{n+1} = w_1^n, \tag{7.10$_2$}$$

or, equivalently,

$$\underset{\sim}{w}^{n+1} = G \underset{\sim}{w}^n, \tag{7.11}$$

where the amplification matrix G is defined by

$$G = \begin{bmatrix} 1 - \dfrac{3}{2}\bar{\lambda}\Delta t & \dfrac{1}{2}\bar{\lambda}\Delta t \\ 1 & 0 \end{bmatrix}. \tag{7.12}$$

The conditions of numerical stability of this scheme for the two spatial discretizations here considered can be determined from (7.12) according to standard methods, using again (6.9) and (6.13) for the advection and diffusion equations, respectively. The stability limits in the diagonal mass approximation are reported in Table 7.1. For the mixed advection-diffusion case, the stability analysis is rather cumbersome. For all our practical purposes we have found it to be adequate to satisfy the two stability conditions of the advection and diffusion equations, as has been indicated in the table.

The solution of (7.11) is $\underset{\sim}{w}^n = \underset{\sim}{\tilde{w}}\, e^{-\tilde{\lambda}n\Delta t}$, with $\underset{\sim}{\tilde{w}}$ and $\tilde{\lambda}$ defined by the eigenvalue/vector problem

$$e^{-\tilde{\lambda}\Delta t}\, \underset{\sim}{\tilde{w}} = G \underset{\sim}{\tilde{w}} \tag{7.13}$$

which provides for the fully discrete response variable $\tilde{\lambda}$ the values

$$\frac{\tilde{\lambda}_{1,2}}{\bar{\lambda}} = -\frac{1}{\bar{\lambda}\Delta t}\, \ell n\, \frac{1}{2}\left\{(1 - \frac{3}{2}\bar{\lambda}\Delta t) \pm \left[1 - \bar{\lambda}\Delta t + \frac{9}{4}(\bar{\lambda}\Delta t)^2\right]^{\frac{1}{2}}\right\}. \tag{7.14}$$

TABLE 7.1 – Numerical stability conditions for the advection–diffusion equation in one dimension. Euler and Adams-Bashforth explicit time-integration schemes combined with linear and quadratic elements (diagonal mass representation).

$$c = v\,\Delta t/h; \quad d = \nu\,\Delta t/h^2 ; \quad Re_c = vh/\nu = c/d .$$

Scheme	Diffusion equation		Advection equation		Advection – diffusion equation	
	L	Q	L	Q	L	Q
Euler	$d \leqslant \dfrac{1}{2}$	$d \leqslant \dfrac{1}{4}$	unconditionally unstable		$2c \leqslant Re_c \leqslant \dfrac{2}{c}$	$4c \leqslant Re_c \leqslant \dfrac{2}{c}$
Adams–Bashforth	$d \leqslant \dfrac{1}{4}$	$d \leqslant \dfrac{1}{8}$	$c \leqslant \dfrac{1}{2}$	$c \leqslant \dfrac{1}{4}$	$d \leqslant \dfrac{1}{4},\ c \leqslant \dfrac{1}{2}$	$d \leqslant \dfrac{1}{8},\ c \leqslant \dfrac{1}{4}$

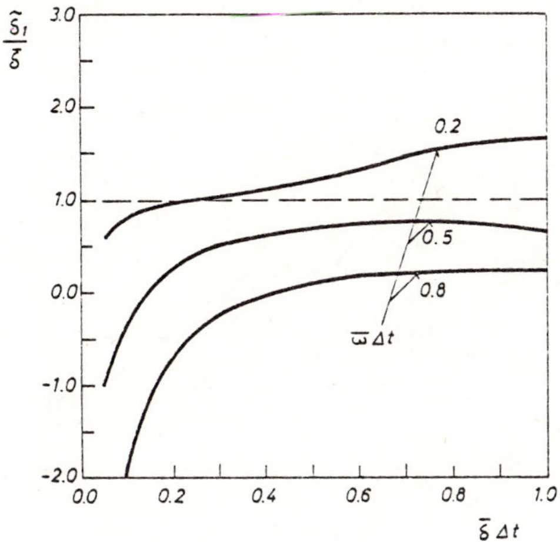

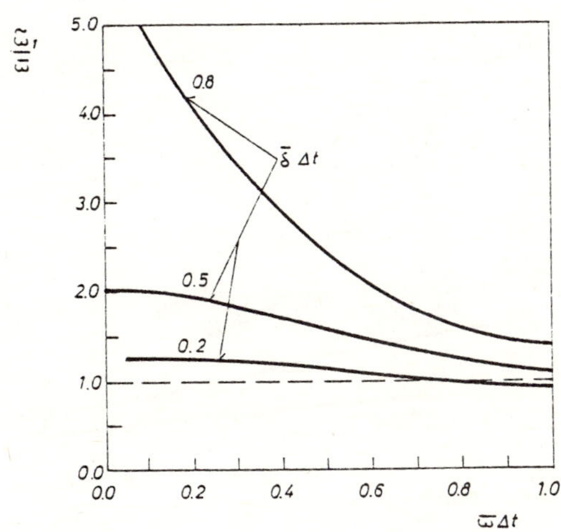

Figure 7.2 - Damping ratio $\tilde{\delta}_1/\bar{\delta}$ and frequency ratio $\tilde{\omega}_1/\bar{\omega}$ of the fundamental mode of Adams-Bashforth integration scheme

E

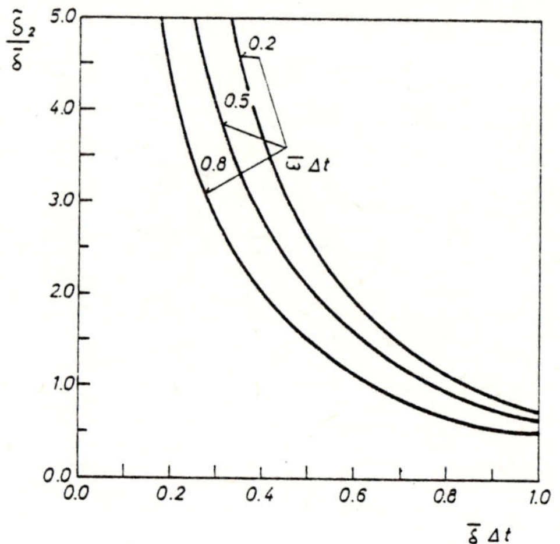

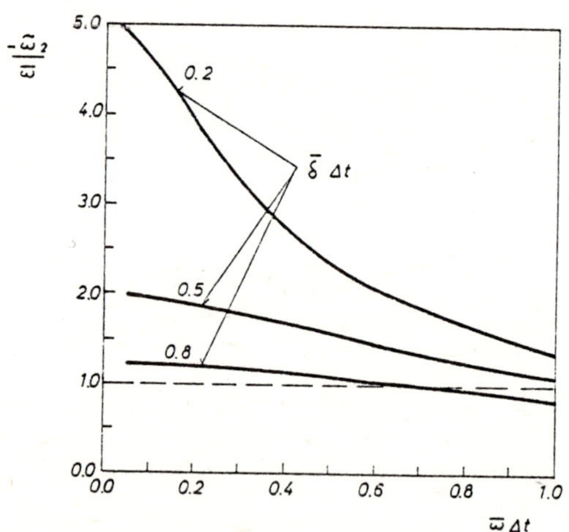

Figure 7.3 – Damping ratio $\tilde{\delta}_2/\bar{\delta}$ and frequency ratio $\tilde{\omega}_2/\bar{\omega}$ of the spurious mode of Adams-Bashforth integration scheme

The solution with the positive sign is the fundamental mode, whereas that with the negative sign corresponds to the spurious mode present in any three-level scheme. Since the separation of the real and imaginary parts of $\tilde{\lambda}$ in (7.14) is not simple to perform analytically, we have calculated numerically the ratios $\tilde{\delta}_i/\bar{\delta}$ and $\tilde{\omega}_i/\bar{\omega}$, i = 1,2, solving (7.14) in the complex field. Figures 7.2 and 7.3 report the solutions corresponding to the fundamental and spurious modes, respectively. By comparing Figs. 7.2 and 7.1 we see that the damping and frequency responses of the Adams-Bashforth scheme are very similar to those of Euler's first-order scheme. Therefore, the considerations developed before for Euler's scheme apply equally well to the three-level scheme, the only advantage of using it being that it does not suffer from the severe stability restriction ($Re_c < 2/c$) of Euler's scheme in advection dominated situations. Finally, Fig. 7.3 shows that the spurious mode is always damped and that this damping increases as the value of Δt decreases.

8. NUMERICAL EXAMPLES

8.1 Unsteady driven cavity flow

As a first application of the method described in this chapter, we consider the flow in a square cavity with a moving wall. The cavity contains a fluid of constant density and viscosity which is set into recirculating motion by applying a constant tangential velocity to the top wall (see Fig. 8.1). The spatial discretizations shown in Fig. 8.2 consist of non-uniform meshes of 20x20 four-node elements and 10x10 nine-node elements involving the same number of degrees of freedom. The solution is calculated for a Reynolds number Re = 1000. A pressure datum, p = 0, is specified at the middle of the bottom wall. The transient calculation is started from initial conditions of zero velocity and continued until a steady-state regime is reached.

In Fig. 8.3 we show contour plots of the pressure and the stream function at the beginning of the motion (t = 5). Figure 8.4 represents the same data for the steady-state solution (t = 40). The pressure fields obtained with both bilinear and biquadratic elements were found to be completely free of numerical oscillations and in substantial agreement with the results of Bercovier and Engelman (1979). The plots of the stream function clearly indicate the presence of two

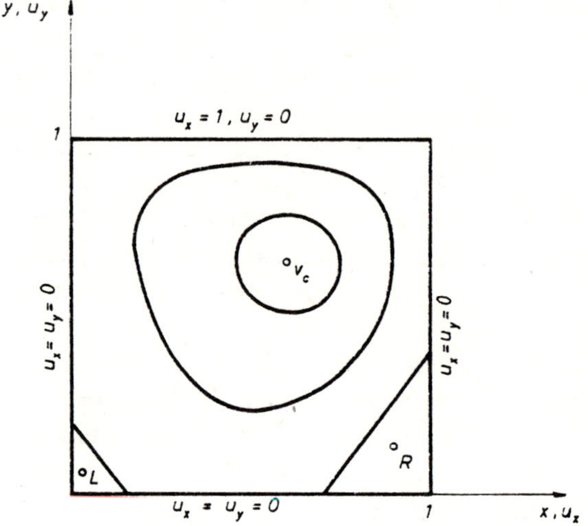

Figure 8.1 - Schematic of the driven cavity problem

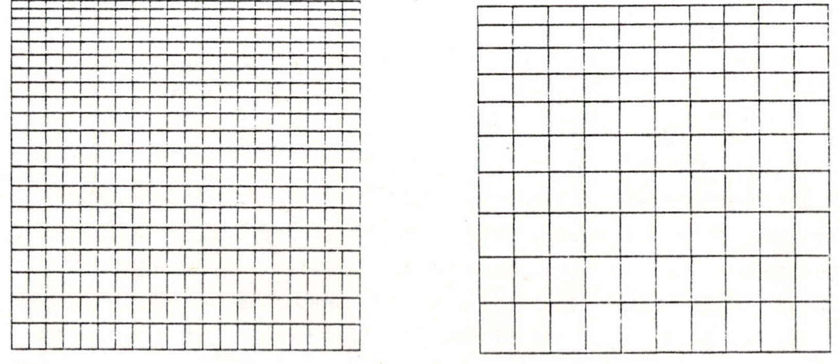

Figure 8.2 - Driven cavity problem: non-uniform meshes of 4-node and 9-node elements

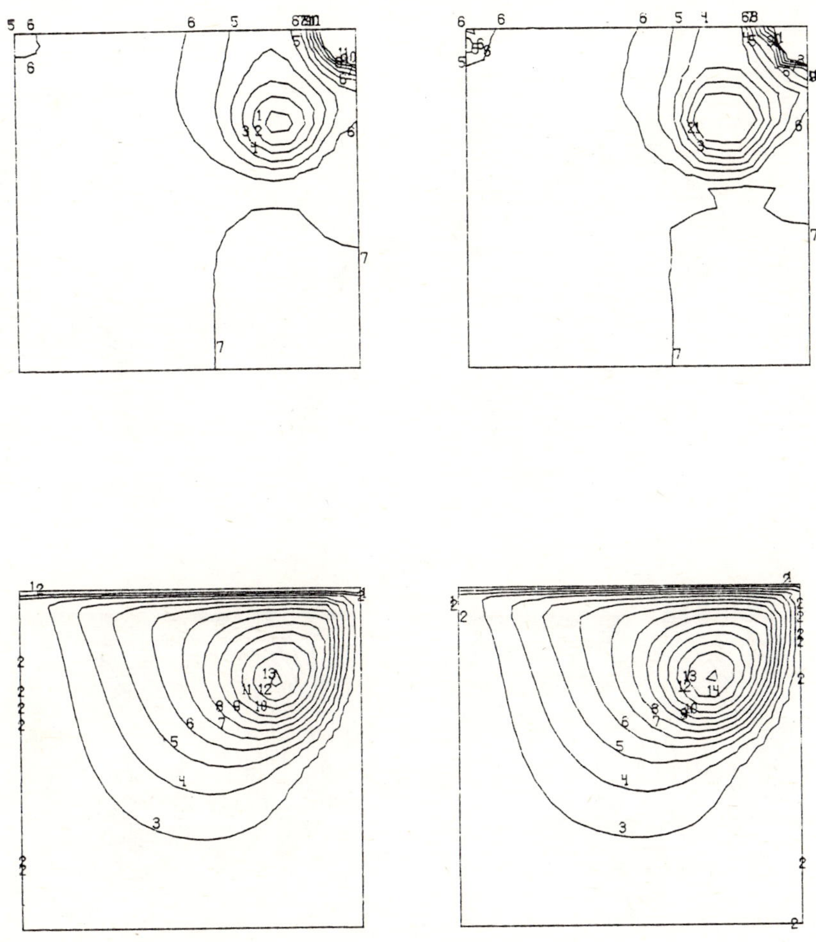

Figure 8.3 - Driven cavity problem (Re = 1000). Pressure contours and streamlines at time t = 5.
(left: bilinear elements; right: biquadratic elements).
Motion started from rest at t = 0.

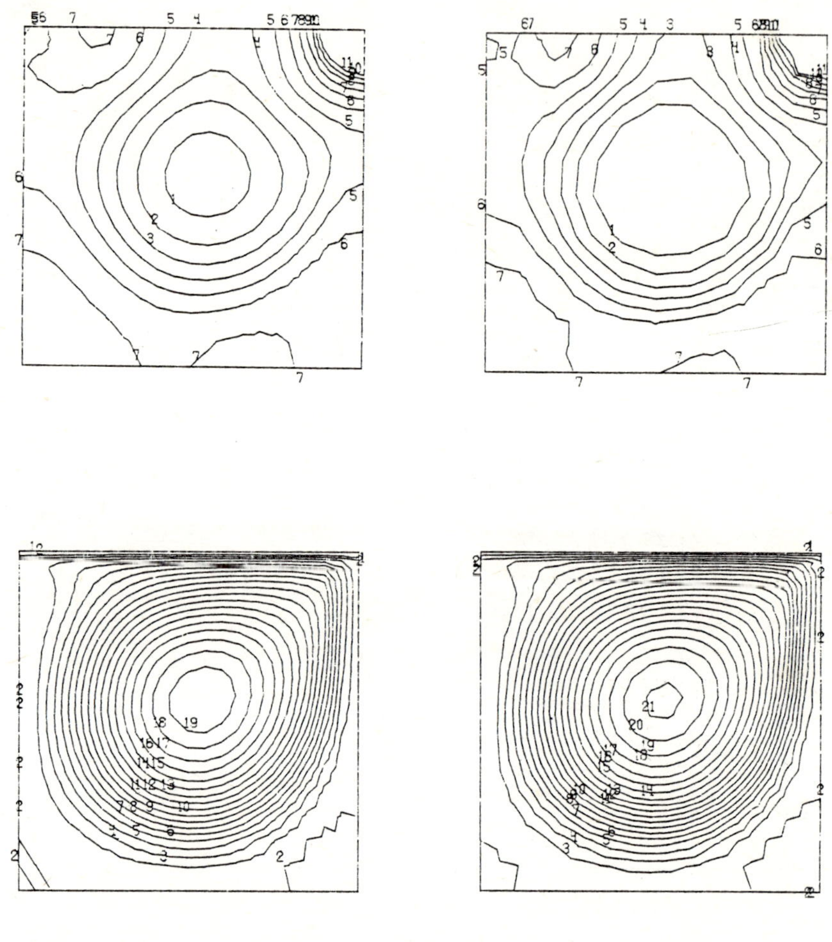

Figure 8.4 - Driven cavity problem (Re = 1000). Pressure contours and streamlines at time t = 40 (left: bilinear elements; right: biquadratic elements)

TABLE 8.1 – Unsteady driven cavity flow at Re = 1000. Comparison of results obtained with bilinear and biquadratic elements during transient motion started from rest at time t = 0. See Fig. 8.1 for definition of symbols.

	t	x_{vc}	y_{vc}	ψ_{vc} $\times 10^{-2}$	$- P_{vc}$ $\times 10^{-2}$	x_R	y_R	ψ_R $\times 10^{-4}$	x_L	y_L	ψ_L $\times 10^{-4}$
bilinear	5	0.750	0.722	5.603	6.429						
	10	0.650	0.626	7.049	6.219	0.900	0.148	− 1.38			
	15	0.600	0.626	7.862	6.255	0.850	0.148	− 4.78			
	40	0.550	0.577	8.913	6.58	0.900	0.148	− 6.41	0.05	0.072	− 0.039
	steady state										
biquadratic	5	0.750	0.724	6.052	8.058						
	10	0.650	0.626	7.682	7.512	0.950	0.34	− 3.0			
	15	0.600	0.626	8.613	7.740	0.850	0.148	− 6.37			
	40	0.550	0.574	9.637	7.796	0.900	0.148	− 9.45	0.05	0.074	− 1.60
	steady state										

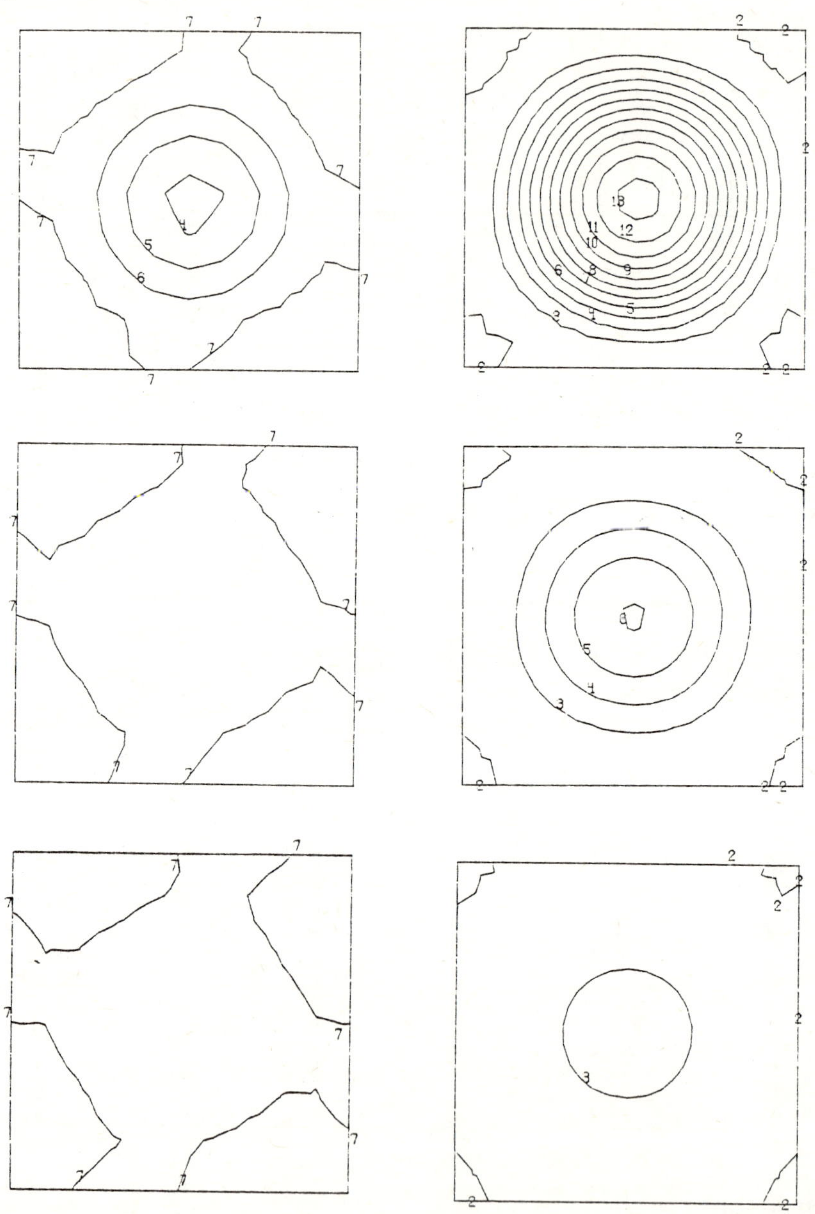

Figure 8.5 - Driven cavity problem (Re = 1000). Pressure contours and streamlines at times t = 60, 80, 100 for the viscous decay of the main vortex following reduction of the wall velocity to zero at time t = 50. Lines labelled by the same number correspond to the same value of the variable.

secondary eddies at the lower corners of the cavity. Some relevant numerical values are given in Table 8.1, which compares the position and strength of the primary and two secondary vortices, and the value of the pressure at the main vortex centre for both discretizations. The steady-state position of the main vortex centre for both discretizations is quite close to the values x = 0.55 and y = 0.56 calculated by Bercovier and Engelman on a uniform 10x10 mesh of nine-node elements.

Although the results obtained with the two discretizations are in general agreement, some discrepancies are nevertheless noted. These may probably be attributed to the fact that the boundary conditions are discontinuous at the upper corners of the cavity. In the present computations, the condition $u_x = 1$ was prescribed along the entire upper boundary including the corner nodes. However, setting this boundary condition in identical fashion in the meshes on 4-node and 9-node elements actually implies different representations along the vertical edges of the upper corner elements.

We have employed the proposed method (using nine-node elements) to model the evolution of the steady-state solution following a sudden reduction to zero of the upper-wall velocity. In Fig. 8.5 we report the solutions at t = 60, 80 and 100. The contour plots clearly show the viscous decay of the main vortex which progressively looses its initially non-symmetric character and tends to take a circular shape in the central region of the cavity. Four identical counter-rotating secondary eddies accompany the decay of the main vortex.

8.2 Axisymmetric flow through a sudden enlargement

A sketch depicting the geometry of the flow region and the boundary conditions is presented in Fig. 8.6. At the channel inlet, we assume fully developed Poiseuille flow with a Reynolds number of 60. At the outlet, we adopt boundary conditions of zero surface stress. The flow region is discretized with 234 four-node elements.

The calculations were started from an initial situation of zero-velocity and continued until a steady-state regime was obtained. Fig. 8.6 shows the streamlines; due to the coarseness of the mesh, no detailed picture of streamlines in the recirculation region could be obtained. The pressure contours are also given on Fig. 8.6; the weak treatment of the tangential velocities along the boundary produced an ex-

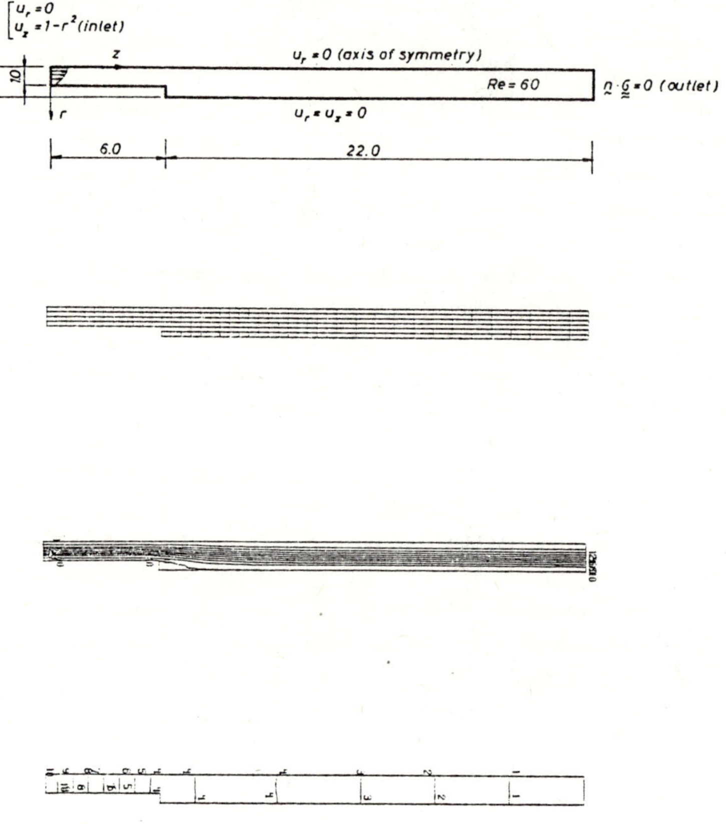

Figure 8.6 - Axisymmetric flow through a sudden elargement. Problem description; mesh of bilinear elements; streamlines and pressure contours (steady-state solution, Re = 60)

tremely regular pressure field. In particular, the pressure
was found to be perfectly uniform in the radial direction at
the inlet section, as required by the assumed inlet boundary
condition of fully developed flow. At the outlet section,
which is far downstream of the enlargement, pressure was again
found to be radially uniform.

8.3 <u>Flow over a circular cylinder</u>

Figure 8.7 shows the domain and the boundary conditions. The
steady-state solution corresponding to Re = 20 was sought
using the mesh of 4-node elements depicted in Fig. 8.7. The
simulation was started from rest and the condition of zero
tangential velocity along internal and external boundaries was
satisfied in a weak sense only. Our primary interest in running
this problem was to check the weak treatment of tangential
velocities in the presence of a curved boundary. The results
were quite satisfactory in the sense that the discrete pressure
field was found free of spurious oscillations; the pressure
contours and the accompanying streamline profiles are shown in
Fig. 8.7.

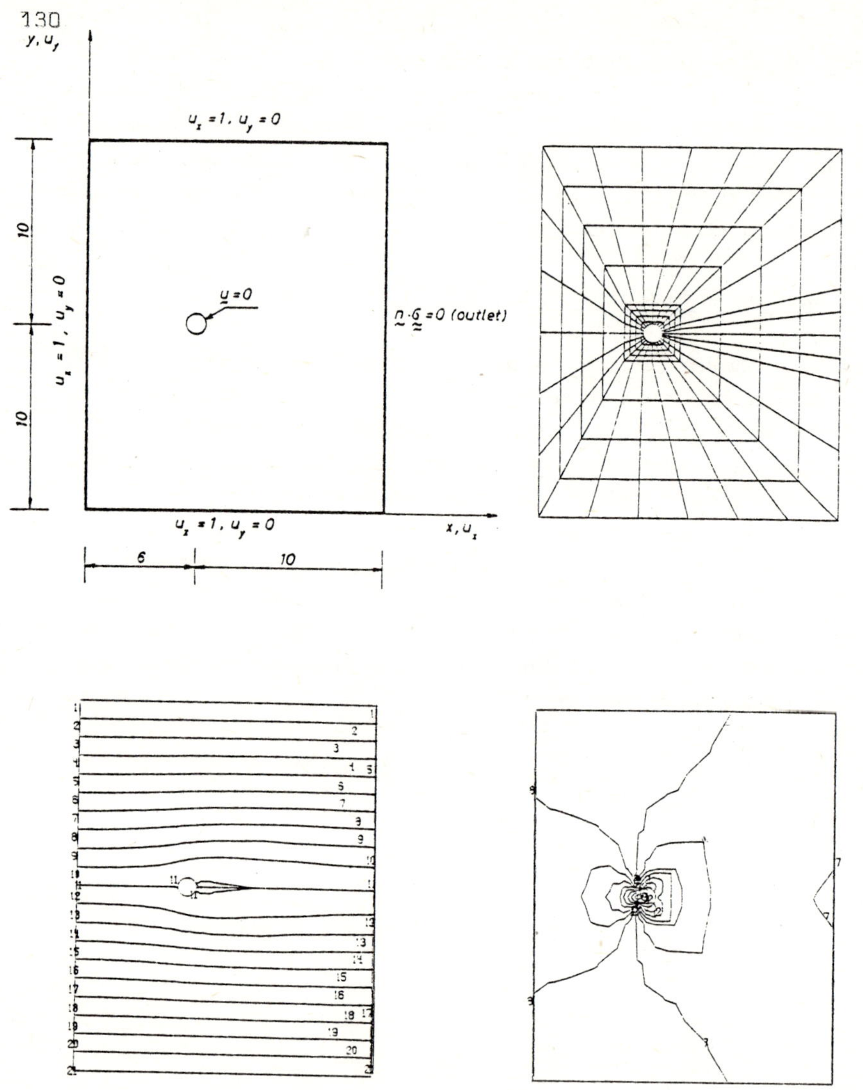

Figure 8.7 - Flow over a circular cylinder. Problem description; mesh of bilinear elements; streamlines and pressure contours (steady-state solution, Re = 20)

REFERENCES

Bercovier, M. and Engelman, M. (1979) A finite element for the numerical solution of viscous incompressible flows. J.Comp.Phys., $\underline{30}$, 181-201

Bristeau, M.O., Pironneau, O., Glowinski, R., Periaux, J. and Perrier, P. (1979) On the numerical solution of non-linear problems in fluid dynamics by least square and finite element methods. (I) Least square formulations and conjugate gradient solution of the continuous problems. Comp.Meth. Appl.Mech.Engng., $\underline{17/18}$, 619-657

Chorin, A.J. (1968) Numerical solution of the Navier-Stokes equations. Math.Comp., $\underline{22}$, 745-762

Chorin, A.J. (1969) On the convergence of discrete approximations of the Navier-Stokes equations. Math.Comp. $\underline{23}$, 341-353

Donea, J., Giuliani, S. and Laval, H. (1979) Accurate explicit finite element schemes for convective-conductive heat transfer problems, in: Finite Element Methods for Convection Dominated Flows, T.J.R. Hughes ed., AMD, Vol. 34, ASME, New York

Fortin, M. and Thomasset, F. (1979) Mixed finite-element methods for incompressible flow problems. J.Comp.Phys., $\underline{31}$, 113-145

Goda, K. (1979) A multistep technique with implicit difference schemes for calculating two- or three-dimensional cavity flows. J.Comp.Phys., $\underline{30}$, 76-95

Gresho, P.M., Lee, R.L., Stullich, T.W. and Sani, R.L. (1978) Solution of the time-dependent Navier-Stokes equations via F.E.M., in: Second Int.Conf. on Finite Elements in Water Resources, London, U.K., July 10-14

Heinrich, J.C., Marshall, R.S. and Zienkiewicz, O.C. (1978) Penalty function solution of coupled convective and conductive heat transfer. Proc. 1st Int.Conf.Num.Meth. in Laminar and Turbulent Flow, Swansea, U.K.

Hirt, C.W. and Cook, J.L. (1972) Calculating three-dimensional flows around structures and over rough terrain. J.Comp.Phys., $\underline{10}$, 324-340

132

Hughes, T.J.R., Liu, W.K. and Brooks, A. (1979) Finite element analysis of incompressible viscous flows by the penalty function formulation. J.Comp.Phys., $\underline{30}$, 1-60

Huyakorn, P., Taylor, C., Lee, R. and Gresho, P. (1978) A comparison of various mixed-interpolation finite elements in the velocity-pressure formulation of the Navier-Stokes equations. Computers & Fluids, $\underline{6}$, 25-35

Ladyzhenskaya, O.A. (1969) The Mathematical Theory of Viscous Incompressible Flow, 2nd Ed., Gordon and Breach, New York

Lambert, J.D. (1973) Computational Methods in Ordinary Differential Equations. Wiley, London

Laval, H. (1980) Etude numérique des courants de convection par la méthode des éléments finis. Dissertation, University of Louvain, Belgium

Olson, M.D. and Tuann, S.Y. (1978) Primitive variables versus stream function finite element solutions of the Navier-Stokes equations. in: Finite Elements in Fluids, Vol. 3, Wiley, London

Rao, C.R. and Mitra, S.K. (1971) Generalized Inverse of Matrices and its Applications. Wiley, New York

Raviart, P.A. (1979) Méthodes d'Eléments Finis en Mécanique des Fluides. Cours de l'Ecole d'Eté d'Analyse Numérique

Richtmyer, R.D. and Morton, K.W. (1967) Difference Methods for Initial-value Problems, 2nd Ed., Interscience Publishers, New York

Takami, H. and Kawahara, K. (1974) Numerical study of three-dimensional flow within a cubic cavity. J.Phys.Soc. Japan, $\underline{37}$, 1695-1698

Taylor, C. and Ijam, A.Z. (1979) A finite element numerical solution of natural convection in enclosed cavities. Comp.Meth.Appl.Mech.Engng., $\underline{19}$, 429-446

Temam, R. (1977) On the Theory and Numerical Analysis of the Navier-Stokes Equations. North-Holland, Amsterdam

Zienkiewicz, O.C. (1977) The Finite Element Method , 3th Ed ., Mc Graw-Hill, London

CHAPTER 5

LARGE SCALE WEATHER FORECASTING USING FINITE ELEMENT METHODS

M.J.P. Cullen

Meteorological Office, Bracknell, Berkshire, UK

1.   INTRODUCTION

This chapter discusses the application of finite element methods to large-scale weather forecasting models. Because these models use a very large amount of computing time, the shortest possible gridlength for a global model with even the most powerful current computers, such as the CRAY1 or CYBER203E, is about 150 km. This is barely sufficient to resolve important features of the atmosphere. There is therefore great interest in establishing which numerical techniques are most accurate and efficient in solving the equations. A large number of techniques have been studied, including the finite element method. This chapter describes the results obtained using it and compares them with those obtained by other techniques.

Most work on this problem has been carried out using finite difference methods, since these can be very simple. However there have been difficulties in obtaining schemes which remain stable over the period of integration, let alone producing accurate forecasts. Extensive reviews of this work can be found in Mesinger and Arakawa (1977) and Kasahara (1979). Some of these difficulties are caused by the nonlinear advective terms. The most popular technique for avoiding nonlinear instability is to require the finite difference scheme to satisfy conservation laws exactly. Part of the incentive for investigating finite element methods is that the representation of fields by functions allows more realistic nonlinear behaviour. The spherical harmonic method, where global functions are used to represent the data, also has this advantage. However, a local representation may be a more correct reflection of the underlying physics.

The forecasting equations are evolutionary and the main errors arise from evolutionary effects. The conventional analysis of the finite element method (e.g. Strang and Fix (1973), Zienkiewicz (1977)) concentrates on the error made in representing a fixed data field. When calculating asymptotic error estimates these approximation errors can mask the errors made in the evolution. It is therefore necessary to analyse the error in a way analogous to finite difference analysis as described in Cullen and Morton (1980).

The following sections of this chapter review these matters in more detail. We first discuss some properties of the forecasting equations. The error analysis is reviewed and it is shown that by appropriate combinations of finite element techniques even very complicated problems can be solved to high order accuracy by the simplest finite element representation. The advantages of the finite element solution are demonstrated in a solution of the shallow water equations on a sphere. Solutions are then presented for complete atmospheric models. It is found in these cases, however, that the forecast error becomes rather insensitive to the numerical scheme and other errors predominate. The complete model has not been a very satisfactory test-bed for numerical schemes. It is thus not clear at present whether finite elements have an important part to play.

2.   NUMERICAL WEATHER PREDICTION MODELS

2.1   The Governing Equations

In this section we briefly review the equations used in numerical weather prediction. A much fuller description of the subject is given in Haltiner (1971) and elsewhere. Nearly all current operational forecasting models use the primitive equations of motion. These can be written as follows:

$$\frac{\partial U}{\partial t} + \frac{1}{a\cos\theta} \left[\frac{\partial}{\partial \lambda}(Uu) + \frac{\partial}{\partial \theta} (Vu\cos\theta) \right] + \frac{\partial}{\partial \sigma} (p_* \dot{\sigma} u)$$

$$-V(f + \frac{u\tan\theta}{a}) + \frac{1}{a\cos\theta} (p_* \frac{\partial \phi}{\partial \lambda} + RT \frac{\partial p_*}{\partial \lambda}) = K\nabla \cdot p_* \nabla u + g\frac{\partial \tau\lambda}{\partial \sigma} \qquad (1)$$

$$\frac{\partial V}{\partial t} + \frac{1}{a\cos\theta} \left[\frac{\partial}{\partial \lambda}(Uv) + \frac{\partial}{\partial \theta} (Vv\cos\theta) \right] + \frac{\partial}{\partial \sigma} (p_* \dot{\sigma} v) + U(f + \frac{u\tan\theta}{a})$$

$$+ \frac{1}{a} (p_* \frac{\partial \phi}{\partial \theta} + RT \frac{\partial p_*}{\partial \theta}) = K\nabla \cdot p_* \nabla v + g \frac{\partial \tau\theta}{\partial \sigma} \qquad (2)$$

$$\frac{\partial \phi}{\partial \sigma} + \frac{RT}{\sigma} = 0 \tag{3}$$

$$\frac{\partial p_*}{\partial t} + \frac{1}{a\cos\theta} \left[\frac{\partial}{\partial \lambda} (U) + \frac{\partial}{\partial \theta} (V\cos\theta) \right] + \frac{\partial}{\partial \sigma} (p_*\dot{\sigma}) = 0 \tag{4}$$

$$\frac{\partial}{\partial t} (p_* T) + \frac{1}{a\cos\theta} \left[\frac{\partial}{\partial \theta}(UT) + \frac{\partial}{\partial \theta}(VT\cos\theta) \right] + \frac{\partial}{\partial \sigma} (p_*\dot{\sigma}T) - \frac{KT_\omega}{\sigma}$$

$$= k\nabla \cdot p_*^{k+1} \nabla \left(\frac{T}{p_* k}\right) + \frac{g}{C_p} \frac{\partial H}{\partial \sigma} + \frac{p_*}{C_p} [\dot{Q}+L\dot{P}] \tag{5}$$

$$\frac{\partial}{\partial t} (p_* q) + \frac{1}{a\cos\theta} \left[\frac{\partial}{\partial \lambda}(Uq) + \frac{\partial}{\partial \theta}(Vq\cos\theta) \right] + \frac{\partial}{\partial \sigma} (p_*\dot{\sigma}q)$$

$$= k\nabla \cdot p_* \nabla_q + g\frac{\partial M}{\partial \sigma} - p_*\dot{P} \tag{6}$$

The coordinates are spherical polars, λ is longitude, θ is latitude and a is the radius of the earth.

u,v	– horizontal wind components
p_*	– surface pressure
U,V	– $p_* u$, $p_* v$ momentum components
p	– pressure
$\sigma = p/p_*$	– vertical coordinate, zero at top of atmosphere, unity at earth's surface
$\dot{\sigma} \equiv \dfrac{D\sigma}{Dt} \equiv \dfrac{\partial \sigma}{\partial t} + \dfrac{u}{a\cos\theta} \dfrac{\partial \sigma}{\partial \lambda} + \dfrac{v}{a\cos\theta} \dfrac{\partial}{\partial \theta} (\sigma\cos\theta)$	– vertical velocity in σ coordinate
$\omega \equiv \dfrac{Dp}{Dt}$, D/Dt	– defined as above; vertical velocity in p coordinate
ϕ	– geopotential
f	– Coriolis parameter = $2\Omega\sin\theta$ where Ω is angular velocity of earth's rotation
g	– acceleration due to gravity
$\tau_\lambda, \tau_\theta$	– surface stress components
T	– temperature
q	– mixing ratio of water vapour
k	– diffusion coefficient
R, C_p	– gas constant and specific hear of air
K	– R/C_p

136

                Q   – sensible heating of air by radiation etc.

                L   – latent heat of evaporation

                P   – amount of condensation of water

              H,M   – vertical fluxes of sensible heat and
                              moisture due to subgridscale effects

It is seens that this system of equations contains essentially first order hyperbolics predicting momentum, heat, moisture and surface pressure. They include diffusion terms, but these are essentially artifical, since the true physical value of the diffusion coefficient is very small. Higher values are usually needed for computational reasons. The equations also include terms representing other effects such as radiation which have to be represented separately. It is difficult in any practical case to decide whether an unsuccessful forecast is due to wrong initial data, poor numerical solution of the equations, poor representation of the extra terms or other reasons. Later in this chapter a solution of (1) to (6) using finite element methods will be presented, but it is difficult to know how fair a test it is of the numerical technique.

2.2   Analysis of the equations.

In developing numerical methods to solve this system it has been common to study solutions of the shallow water equations instead (e.g. Grammeltvedt (1969)). These can be set out as follows:

$$\frac{\partial U}{\partial t} + \frac{1}{a\cos\theta} \left[\frac{\partial}{\partial \lambda} (Uu) + \frac{\partial}{\partial \theta} (Vu\cos\theta) \right]$$

$$- V(f + \frac{u\tan\theta}{a}) + \frac{\phi}{a\cos\theta} \frac{\partial \phi}{\partial \lambda} = 0 \tag{7}$$

$$\frac{\partial V}{\partial t} + \frac{1}{a\cos\theta} \left[\frac{\partial}{\partial \lambda}(Uv) + \frac{\partial}{\partial \theta} (Vv\cos\theta) \right]$$

$$+ U(f + \frac{u\tan\theta}{a}) + \frac{\phi}{a} \frac{\partial \phi}{\partial \theta} = 0 \tag{8}$$

$$\frac{\partial \phi}{\partial t} + \frac{1}{a\cos\theta} \left[\frac{\partial}{\partial \lambda} (U) + \frac{\partial}{\partial \theta} (V\cos\theta) \right] = 0 \tag{9}$$

where the notation is the same as before, except that now
$U=\phi u$, $V=\phi v$. This is a system of first order hyperbolics in
three variables. Its behaviour can be illustrated by the
following standard analysis (e.g. Temperton (1976)).
Linearise equations (7) to (9) about a basic state of rest
with $\phi=\phi_o$.

To avoid complications consider only Cartesian
geometry with $x=a\lambda\cos\theta$, $y=a\theta$ and regard f as a constant.
This gives

$$\frac{\partial u}{\partial t} - fv + \frac{\partial \phi}{\partial x} = 0 \tag{10}$$

$$\frac{\partial v}{\partial t} + fu + \frac{\partial \phi}{\partial y} = 0 \tag{11}$$

$$\frac{\partial \phi}{\partial t} + \phi_o \left(\frac{\partial u}{\partial x} + \frac{\partial v}{\partial y}\right) = 0 \tag{12}$$

This has normal mode solutions proportional to
$\exp(\,i(kx-\omega t))$ where

$$\begin{vmatrix} i\omega & -f & ik \\ f & i\omega & 0 \\ \phi_o ik & 0 & i\omega \end{vmatrix} = 0$$

so that $\omega = 0$ or $\pm \sqrt{k^2\phi_o + f^2}$. The associated eigenfunctions
have the form

$$\begin{pmatrix} u \\ v \\ \phi \end{pmatrix} = \begin{pmatrix} 0 \\ ik/f \\ 1 \end{pmatrix} \quad \text{and} \quad \begin{pmatrix} \dfrac{KC}{f^2-C^2} \\ \dfrac{ikf}{f^2-C^2} \\ 1 \end{pmatrix} \quad \text{with } C = \pm \sqrt{k^2\phi_o + f^2}$$

The first solution is a steady flow satisfying

$$-fv + \frac{\partial \phi}{\partial x} = 0, \quad fu + \frac{\partial \phi}{\partial y} = 0 \tag{13}$$

These are the geostrophic relations. The other solutions are gravity waves.

In spherical geometry the geostrophic solution is no longer steady and it forms the basis of the motions of interest to weather forecasters. The problem in numerical forecasting is to forecast the evolution of this mode accurately in the presence of the others. A numerical method can be designed to approximate the individual terms of (1) to (6) or (7) to (9) accurately. However, a normal mode analysis of it may show that not all of the modes are well approximated. This problem is discussed in detail by Kreiss (1979) and elsewhere. Early forecasting models filtered out the gravity waves analytically, but to do so requires making approximations which are no longer considered acceptable.

2.3   <u>Properties of the equations.</u>

The system (1) to (6) has several properties which influence the choice of numerical method. These are briefly reviewed here, but are set out fully in Mesinger and Arakawa (1977) and Kasahara (1979). The main features are as follows:

(i)   The equations describe three-dimensional flow, but because of the hydrostatic assumption the behaviour in the vertical direction is quite different from that in the horizontal directions. For instance, sound waves can only propagate horizontally. The initial-boundary value problem becomes mathematically illposed (Oliger (1979)).

(ii)   The equations contain a wide variation of signal speeds ranging from 300 msec^{-1} to the advective speed of typically 10 msec^{-1}. Thus implicit methods are required for at least parts of the solutions; since the evolution of meteorological systems is on the advective time scale.

(iii)   The equations are nonlinear, but the nonlinearity is in the advective terms. All the equations contain terms of the form $u\, \partial/\partial x + v\, \partial/\partial y$. It is possible to make savings by evaluating parts of these terms once only for all the

equations. This form of nonlinearity
can lead to shocks and the numerical
scheme must be able to handle them.

(iv)   The equations have a high Reynolds number,
         even if the effective viscosity is a
         turbulent eddy viscosity rather than a
         molecular viscosity. The required
         solution can be thought of as the
         limit as $\nu \to 0$ of the viscous solution
         and will contain internal boundary layers
         (fronts) where the flow is turbulent.

(v)   The flow of interest to meteorologists
         is contained in the troposphere. This
         is bounded below by the earth's surface,
         which is very rough, and above by a
         strongly stably stratified layer
         (the stratosphere). It is difficult to
         know what upper boundary condition to
         impose and how well the lower boundary
         layer has to be represented.

(vi)   The computational demands of such a
         model are large and the efficiency of
         the numerical scheme is very important.
         It is not possible to refine the mesh
         sufficiently for convergence of the
         solutions and it is important to
         extract as much information as
         possible from the computations that
         are practicable.

(vii)   The solution consists of travelling
         disturbances of widely varying horizontal
         scale. If the mesh does not actually
         follow the disturbances in time it is
         necessary to use a uniform mesh. This is
         because the regions where mesh refinement
         is needed are not known in advance and
         because disturbances propagating through
         a variable mesh will be partially reflected
         or refracted wherever the characteristics
         of the mesh change. This argument does not
         apply in the vertical. Mesh refinement
         is routinely used to model the boundary
         layers at the earth's surface and the top
         of the troposphere.

(viii)   Since integrations have to be carried out
         over long time periods, methods with implicit
         damping can only be used if the damping time

for a given disturbance is much larger
than its lifetime.

3. GENERALISED ERROR ANALYSIS FOR FINITE ELEMENT METHODS APPLIED TO NUMERICAL WEATHER PREDICTION

3.1 General Principles.

At the present time nearly all forecasting models use
a spatial grid which does not vary with time. The analysis
of the finite element method given here will assume this.
Some attempts have been made to use moving grids and it is
possible that finite element methods could be used this way
in the future. The analysis here is set out more formally
in Cullen and Morton (1980).

We consider the time dependent problem

$$\frac{\partial \underline{u}}{\partial t} = L\underline{u}$$

$$\underline{u} = \underline{u}^o \text{ at } t = 0 \tag{14}$$

where $\underline{u}$ is a vector values function on $\mathbb{R}^n$ and L is a
possibly nonlinear operator which has real coefficients
not depending explicitly on time. The systems (1) to (6)
and (7) to (9) are both of this form. Suppose that for each
$t \varepsilon [0,T]$ $\underline{u}$ lies in a Hilbert space V which is called the
solution space. Then any procedure for solving (14) can
be broken down into the following stages:

(i)   Representation of the initial data $\underline{u}^o$

(ii)   Forward integration to a time $t = T$

(iii) Recovery of as much information as possible
       from the numerical solution at $t = T$.

The finite element method is a natural choice for
stage (i). Following Aubin (1972) we associate the triplet
(V_h, P_h, r_h) with any procedure for approximating members of
V on a discrete mesh in $\mathbb{R}^n$ characterised by a positive
mesh length h. V_h is the space of discrete parameter values
defining an approximation and r_h is a restriction operator
from V to V_h which associates such values with a member of V.
The third element is a prolongation operator $P_h = V_h \rightarrow V$, which
creates the approximation in V from the discrete parameter
values. We assume that $r_h P_h$ is the identity on V_h so that
$P_h r_h$ is a projection operator, $(P_h r_h)^2 = P_h r_h$. The initial
data in V_h is then $\underline{u}_{ho} = r_h \underline{u}_o$. The error made at this stage

is the subject of standard analysis. Of particular interest
is the case where $P_h r_h$ is a least squares fit, so that

$$\int_{\mathbb{R}^n} (P_h r_h u - u)^2 \quad \text{is minimised}$$

This is very useful in the recovery stage. Suppose
that the forward integration (ii) is carried out perfectly
in the sense that the numerical solution u_{hT} equals the
best fit to the true solution, $P_h r_h u_T$. Then information
about u_T can be recovered to a much higher order of
accuracy than the basic approximation error $||u - P_h r_h u||$.
This also applies to derivatives of u_T, see Chandler (1980).
Therefore use of a best fit to approximate u will give very
accurate information if stage (ii) can be carried out
successfully.

From the foregoing argument we see that the important
error in the forward integration stage can be written as
either

$$e = ||P_h r_h u_T - P_h u_{hT}|| \quad \text{or} \quad e_h = r_h e \tag{15}$$

The quantity e or e_h is called the evolutionary
error. To estimate it, suppose that (14) is approximated
in space by the equation

$$\frac{\partial u_h}{\partial t} = L_h u_h \tag{16}$$

where $L_h = V_h \to V_h$ is an operator in V_h which in some sense
approximates L. Then subtract (16) from the restriction of
(14) to obtain

$$\frac{\partial e}{\partial t} h = r_h L_u - L_h u_h$$

$$= (r_h L - L_h r_h) u + (L_h r_h u - L_h u_h) \tag{17}$$

If L and L_h are linear, the second term on the right
of (17) is just $L_h e_h$. The first term on the right is the
forcing term for the evolutionary error and is called the
truncation error (T.E.). It represents the difference between
operating with L on u before then applying the restriction,
and taking the restriction of u before operating on it with
L_h, as shown diagrammatically in Figure 1.

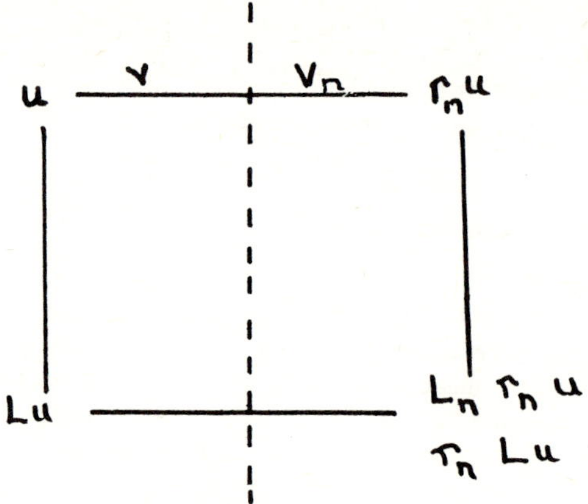

Figure 1   Diagrammatic representation of truncation error.

The finite element method will give an accurate forward integration if the T.E. is small and the time integration is accurately carried out. The T.E. depends on the choice of L_h. If the Galerkin method is used with the finite element functions χ_n as the test functions:

$$\int (LP_h u_h - P_h L_h u_h)\, \chi_n = 0$$

$$L_h = r_h L P_h \tag{18}$$

For an evolutionary problem we shall wish to choose an L_h to give a low T.E. The subsequent sections will show when this can be done using a Galerkin method and describe how to optimise it in general circumstances. For complicated operators L it is sufficient to reduce the T.E. of each component. Thus write $Lu = L^{(q)} L^{(q-1)} \ldots L^{(1)} u$. Approximate each $L^{(i)}$ by $L^{(i)}_h$ with T.E.

$O(h^{\mu(i)})$ and choose the approximation $L_h = L_h^{(q)} L_h^{(q-1)} \ldots L^{(1)}_h u$.

$$\tag{19}$$

We can write the total T.E. as

$(r_h L - L_h r_h) u$

$$= \sum_{i=1}^{q} L_h^{(q)} \ldots L_h^{(i+1)} (r_h L^{(i)} - L_h^{(i)} r_h) L^{(i-1)} \ldots L^{(1)} u \tag{20}$$

Thus the order of accuracy of the approximation to L is $\min_{i=1,q}(\mu^{(i)})$ and the total T.E. is the sum of the T.E. from

each component of L operated on by appropriate L and L_h terms.
A similar argument holds if $L=L_1+L_2$, and therefore we can
estimate the T.E. of an approximation to any operator L consisting of sums, products and derivatives by estimating the
T.E. of the approximations to a simple product and a simple
derivative given by the L_h of the form (19). In many cases
L is not broken down into the simplest possible components
but the same principles of analysis apply.

3.2 Estimation of truncation error for derivatives and products using linear elements.

The final remark in section 3.1 suggests that it is most
important to analyse the T.E. of the finite element approximations to simple operators L. We therefore present here the
truncation errors calculated for the one-dimensional
Galerkin approximation to derivatives and products using
linear and quadratic elements on a regular mesh. The analysis
is set out in more detail in Cullen and Morton (1980).

The analysis is carried out by writing out the finite
element algorithm at a particular mesh point and using
Fourier analysis. The simplest example is the linear element
Galerkin approximation to

$$\frac{\partial u}{\partial t} = \frac{\partial u}{\partial x} \tag{21}$$

on a regular grid with grid-points x_n, x=nh. The suffix n
indicates a discrete value at x=nh.

$$\frac{1}{6} \left(\frac{\partial u}{\partial t}\right)_{n-1} + \frac{2}{3} \left(\frac{\partial u}{\partial t}\right)_{n} + \frac{1}{6} \left(\frac{\partial u}{\partial t}\right)_{n+1} = \frac{1}{2h} (u_{n+1} - u_{n-1}) \tag{22}$$

Equation (21) defines L and (22) L_h in the general
definition (17). We define r_h by making $L_h r_h u$ the least
squares fit to u using piecewise linear functions. Write

$$(r_h u)_n = \alpha(\xi) e^{in\xi}$$

where
$$\xi = kh$$

Substituting the function $u = e^{ikx}$ into the formula
for T.E. gives

$$(r_h L - L_h r_h) e^{ikx}$$

$$= ik r_h (e^{ikx}) - L_h \alpha(\xi) e^{in\xi}$$

144

$$= i\alpha(\xi)(k-\beta(\xi))\,e^{in\xi} \qquad (23)$$

Fourier analysis of (22) gives $\beta(\xi)= \dfrac{3\sin\xi}{2+\cos\xi}$

Calculating the best least squares fit to e^{ikx} gives

$$\alpha(\xi) = \frac{6(1-\cos\xi)}{\xi^2(2+\cos\xi)} \qquad (24)$$

For small values of $\xi, \beta(\xi)$ has the expansion $\xi(1-\dfrac{(\xi)^4}{180} + 0((\xi)^6)$ and so (23) is $0((\xi)^4)$.

To obtain the truncation error for a product we must use two functions $u=e^{ikx}$ and $v=e^{i\ell x}$. Define $\eta=\ell h$. Then the T.E. is

$$r_h(e^{ikx}e^{i1x}) - L_h(\alpha(\xi)\alpha(\eta)e^{in\xi}e^{in\eta})$$

The Galerkin approximation to the product is

$$\frac{1}{6}(uv)_{n-1} + \frac{2}{3}(uv)_n + \frac{1}{6}(uv)_{n+1}$$

$$= \frac{1}{2}u_n v_n + \frac{1}{12}(u_n v_{n+1} + u_{n+1} v_n + u_{n+1} v_{n+1}$$

$$+ u_{n-1} v_n + u_n v_{n-1} + u_{n-1} v_{n-1}) \qquad (25)$$

The Fourier transform of (25) gives

$$L_h(e^{in\xi}, e^{in\eta}) = \frac{3+\cos\xi+\cos\eta+\cos(\xi+\eta)}{4+2\cos(\xi+\eta)}\, e^{i(\xi+\eta)n}$$

$$\equiv \gamma(\xi,\eta)e^{i(\xi+\eta)n} \qquad (26)$$

so that the T.E. is proportional to

$$\alpha(\xi+\eta) - \alpha(\xi)\,\alpha(\eta)\,\gamma(\xi,\eta)$$

As $\xi, \eta \to 0$ this is

$$-\frac{1}{720}(2\xi^3\eta + 3\xi^2\eta^2 + 2\xi\eta^3) + 0((\xi,\eta)^6)$$

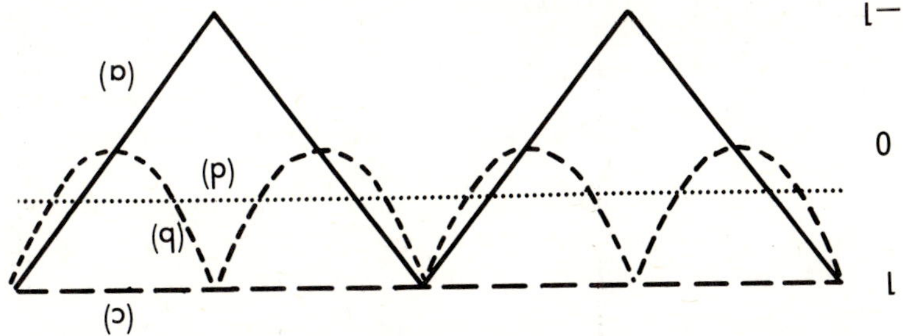

Figure 2   Piecewise linear approximations to the square
            of a function. For explanation see text.

The effect of this scheme is illustrated in Fig. 2 for
the short wave with $\xi=\eta=\pi$. Curve (a) is the least squares
fit to u. Curve (b) is the piecewise quadratic obtained
by squaring u and (c) is the best fit to it with a piecewise
linear function. It has a constant value of $\frac{1}{3}$. Squaring
the nodal values would have given a constant value of 1,
shown in (d). Thus (25) is a fourth order approximation to
a product with the extra property of damping short wave
interactions. As discussed in Mesinger and Arakawa (1977),
this is desirable for successful long-term integrations of
the weather forecasting equations.

3.3. <u>Estimates of truncation error for other operators</u>.

Truncation error results for other operators using
these elements can be derived in the same way. For the
advective opertor $L(u,v)=u\,\frac{\partial v}{\partial x}$ consider two methods:

   (i)   Direct Galerkin approximation. Using the
same functions and notation as above, this
gives an algorithm with Fourier transform

$$\frac{\sin\eta+4\sin\tfrac{1}{2}\eta\cos\tfrac{1}{2}\xi\cos\tfrac{1}{2}(\xi+\eta)}{h(2+\cos(\xi+\eta))} \tag{27}$$

and $\quad |\text{T.E.}| \sim \dfrac{L[4\eta^{4}+8\eta^{3}\xi+7\eta^{2}\xi^{2}-2\eta\xi^{3}]}{720} \tag{28}$

   (ii)   Two stage Galerkin approximation. Calculate
the Galerkin approximation to $\frac{\partial v}{\partial x}$, call this

ω. Then calculate the product $u\omega$. This gives
an algorithm with Fourier transform

$$\frac{3\sin\eta[3+\cos\xi+\cos\eta+\cos(\xi+\eta)]}{h(2+\cos\eta)(4+2\cos(\xi+\eta))} \tag{29}$$

and

$$|\text{T.E}| \sim \frac{L[2\xi^3\eta + 3\xi^2\eta^2 + 2\xi\eta^3 - 4\eta^4]}{720} \qquad (30)$$

In the case $\xi = \eta$ (30) gives much lower error than (28).

The effect of the two stage method is illustrated in Fig. 3. Diagram (a) is the projected function $p_h r_h u$ and (b) is $p_h r_h v$. We have chosen $\xi = \pi$ and $\eta = \frac{2\pi}{3}$.
Diagram (c) shows, (i), the piecewise constant derivative

$$\frac{\partial}{\partial x}(p_h r_h v)$$

and (ii), the piecewise linear approximation to it, $p_h r_h \omega$. Diagram (d) shows, (i), the product $(p_h r_h u)\frac{\partial}{\partial x}(p_h r_h v)$

and, (ii), the piecewise linear approximation to it. This is the result obtained using the single stage algorithm (27). Diagram (e) shows, (i), the product $(p_h r_h u)(p_h r_h \omega)$ and, (ii), the piecewise linear approximation to it. This is the result obtained using (29). It can be seen that (29) has the advantage of using a better representation of the function at all stages of the calculation, rather than using the discontinuous function shown at (c) (i) in the product calculation.

A second example of the two stage method is in approximating $Lu = \frac{\partial^2 u}{\partial x^2}$. The simple Galerkin method gives a T.E. asymptotically $(\xi^2)/12$. The two stage method, applying (24) twice, gives $\xi^4/90$. This method, however, gives zero response for $\xi = \pi$ while the single stage method does not. Thus a two grid length wave will not be diffused which is unlikely to be acceptable. A fourth order scheme of the same form as the algorithm given by the simple Galerkin method can be derived and used instead.

3.4. Estimates of truncation error for derivative using quadratic elements.

Now consider the finite element Galerkin method using higher order elements. Thomee and Wendroff (1974) show that the Galerkin method for (21) using splines of order μ on a regular mesh, gives an order of accuracy $O(h^{2\mu})$ and, for

$$\frac{\partial u}{\partial t} = \frac{\partial^2 u}{\partial x^2} , O(h^{2\mu-2}).$$

The results with linear elements given above are the case $\mu = 2$. Cullen (1975) shows that the spline Galerkin method gives errors $O(h^{2\mu})$ for products. The spline method has the characteristic that there is only one independent parameter per mesh interval, the other coefficients of the polynomials

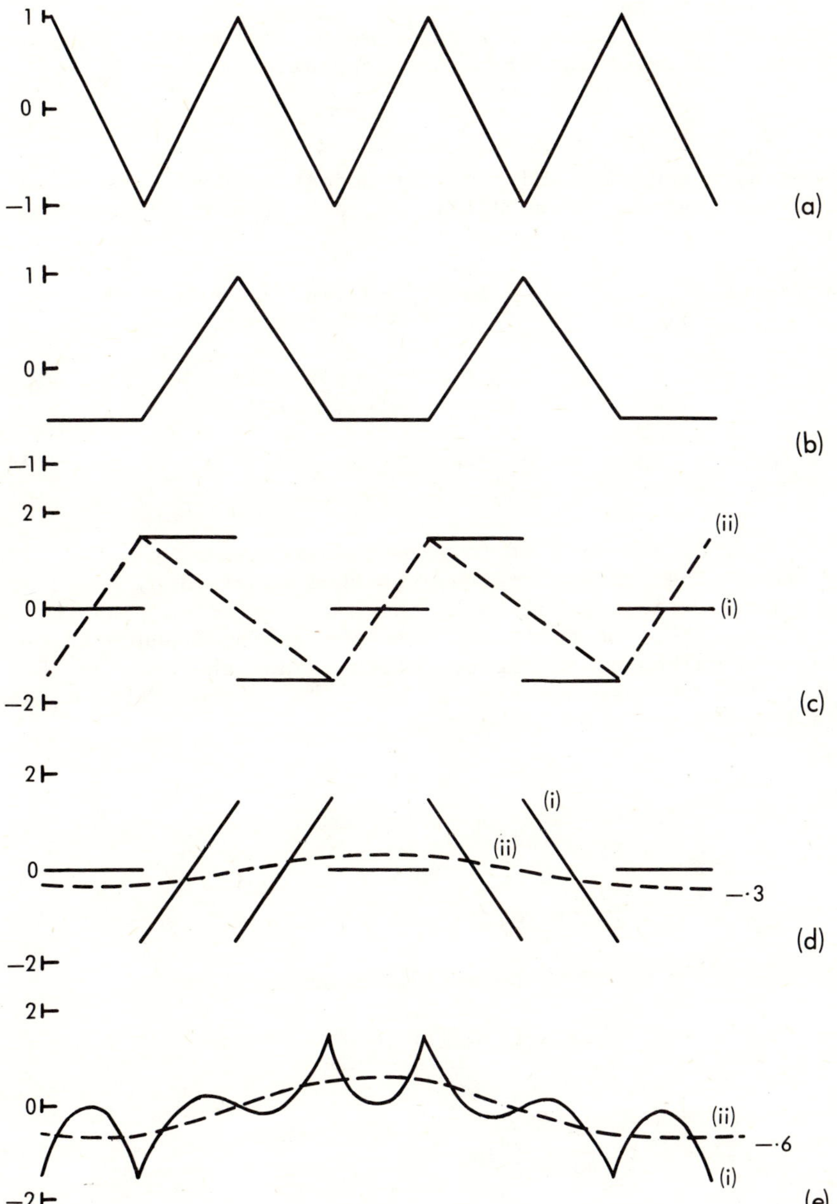

Figure 3   Approximation to nonlinear advective term
For explanation see text.

are determined by continuity constraints. Thus the resulting
algorithms look like generalised finite difference schemes,
which is how they were regarded by Thomee and Wendroff.

As an example of a different higher order element
consider quadratics on a regular mesh. We use the Galerkin
method with exact integration to approximate

$$\frac{\partial u}{\partial t} = \frac{\partial u}{\partial x}$$

These have endpoint and midpoint nodes, so there are three
values to define the quadratic on each element. The scheme
is

$$(V_t)_j + 8(V_t)_{j+\frac{1}{2}} + (V_t)_{j+1} = 10(U_{j+1} - U_j)/h$$

$$-(U_t)_{j-1} + 2(U_t)_{j-\frac{1}{2}} + 8(U_t)_j + 2(u_t)_{j+\frac{1}{2}} - (U_t)_{j+1}$$

$$= 10[2(U_{j+\frac{1}{2}} - U_{j-\frac{1}{2}}) - \tfrac{1}{2}(U_{j+1} - U_{j-1})]/h \tag{31}$$

Fourier analysis of (31) is complicated by the two types
of node. The fairest analysis is that obtained by calculating
an eigensolution of (31) with the same periodicity as the
function e^{ikx}. It turns out that the ratio of endpoint to
midpoint values of an eigensolution is not the same as that
for the best least squares fit to e^{ikx}. If we define $\alpha_o, \alpha_{\frac{1}{2}}$
and β so that

$$(U)_j = \alpha_o(\xi) e^{ij\xi}$$

$$(U)_{j+\frac{1}{2}} = \alpha_{\frac{1}{2}}(\xi) e^{i(j+\frac{1}{2})\xi}$$

$$L_h r_h e^{ikx} = ih^{-1}\beta_o(\xi) e^{ij\xi}$$

the condition for U to be an eigenfunction gives

$$\alpha_{\frac{1}{2}}(\xi) = \alpha_o(\xi)(1 + \frac{1}{48}\xi^2 + O(\xi^4))$$

and

$$\beta_o(\xi) = \xi(1 + \frac{1}{4320}\xi^4 + O(\xi^6))$$

The comparative relative phase errors for quadratic and
linear elements with, first, the same number of elements
and, second, the same number of degrees of freedom is shown
in Table 1.

TABLE 1

ξ	$\pi/4$	$\pi/2$	$3\pi/4$	π	$3\pi/2$	2π
Quadratic	1.0001	1.0011	1.0043	1.007	0.900	0.0
Linear, same number elements	0.9977	0.955	0.696	0.0	–	–
Linear, same number degrees freedom	0.9999	0.9977	0.990	0.955	0.696	0.0

This table suggests that quadratic elements used in this
way may be competitive with linear elements, despite
considering the computational complexity of (31).

We now consider how higher order accuracy can usefully
be achieved with quadratic elements. Douglas and Dupont (1973)
show that fourth order accuracy can be achieved by appropriate
collocation procedures. The band width of the approximate
equations is determined by the support of the finite element
basis functions whichever weighted residual method is used.
For a given band width we can choose the coefficients to
maximise the order of accuracy in the manner of Collatz (1964).
To obtain a useful scheme we want the T.E. to be reduced where
r_h is the least squares projection so that full advantage
can be taken of recovery theorems. This is not possible in
the case just analysed because the eigensolutions of (31)
were not the least squares fit to the eigensolutions of the
continuous equations.

Consider a generalised weighted residual scheme for (21).
Since there are two unknowns per element this scheme must
give two equations associated with each element. Assume that
the scheme is symmetrical, not upwinded, as is the normal
Galerkin scheme or the Gausss point collocation scheme. Then
it can be written in the form either:

$$(U_t)_j + A(U_t)_{j+\frac{1}{2}} + (U_t)_{j+1} = (A+2)(U_{j+1} - U_j)/h$$

$$-(U_t)_{j-1} + B(U_t)_{j-\frac{1}{2}} + C(U_t)_j - (U_t)_{j+1} + B(U_t)_{j+\frac{1}{2}}$$

$$= \frac{C+2B-2}{D-1} \left[D(U_{j+\frac{1}{2}} - U_{j-\frac{1}{2}}) - \tfrac{1}{2}(U_{j+1} - U_{j-1}) \right]/h \tag{32}$$

or

$$A(U_t)_j + (U_t)_{j+\frac{1}{2}} + B(U_t)_{j+1}$$

$$= \frac{2(A+B+1)}{h}[\, C(U_{j+1}-U_{j+\frac{1}{2}}) + (1-C)(U_{j+\frac{1}{2}}-U_j)\,] \tag{33}$$

$$B(U_t)_j + (U_t)_{j+\frac{1}{2}} + A(U_t)_{j+1}$$

$$= \frac{2(A+B+1)}{h}[\,(1-C)(U_{j+1}-U_{j+\frac{1}{2}}) + C(U_{j+\frac{1}{2}}-U_j)\,] \tag{33}$$

Some arbitrary coefficients have been eliminated by symmetry and by requiring the schemes to be a consistent approximation to (21).

To analyse the effect of either of these schemes using the least squares restriction r_h, we write

$$(r_h e^{ikx})_j = \alpha_o(\xi)e^{ij\xi}$$

$$(r_h e^{ikx})_{j+\frac{1}{2}} = \alpha_{\frac{1}{2}}(\xi)e^{i(j+\frac{1}{2})\xi} \tag{34}$$

Fourier analysis of this restriction shows that α_o and $\alpha_{\frac{1}{2}}$ are both $1 + O(\xi)^4$. The condition for (34) to be an eigenfunction of (32) or (33) is that

$$\frac{(U_t)_{j+\frac{1}{2}}}{(U_t)_j} = \frac{\alpha_{\frac{1}{2}}(\xi)}{\alpha_o(\xi)}e^{i\xi/2} \quad \text{for all } \xi \tag{35}$$

given that

$$U_{j+\frac{1}{2}}/U_j = \alpha_{\frac{1}{2}}(\xi)/\alpha_o(\xi)e^{i\xi/2}$$

To get a T.E. of order h^μ we then require

$$(U_t)_j = \frac{1}{h}(\xi + O(\xi^{\mu+1}))U_j \tag{36}$$

However, since A, B, C and D are not allowed to depend on ξ we cannot take advantage of all the arbitrary constants to obtain a very high order of accuracy. In the case of (32), once $\alpha_{\frac{1}{2}}/\alpha_o$ is fixed, the first equation only has the arbitrary constant A which can be chosen to give fourth order accuracy. In the case of (33), the sum of the two equations is of the same form as the first equation of (32). Again only fourth order accuracy is possible. If we did not use the eigenfunction condition but allowed $\alpha_{\frac{1}{2}}/\alpha_o$ to be arbitrary, then each of the constants A, B, C and D could be used to eliminate a term of the truncation error, giving tenth order

accuracy for (32) and eighth for (33). The loss of the least
squares projection makes this achievement useless in practice.

Further examples of using this approach are for a
generalised quadratic spline Galerkin scheme. The general
form of this is

$$((U_t)_{j-2} + (U_t)_{j+2}) + A((U_t)_{j-1} + (U_t)_{j+1}) + B(U_t)_j$$

$$= \frac{(B+2A+2)}{h} (\frac{C}{4}(U_{j+2} - U_{j-2}) \qquad\qquad (37)$$

$$+ (1-C)(U_{j+1} - U_{j-1}))$$

This scheme can be made eighth order accurate, while the
standard Galerkin coefficients only give sixth order accuracy.
A final example is the linear element approximation to the
second derivative. The Galerkin method gives an approximation
of the form

$$(U_t)_{j-1} + A(U_t)_j + (U_t)_{j+1} = \frac{A+2}{h^2} (U_{j-1} - 2U_j + U_{j+1}) \qquad (38)$$

and A can be chosen to give fourth order accuracy. This
scheme can be regarded as 'half-lumping' the mass matrix.

It is clear from these examples that there is no clear
relation between finite element arguments and optimal
truncation error for a given bandwidth. The eigenvalue
condition for higher order elements also restricts available
accuracy. The possible advantages of finite elements seem
to be confined to the nonlinear representation and these
empirical techniques should be used for approximating derivatives.

4.   SOLUTION OF THE FORECASTING EQUATIONS BY FINITE ELEMENT
METHODS.

4.1. Construction of the model.

The previous sections have reviewed the theory of finite
element methods as applied to evolution equations. In this
section we discuss the practical details of the solution of
(1) to (6) or (7) to (9). Only finite element approximation
in the horizontal is discussed here. Because of the nature of
the equations, as discussed earlier, it is reasonable to use
different types of numerical method in the horizontal and the
vertical. Finite element methods have been used in the
vertical by Staniforth and Daley (1977). There is little
incentive to use finite elements in time because of the wide
variation of signal speed in the equations, the problem of
time integration is fully discussed in Mesinger and Arakawa

F

152

(1977).

     The solutions presented in this chapter are on a sphere.
No regular grid can be placed on a sphere directly, though
a regular grid can be placed on a map projection. The error
estimates of section 4 are degraded on an irregular mesh to
$O(h)$ for linear elements and $O(h^2)$ for quadratics, but the
practical amount of degradation depends on how rapidly the
mesh varies. In these experiments we use a quasi-regular
grid obtained by subdividing the faces of an icosahedron
inscribed in the sphere. The resulting grid is shown in Fig.
4.

     The solution of these equations is very expensive and, in
view of the high accuracy attainable using linear elements,
we use linear interpolation on the triangles. The nonlinear
terms are calculated using the two stage Galerkin method.
Because of the form of the advective terms in (1) to (6)
this is computationally efficient. Equations (1), (2), (4),
(5) and (6), when the derivatives are expanded, all include
terms of the form

$$\frac{u}{a\cos\theta}\frac{\partial\xi}{\partial\lambda}, \quad \frac{v}{a}\frac{\partial\xi}{\partial\theta} \tag{39}$$

where ξ is successively U, V, P_*, T and q. The two stage
algorithm first calculates the gradients, and then multiplies
them by u and v. The Galerkin approximation to the product
can be written in the form

$$\int (\sum (\frac{\partial u}{\partial t})_n \chi_h)\chi_m = \int (\sum_h u_h \chi_h)(\sum (\frac{\partial\xi}{\partial\lambda})_L \chi_L)\chi_m \tag{40}$$

for all m, where the χ_m are the finite element basis functions.
(40) can be written for each node m as

$$\sum \alpha_{i,m}(\frac{\partial u}{\partial t})_{i,m} = \sum(\sum \beta_{i,j,m} u_{j,m})(\frac{\partial\xi}{\partial\lambda})_{i,m} \tag{41}$$

The suffices i,m range over the nodes adjacent to node m.
Then the term $(\sum \beta_{i,j,m} u_{j,m})$ can be evaluated once for each
i and m and then used in all the equations (1) to (6).

     The complete solution of (1) to (6), or (7) to (9), also
involves representation of physical processes and artificial
smoothing. These aspects are fully discussed for equations
(7) to (9) in Cullen (1974) and for equations (1) to (6) in
Cullen and Hall (1979).

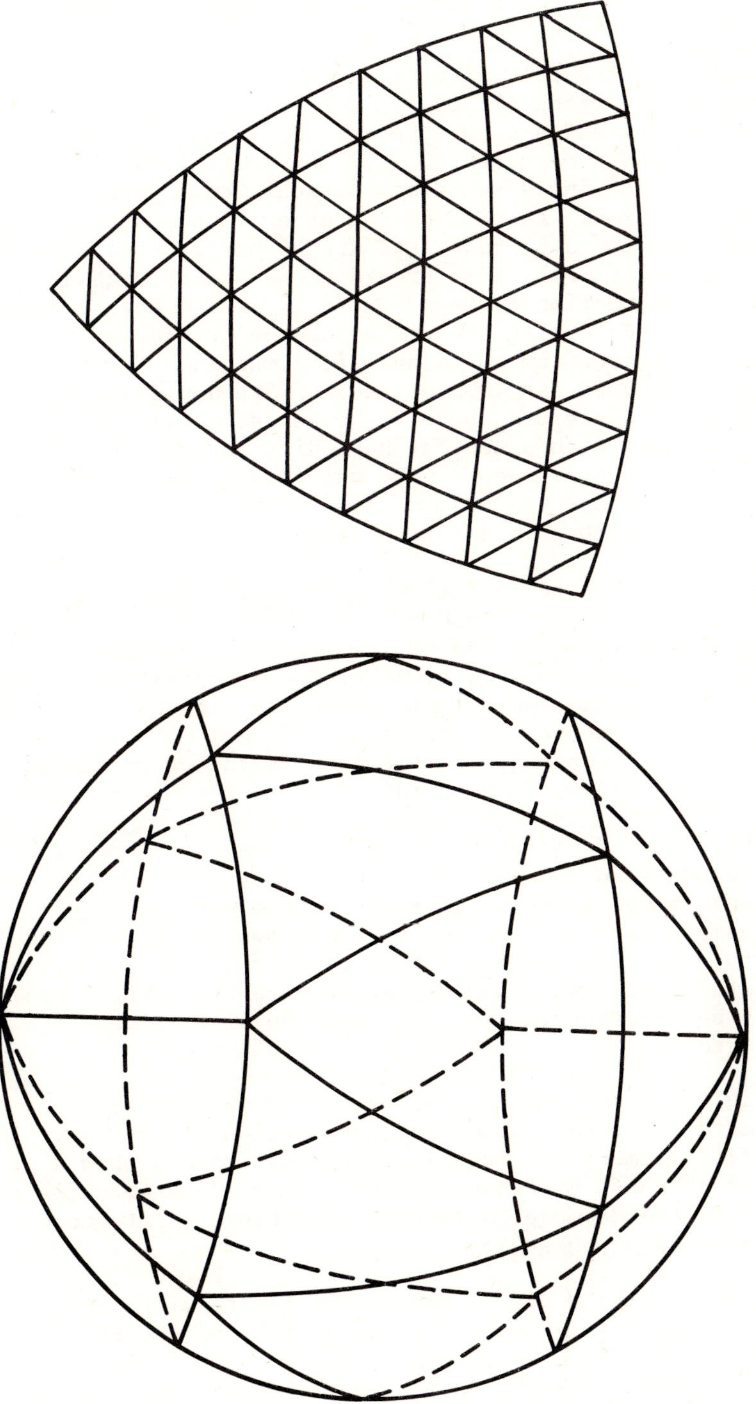

Figure 4   View of spherical grid used for forecasts.

154

4.2 <u>Solution of the shallow water equations on a sphere.</u>

We conclude by presenting two solutions discussed in
more detail in the preceding references. The first is a
solution of equations (7) to (9) for initial data consisting
of a zonal flow and a single spherical harmonic perturbation
of wavenumber 8 superposed on it. The initial vorticity
thus contains a single spherical harmonic component, ζ_9^8 in
standard notation. As time passes other components are
generated and the amplitude $|\zeta_9^8|$ drops sharply. As discussed
by Doron et al (1974), this breakdown occurs through a
nonlinear interaction between ζ_9^8 and ζ_{11}^8 which have a constant
phase difference during the period of the breakdown.
Results are shown from three numerical models. The finite
element model used has 1002 degrees of freedom and is
formulated as described above. The second model used a
Galerkin method with spherical harmonics as basic functions
and 640 degrees of freedom. The third uses second order
finite differences with 4032 degrees of freedom. These two
models are described in detail by Doron et al (1974).

Fig. 5 shows the evolution in time of some of the
vorticity components. These are ζ_9^8, the only component
present initially, ζ_{11}^8, one of the principal subharmonics
and ζ_5^0, one of the zonally symmetric components. In addition,
because the finite element grid possesses a five-fold
symmetry, ζ_6^5 is shown for the finite element model. The
growth of this component is a measure of the error introduced
by the irregularity of the grid.

The curves show that the finite element and spherical
harmonic model follow each other up to day 5 while the finite
difference model behaves differently from very early on
and does not forecast the sudden breakdown of ζ_9^8. The amplitude of ζ_6^5 is significant in this experiment. However, in a
real data problem with non-symmetric data this effect would
have to be re-assessed since it might be smaller than other
model errors.

Fig. 6 shows the phase speed of the wavenumber 8
components and the phase difference between ζ_9^8 and ζ_{11}^8. Both
the finite element and spherical harmonic models show a period
of constant phase difference between days 1 and 4, while the
finite difference model does not. The finite difference
model also shows phase lag.

Results with a higher resolution finite difference model
with 14592 points agree very closely with the spherical
harmonic model solution (Doron et al (1974)). This suggests
that the sudden breakdown of ζ_9^8 is the correct solution and
that the spherical harmonic solution is more accurate than the
finite element solution. It also suggests that the finite
element model is intermediate in accuracy between the two

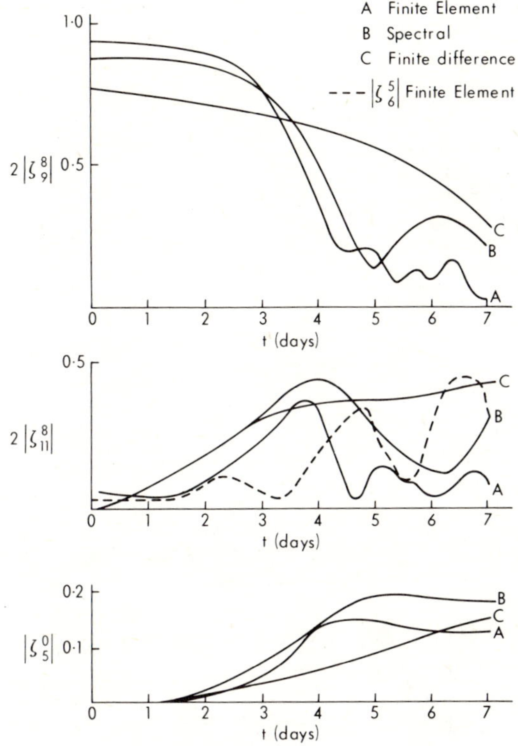

Figure 5   The evolution with time of some of the
principal spectral components.

finite difference models and similar to a second order finite
difference scheme with 10 times the degrees of freedom in two
dimensions.

4.3 Solution of the forecasting equations in a real case

The second experiment uses four different models to
carry out a solution of (1) to (6) for real atmospheric data
for 12Z on 1 March 1965. The forecasts are for the height of
the 500 mb constant pressure surface, (about half the surface
pressure). This is of the order of 5.5 km. The models used
are as follows.

A   Finite element global model with 1002 degrees
     of freedom.

B   As A with 2252 degrees of freedom.

C   As B, but instead of representing the
     velocity components u and v by piecewise

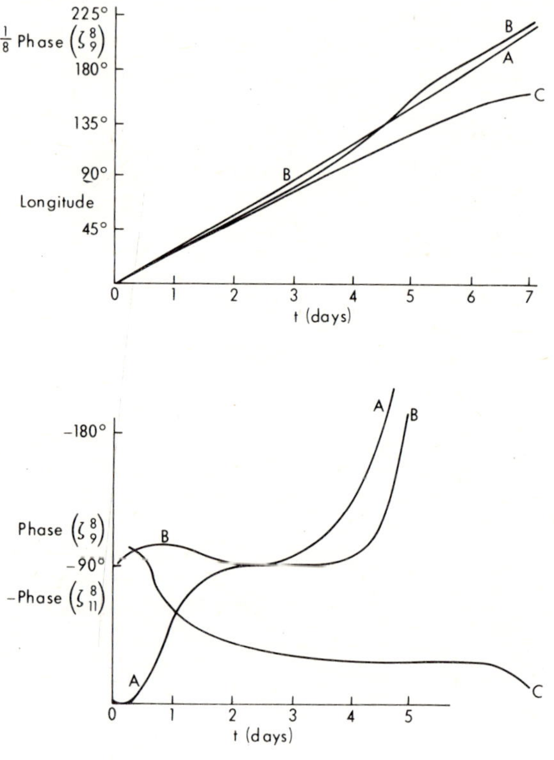

Figure 6   Graph of the variation of phase with time of
             some of the principal spectral components.

       linear functions, the velocity potential
       and stream function are approximated
       instead.

D   A spherical harmonic model with 1848 degrees
       of freedom.

     All these models used finite differences in the vertical
with five equally spaced layers, and the same representation
of physical effects.   Full details can be found in Cullen
and Hall (1979).

     The results are shown in Figs. 7 to 12.   Fig. 7 shows
the 500 mb analysis for 1 March 1965 and Fig. 8 the analysis
three days later.   Figs. 9 and 12 show the forecasts
produced by the four models. Because of the geostrophic
relation, (13), the height field closely resembles the stream-
function at the same level.   The actual developments are as

follows. The high over Greenland breaks down, leaving a
ridge over the North Atlantic. An intense new low forms over
the British Isles, reinforcing a trough originally over western
Europe. The low originally over the western Atlantic becomes
an intense cut-off feature. The trough over the western USA
also cuts-off and the main flow is established over northern
Canada.   The flow over the Pacific remains zonal throughout.
The ridge at 90°E moves to 120°E and further ridging occurs
at about 60°E.

Models A and B give very similar forecasts. Though both
models break down over Greenland, they completely fail to
produce the new low over the British Isles and the associated
northerly flow over the eastern Atlantic. A west to northwest
flow is produced instead. The cut-off low over the eastern
USA is too weak and is only slightly improved by higher
resolution; similarly the low over the western USA is not
cut-off sufficiently.   The development of the ridge over the
western USA is correct though the intensity is too low. The
high latitude trough north of Canada is better forecast by
A than by B. The flow over the Pacific is correct.

The orientation of the ridge originally at 90°E is not
predicted correctly. The weak ridge at 60°E is predicted by
B but not by A. Overall, the low resolution model A predicts
all the main features with some skill but is deficient in
detail. The higher resolution version B is disappointing in
that the improvements are very slight.

The model using stream function and velocity potential,
model C, gives a forecast difference in several respects
from those given by A and B.   The orientation of the ridge at
90°E is well predicted. The ridge at 60°E is too strong.
There is a hint of a trough over the British Isles, but the
flow over the North Atlantic is still northwesterly, not
northerly. The trough in the high latitude flow over
Canada is stronger than in B but is not in the right place,
whereas it was correctly placed in A. In general the model
is more active than A and B but the extra detail is not
always correct.

The spherical harmonic model D gives overdevelopment
at 60°E in high latitudes and also in the ridge at 80°W over
the USA.   The troughing over the British Isles is improved
but the northern Atlantic flow is still not forecast correctly.
The cut-off low centres over the USA are too weak, the
western low is handled worse than in C. The troughing over
Canada is differently forecast but not accurately.

Forecasts from this data using the largest currently
available models, e.g. see Umscheid and Bannon (1977), are
only slightly superior overall to those from models C and D,

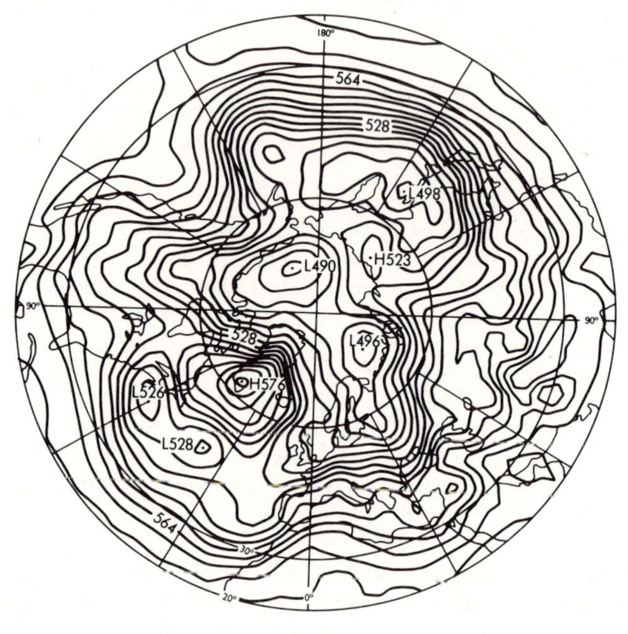

Figure 7   500 mb analysis valid 12Z 1 March 1965.

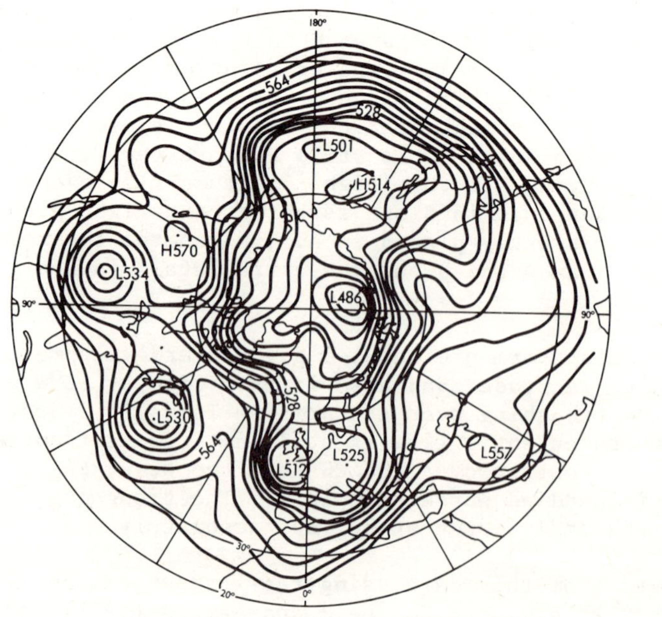

Figure 8   500 mb analysis valid 12Z 4 March 1965.

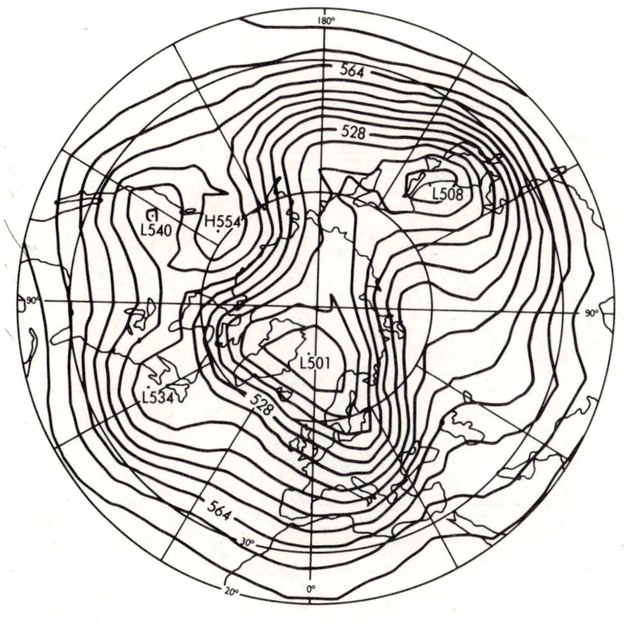

Figure 9   500 mb 72 hour forecast valid 12Z 4 March 1965 – Model A.

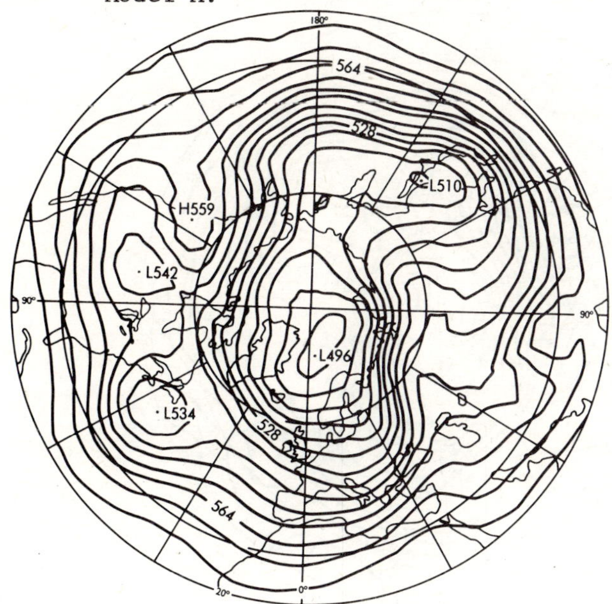

Figure 10   As Figure 9 – Model B.

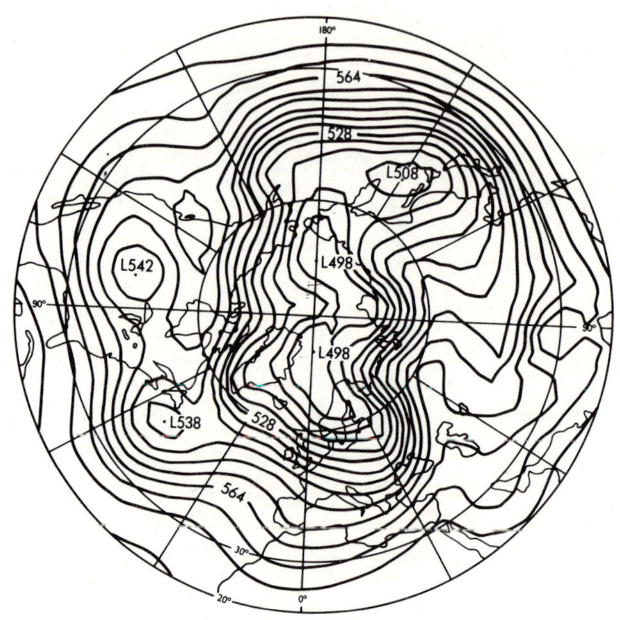

Figure 11   As Figure 9 – Model C.

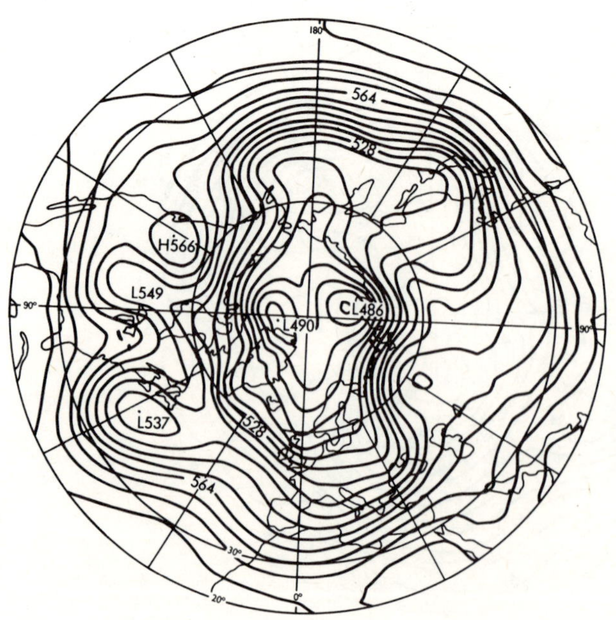

Figure 12   As Figure 9 – Model D.

though they are better in the neighbourhood of the British
Isles.

5.   DISCUSSION

The theoretical analysis of the finite element method
applied to this type of problem indicates that its success in
modelling the advective and wave processes depends on how
good a finite difference approximation is implied by the
scheme. Finite element ideas are not very useful in
constructing the best scheme for this. The Galerkin
approximation to products, however, should give a better
description of nonlinear processes as illustrated by the
results in section 4.2. This applies both to spherical
harmonic and finite element representation. The best
strategy seems to be to combine the finite element method
with a least squares restriction operator, so that the
recovery theorems can be used on the results, and techniques
such as those of Collatz, or Fourier methods, to construct
the approximations to derivatives.

The results from the simple model in section 4.2. support
the predictions from the theoretical analysis. However, the
real case results do not indicate convergence to the observed
solution with increased resolution. Other results in this
field indicate similar difficulties, suggesting that the
accuracy of numerical weather prediction is restricted by
deficiencies of the models other than the numerical
approximation in the horizontal.

ACKNOWLEDGEMENTS

Parts of this work were carried out in collaboration with
Professor K.W. Morton, University of Reading, UK and Mr. C.D.
Hall, UK Meteorological Office.

REFERENCES

AUBIN, J.P. (1972) Approximation of elliptic boundary-value
        problems. Wiley, New York.

CHANDLER, G.A. (1980) Superconvergence for integral equations
        Integral Equations and Applications (ed. R.S.
        Anderssen) Walters Noordhoff (Australia).

COLLATZ, L. (1964)   Numerical treatment of differential
        equations. 3rd ed. Springer-Verlag, New York/
        Berlin.

CULLEN, M.J.P. (1974)   Integration of the primitive equations
        on a sphere using the finite element method.
        Quart. J. Roy. Meteor. Soc., 100: 555-562.

CULLEN , M.J.P. (1975)   Application of the finite element
        method to numerical weather predicition. Ph.D.
        Thesis, University of Reading, UK.

CULLEN, M.J.P. and HALL, C.D. (1979) Forecasting and
        general circulation results from finite element
        models.
        Quart. J. Roy. Meteor. Soc. 105: 571-592.

CULLEN, M.J.P. and MORTON, K.W. (1980)   Analysis of
        evolutionary error in finite element and other
        methods.
        J. Comput. Phys. 34: 245-267.

DORON, E., HOLLINGSWORTH, A., HOSKINS, B.J. and SIMMONS, A.J.
        (1974)   A comparison of grid-point and spectral
        methods in a meteorological problem.
        Quart. J. Roy. Meteor. Soc. 100: 371-383.

DOUGLAS, J. and DUPONT, T. (1973)   A finite element
        collocation method for quasi-linear parabolic
        equations.
        Math. Comp. 27: 17-28.

GRAMMELTVEDT, A. (1969) A survey of finite difference schemes
        for the primitive equations for a barotropic
        fluid. US Mon. Weath. Rev. 97: 384-404.

HALTINER, G.J. (1971)   Numerical weather prediction.
Wiley, New York.

KASAHARA, A. (ed.) (1979)   Numerical methods used in
atmospheric models. Vol. 2 ICSU/WMO GARP
Publication Series No. 17.

KREISS, H.O. (1979)   Difference methods for stiff ordinary
differential equations. SIAM J. Numer. Anal.
$\underline{15}$: 21-58.

MESINGER, F. and ARAKAWA, A. (1977)   Numerical methods in
atmospheric models. Vol. 1. ICSU/WMO GARP
Publication Series No. 17.

OLIGER, J. (1979)   Approximate methods for atmospheric and
oceanographic circulation problems. Springer
.lecture Notes in Physics, $\underline{91}$: 171-184.

STANIFORTH, A.N. and DALEY, R.W. (1977)   A finite element
formulation for the verical discretisation of
sigma coordinate primitive equation models.
US Mon. Weath. Rev., $\underline{105}$: 1108-1118.

STRANG, G. and FIX, G.J. (1973)   An analysis of the finite
element method. Prentice Hall, Englewood Cliffs,
N.J.

TEMPERTON, C.   (1976)   Dynamic initialisation for
barotropic and multi-level models. Quart. J.
Roy. Meteor. Soc., $\underline{102}$: 297-312.

THOMEE, V. and WENDROFF, B. (1974)   Convergence estimates
for Galerkin methods for variable coefficient
initial-value problems. SIAM J. Numer. Anal. $\underline{11}$:
1059-1068.

UMSCHEID, L. and BANNON, P.R. (1977)   A comparison of
three global grids used in numerical prediction
models. US Mon. Weath. Rev. $\underline{105}$: 618-635.

ZIENKIEWICZ, O.C. (1977)   The finite element method. (3rd.
Ed.) McGraw Hill, London.

CHAPTER 6

VARIATIONAL FORMULATION OF FREE SURFACE GRAVITY FLOW INCLUDING
GATED FLOW USING ADAPTIVE FINITE ELEMENTS

W.D. Liam Finn and Erol Varoḡlu

University of British Columbia

ABSTRACT

A generalization of the finite element method for solving
directly gravity flow problems involving one or more free surfaces is described. The global finite element equations
defining the flow are derived from a variational formulation
that recognises that both the dependent variables and the
domain of integration undergo variations. This generality
ensures that the finite element equations simultaneously yield
the values of the dependent variables and the free surfaces.
The usual instabilities and programming difficulties associated with trial free surface methods are eliminated. The
method is presented for free flow over a spillway, gated flow
over a spillway and gated flow in a channel. The latter
examples, a recent extension of the method, involve two free
surfaces, one upstream and one downstream of the gate, joined
by an impenetrable boundary segment.

INTRODUCTION

A classic example of gravity flow with a free surface is flow
over a spillway crest. This problem has continued to engage
the attention of numerical analysts since Southwell and Vaisey
(1946) attempted a solution by finite differences using relaxation methods.

When steady-state conditions are reached in free surface
gravity flow, the flow is governed by an elliptic partial
differential equation. On the known part of the boundary, one
boundary condition must be satisfied; on the free surface two
boundary conditions must be satisfied to obtain a unique solution. The existence of the free surface makes the gravity
flow problem intrinsically nonlinear. A simple example to
illustrate how boundary conditions on a free surface introduce
nonlinearity into a solution of a linear differential equation
is given in Appendix I (Finn and Varoḡlu, 1976).

Current finite element methods for fluid mechanics problems cannot be applied directly to solve flow problems with free surfaces because they are all applicable to domains with fixed boundaries (Zienkiewicz, 1971). Therefore, in applying such methods, a tentative location of the free surface is assumed and the distribution of the dependent variable is determined for the resulting fixed domain. Two boundary conditions must be satisfied on free surfaces; e.g., there is no flow across the surface and the pressure is atmospheric. The trial fixed domain solution will not usually satisfy both boundary conditions simultaneously and, therefore, the trial free surface is adjusted and a solution obtained for the revised fixed domain. This process is repeated until the required adjustment in the free surface is considered negligible. Thus, the solution of the nonlinear free surface problem is achieved by a series of solutions to linear problems with prescribed boundaries.

Details of the trial free surface method for gravity flow through porous media have been given by Finn (1967) and Taylor and Brown (1967). The procedure works well for the inherently stable problems of flow through porous media. However, during the iteration process, addition and deletion of nodes and elements may be required which create complexities in programming and are inefficient processes. In sharp contrast to experience with porous flow problems, Southwell and Vaisey (1946) found that flow problems characterized by free streamlines were among the hardest they had attempted and concluded that the existence of an essential instability makes any tentative solution liable to diverge. More recent attempts using finite elements have also encountered divergent behaviour (Tu, 1971).

The severe difficulties with the trial free surface method have led to attempts to find a workable alternative to the solution based on a sequence of fixed domain solutions. In fluid mechanics problems, the global finite element equations are often obtained by a variational formulation over a fixed domain. The free surface problem, however, is characterized by a variable domain. Luke (1967) has presented a variational principle for free surface fluid problems. Ikegawa and Washizu (1973) used a special case of this principle to analyse flow over a spillway. In their analysis, the unknowns of the problem are taken as the discrete values of the stream function at internal nodes and one coordinate of the nodes on the free surface. This method may require, in general, addition or deletion of nodes and elements during iterations. The actual flow rate, Q, cannot be found directly from the analysis. Instead, for several values of flow rate Q, free surfaces are evaluated and the exact Q is assumed to be the one associated with the smoothest free surface. The use of the variational principle did not give the free surface directly. The variational principle of Ikegawa and Washizu (1973) was

used by Aitchison (1980) for analysing the flow over a submerged weir and by Betts (1978) for the problem of flow downstream of a two-dimensional contraction. Both used finite
element methods of analysis. Aitchison (1980) found that most
solutions for critical flow exhibited waves on the upstream
surface. This is physically incorrect. Betts (1978) found
for subcritical flow that a misplaced boundary results in the
appearance of an additional train of waves over the whole flow
including the upstream surface. The downstream boundary was
moved until the upstream waves were eliminated. Both authors
adopted the Ikegawa and Washizu (1973) criterion that the flow
corresponding to the smoothest free surface was the correct
free surface. Aitchison (1979) found that only a few iterations of the boundary position were needed in those cases
where convergence occurred and that non-convergence after 8-10
iterations indicated that convergence would not be achieved
after many more iterations. All these results support the
conclusion by Southwell and Vaisey (1946) about the essential
instability of the problem. It will be shown later that the
approaches based on Ikewage and Washizu's variational principle cannot determine the free surface directly and have
difficulties with convergence because the variational principle is not general enough and the variations based on it are
too restrictive.

A new variational principle for free surface gravity flow
problems is presented here. The principle is used to generate
global finite element equations which permit the direct and
simultaneous determination of both the flow field of the
problem and the free surface. In implementing the variational
principle a variable domain of integration is used which is
compatible with the concept of the free surface and allows the
simultaneous satisfaction of both boundary conditions on the
free surface. A novel feature of the finite element formulation is the automatic generation of a completely adaptive
finite element mesh. Therefore, no restructuring of the finite element mesh is required during variations in the region
of flow.

The variational principle and the associated solution
technique were developed for flow through porous media by Finn
and Varoḡlu (1975) and extended to free streamline flow by
Varoḡlu and Finn (1977,1978). The method will be developed
for free gravity flow over a spillway and extended to gated
flow in a channel and over a spillway. Each of the latter two
cases involves two different free surfaces joined by a section
of impenetrable boundary.

STATEMENT OF THE PROBLEM

Free gravity flow over a spillway is illustrated in Figure 1.
The geometries of the reservoir bed and of the spillway are
given and the far upstream and downstream boundaries of the

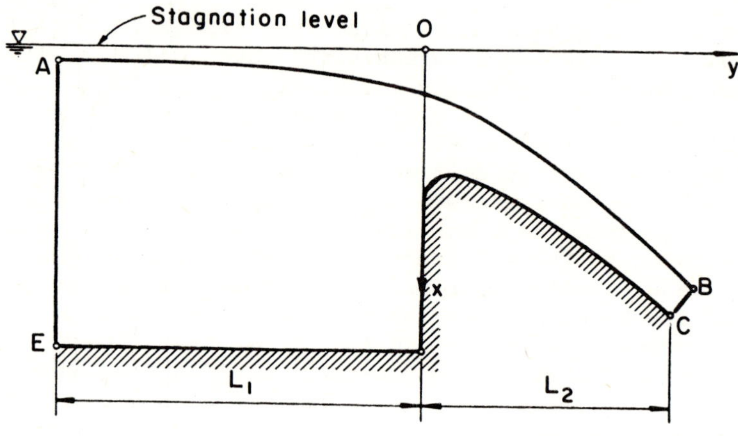

FIGURE 1   Flow over a Spillway

flow region are assumed to be normal to the reservoir bed and
the spillway respectively. The end points E and C of these
boundaries are taken at prescribed distances L_1 and L_2 from
the vertical upstream face of the spillway. The free surface,
including the end points A and B, is not known a priori.
Either the stagnation level of the free surface is given and
the corresponding rate of flow Q for unit width of the spillway and the flow field are sought or the flow rate Q is given
and the stagnation level and the flow field are sought. In
the following analysis Q is assumed to be unknown and the
stagnation level to be given. The origin of the coordinate
system is on the stagnation level with the x-axis vertical,
positive downwards and the y-axis horizontal. Flow is assumed
to be two-dimensional, steady, incompressible, non-viscous and
irrotational. The problem is formulated in terms of the
stream function $\psi(x,y)$. Flow in ABCDEA (Figure 1) is governed
by the Laplace equation

$$\nabla^2 \psi = 0. \tag{1}$$

The boundary conditions are

$$\psi = \text{constant} > 0, \quad \text{on AB}, \tag{2}$$

$$\partial\psi/\partial n = \sqrt{2gx} \quad \text{on AB}, \tag{3}$$

$$\psi = 0 \quad \text{on CDE}, \tag{4}$$

$$\partial\psi/\partial n = 0 \quad \text{on AE and BC} \tag{5}$$

in which n is the outward normal to the boundary of the region

of flow and g is the acceleration of gravity. On the far
upstream and downstream boundaries flow is taken normal to the
boundary, which is an approximation to the actual flow condition at these boundaries. The value of the constant ψ on AB
is the unknown flow rate Q corresponding to the given stagnation level from which the x-coordinate of the free surface is
measured.

VARIATIONAL FORMULATION

The global finite element equations governing fluid mechanics
problems are often derived from the minimization of an appropriate functional. This minimization procedure involves the
vanishing of the first variation subject to the appropriate
boundary conditions. In the procedure which follows the first
variation will be constructed directly without recourse to the
governing functional of the free surface problem.

Consider the functional I

$$I = \iint_{ABCDEA} [(\frac{\partial \psi}{\partial x})^2 + (\frac{\partial \psi}{\partial y})^2] dxdy \qquad (6)$$

in which $\psi(x,y)$ and the domain of integration both vary. This
is not the appropriate functional for the free surface problem
but from its first variation δI the correct first variation
δU, from which the global finite element equations may be
derived, will be constructed.

The domain of integration ABCDEA varies because of the
variation of the free surface AB (Figure 1). The points A and
B are constrained to vary on the far upstream and downstream
boundaries respectively. The first variation of I when the
domain of integration is variable is given by Courant and
Hilbert (1953).

$$\delta I = -2 \iint_R \nabla^2 \psi \bar{\delta}\psi \, dxdy + 2 \int_\Gamma (\frac{\partial \psi}{\partial x} \frac{\partial x}{\partial n} + \frac{\partial \psi}{\partial y} \frac{\partial y}{\partial n}) \bar{\delta}\psi ds$$

$$+ \int_\Gamma [(\frac{\partial \psi}{\partial x})^2 + (\frac{\partial \psi}{\partial y})^2] (\delta x \frac{\partial x}{\partial n} + \delta y \frac{\partial y}{\partial n}) ds \qquad (7)$$

in which R denotes the region, Γ the boundary, n the outward
normal, s the arc length along Γ and $\bar{\delta}\psi$ the variation at a
fixed argument point given by

$$\bar{\delta}\psi = \delta\psi - \frac{\partial \psi}{\partial x} \delta x - \frac{\partial \psi}{\partial y} \delta y. \qquad (8)$$

Admissible functions $\psi(x,y)$ are chosen to satisfy the boundary

170

conditions, Equations (2) and (4). It is important to note that although Equation (2) specifies that ψ must have a constant value along AB it does __not__ specify the value of the constant. The constant is the total flow rate Q which is unknown. Therefore, ψ may have a total variation $\delta\psi = \delta Q$ on AB. In view of these boundary conditions and the admissible variations of the domain of integration Equation (7) can be reduced to

$$\delta I \;=\; -\,2 \iint\limits_{R} \nabla^2 \psi\, \overline{\delta}\psi \; dxdy + 2 \int\limits_{AE} \frac{\partial\psi}{\partial n}\, \overline{\delta}\psi \; ds + 2 \int\limits_{CB} \frac{\partial\psi}{\partial n}\, \overline{\delta}\psi \; ds$$

$$+\, 2 \int\limits_{BA} \frac{\partial\psi}{\partial n}\, \overline{\delta}\psi \; ds + \int\limits_{BA} \left(\frac{\partial\psi}{\partial n}\right)^2 \left(\delta x\, \frac{\partial x}{\partial n} + \delta y\, \frac{\partial y}{\partial n}\right) ds \tag{9}$$

Using Equation (8), we obtain

$$\int\limits_{BA} \frac{\partial\psi}{\partial n}\, \overline{\delta}\psi \; ds \;=\; \int\limits_{BA} \frac{\partial\psi}{\partial n}\, \delta\psi \; ds - \int\limits_{BA} \left(\frac{\partial\psi}{\partial n}\right)^2 (\delta x\, dy - \delta y\, dx) \tag{10}$$

Then

$$\delta I \;=\; -\,2 \iint\limits_{R} \nabla^2 \psi\, \overline{\delta}\psi \; dxdy + 2 \int\limits_{AE} \frac{\partial\psi}{\partial n}\, \overline{\delta}\psi \; ds + 2 \int\limits_{CB} \frac{\partial\psi}{\partial n}\, \overline{\delta}\psi \; ds$$

$$+\, 2 \int\limits_{BA} \frac{\partial\psi}{\partial n}\, \delta Q \; ds - \int\limits_{BA} \left(\frac{\partial\psi}{\partial n}\right)^2 (\delta x\, dy - \delta y\, dx) \tag{11}$$

Now consider the first variation δU given by

$$\delta U \;=\; \delta I - 2 \int\limits_{BA} \sqrt{2gx}\; \delta Q \; ds + \int\limits_{BA} 2gx\, (\delta x\, dy - \delta y\, dx) \tag{12}$$

Then

$$\delta U \;=\; -\,2 \iint\limits_{R} \nabla^2 \psi\, \overline{\delta}\psi \; dxdy + 2 \int\limits_{AE} \frac{\partial\psi}{\partial n}\, \overline{\delta}\psi \; ds + 2 \int\limits_{CB} \frac{\partial\psi}{\partial n}\, \overline{\delta}\psi \; ds$$

$$+\, 2 \int\limits_{BA} \left(\frac{\partial\psi}{\partial n} - \sqrt{2gx}\right) \delta Q \; ds$$

$$-\, \int\limits_{BA} \left[\left(\frac{\partial\psi}{\partial n}\right)^2 - 2gx\right](\delta x\, dy - \delta y\, dx) \tag{13}$$

It can be shown that the function $\psi(x,y)$ which satisfies the
boundary conditions specified by Equations (2) and (4) and for
which

$$\delta U = 0 \qquad\qquad (14)$$

for all admissible variations $\delta x, \delta y, \overline{\delta\psi}$ also satisfies Equations
(1),(3) and (5). Therefore, the variational formulation,
Equation (14), is equivalent to the boundary value problem
specified by Equations (1)-(5). The constant value of the
stream function ψ on the free surface gives the rate of flow Q.

IKEGAWA-WASHIZU VARIATIONAL PRINCIPLE

A variational principle for free surface gravity flow was formulated by Ikegawa and Washizu (1973) and applied by them to
the problem of free flow over a spillway. They assumed that
during steady state flow, as illustrated in Figure 1, the
functional χ defined by

$$\chi = \frac{1}{2} \iint_R [(\frac{\partial\psi}{\partial x})^2 + (\frac{\partial\psi}{\partial y})^2]dxdy - \frac{1}{2}\int_{AB} gx^2dy \qquad (15)$$

has an extremum. Equation (15) contains a sign correction due
to Betts (1976). Ikegawa and Washizu (1973), Betts (1978)
and Aitchison (1979,1980) obtain the first variation $\delta\chi$ as

$$\delta\chi = -\iint_R (\frac{\partial^2\psi}{\partial x^2} + \frac{\partial^2\psi}{\partial y^2})\,\delta\psi\,dxdy + \int_{AB}[gx - \frac{1}{2}(\frac{\partial\psi}{\partial n})^2]\delta xdy \qquad (16)$$

They obtain their global finite element equations for the flow
field by discretizing Equation (16) using the usual procedures
of finite element analysis. It is clear from a comparison of
Equations (13) and (16) that the first variation $\delta\chi$ is quite
different from the first variation δU. The most significant
difference is that the surface integral containing the variation δQ is missing from $\delta\chi$. Without this integral it is
impossible to find the flow Q directly from the global finite
element equations, i.e., from $\delta\chi = 0$.

For a given stagnation level, solutions of the flow field
using $\delta\chi = 0$ are obtained by computing associated free surfaces
for several given Q and from this set of solutions selecting
the "correct solution" by some ancillary criterion unrelated
to the variational principle. Ikegawa and Washizu (1973)
assumed that the flow Q with the smoothest surface is the
solution. Similar criteria are used by Betts (1978).
Aitchison (1979,1980) notes that for the problem of flow over
a submerged weir for a given stagnation level only one value
of Q produced a flow without waves and assumed this to be the

solution.   She recommended finding this Q by an interpolation procedure.   Aitchison also noted that for this procedure to work Q had to lie within a certain range (unspecified).

The difficulties arise for two reasons: (1) the functional in Equation (15) is not complete, and   (2) variations are carried out subject to the constraint the ψ = Q (specified) on the free surface.   It can be seen with reference to Equation (8) that this implies that the total variation $\delta\psi$ = 0 on AB or δQ = 0 on AB.   Therefore

$$\bar{\delta\psi} \;\; = \;\; - \frac{\partial\psi}{\partial x} \, \delta x - \frac{\partial\psi}{\partial y} \, \delta y \tag{17}$$

This equation shows that the variation in ψ due to variation of the free surface AB is equal and opposite to the fixed argument point variation $\bar{\delta\psi}$ to ensure a total variation $\delta\psi$ = 0. This lack of a variation $\delta\psi$ (=δQ) on AB removes an important control on the location of AB to ensure the minimization of the functional governing the flow field and is directly responsible for flow solutions incompatible with the specified stagnation level.

The first variation δU does contain a boundary integral involving total variations in Q on the free surface.   The presence of this integral ensures that the resulting global finite element equations will yield a solution compatible with the stagnation level.

There are other differences between $\delta\chi$ and δU.   Note that δU also contains integrals relating to the upstream and downstream boundaries and that the variation in ψ, $\bar{\delta\psi}$, in the area integral is a variation with respect to a fixed argument point.   It was earlier asserted that the functional defined by Equation (16) is not complete.   The basis of this assertion is that the first variation δU cannot be derived from χ even using the completely general variational process described by Varo$\bar{\text{g}}$lu and Finn (1977,1978) and only the condition δU = 0, which has been shown to be an equivalent formulation of the flow problem, yields stable unique solutions for each specified stagnation level.   The variational principle $\delta\chi$ = 0 can only be shown to be an equivalent formulation if Q is assumed to be known.   In fact, of course, Q is the solution we are seeking.

FINITE ELEMENT FORMULATION

The variational formulation will be used in discretized form. The nodes in the x-y plane are defined as the points of intersection of the stream lines

$$\psi(x,y) \;\; = \;\; \alpha_i Q, \qquad i=1,2,\ldots,p \tag{18}$$

with the fixed vertical lines

$$y = y_j, \qquad j=1,2,\ldots,q-1 \tag{19}$$

and the fixed far downstream boundary $y = r_1x+r_2$. As one of the nodal coordinates, ψ, is not a spatial coordinate, the finite element mesh is not fixed in the space of the flow field. The boundary of the flow region is defined by ψ_{AB} and, therefore, as the domain varies the stream lines vary and the mesh automatically adjusts to the change to fill the new domain. Meshes with nodes defined by the spatial coordinates do not, in general, have the capacity to adapt except in a limited region near the boundary. This often leads to severe mesh distortion along the important boundary elements. Also, in many cases with a fixed mesh, boundary movements eliminate elements and require a restructuring of the finite element mesh.

The reservoir bed and spillway EDC and the free surface AB correspond to the streamlines $\psi=0$ ($\alpha_1=0$) and $\psi=Q$ ($\alpha_p=1$), respectively. The x-coordinates of the nodes on the boundary segment EDC are known. Let the number of nodes with unknown x-coordinates be denoted by n. Then the n+1 discretized unknowns of the problem are (i) the total rate of flow Q for given stagnation level; (ii) the x-coordinates x_i (i=1,2,..., n) of the nodes inside the flow region and on the boundaries excluding the reservoir bed and the spillway.

By prescribing the parameters

$$0 = \alpha_1 < \alpha_2 < \ldots < \alpha_p = 1 \tag{20}$$

and

$$-1 = \beta_1 < \beta_2 < \ldots < \beta_q = 1 \tag{21}$$

which are defined as

$$\beta_j = \begin{cases} y_j/L_1, & y_j \le 0, \\ y_j/L_2, & y_j > 0, \end{cases} \tag{22}$$

a mesh can be defined in the x-y plane which _adapts_ to the variable flow region depending on the values of the unknowns Q and x_i (i=1,2,...,n). The discretization of the flow region in the x-y plane is shown in Figure 2.

For the discretized region, δU can be written as

$$\delta U = \delta \sum_{ijk} I_{ijk} + \sum_{ij} \delta \Gamma_{ij} \tag{23}$$

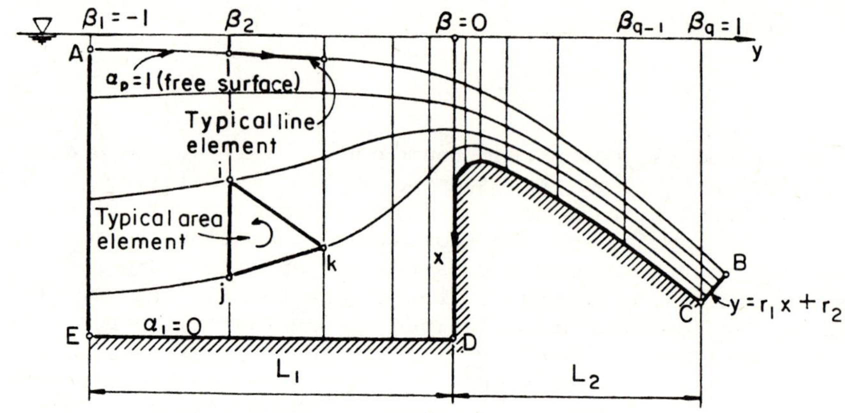

FIGURE 2   Adaptive Finite Element Mesh

where

$$I_{ijk} = \iint_{ijk} [(\frac{\partial \psi}{\partial x})^2 + (\frac{\partial \psi}{\partial y})^2] \, dxdy \qquad (24)$$

and

$$\delta\Gamma_{ij} = -2 \int_{ij} \sqrt{2gx} \; ds\delta Q - \int_{ij} 2gx(\delta xdy - \delta ydx). \qquad (25)$$

The change in sign of the second integral from that in
Equation (12) is due to the change in the direction of integration. The first summation in Equation (23) is performed
over all the triangular area elements in the flow region; the
second summation is over all the line elements in the free
surface AB. In a typical triangular area element ijk the
unknown function $x(\psi,y)$ is approximated by a linear function
in ψ and y. This is accomplished most conveniently by speci-
fying $\psi(x,y)$ as a linear function in x,y by

$$\psi(x,y) = \frac{1}{2A_{xy}} \{a_i + Y_i x + X_i y, \; a_j + Y_j x + X_j y, \; a_k + Y_k x + X_k y\} \begin{Bmatrix} \psi_i \\ \psi_j \\ \psi_k \end{Bmatrix}$$

$$(26)$$

The x and y coordinates of the node i are denoted by x_i and y_i,
respectively. The value of the stream function at node i is
ψ_i. The rest of the symbols appearing in Equation (26) are
defined in Table 1.

The differentiation of $\psi(x,y)$ with respect to x and y
yields

$$\frac{\partial \psi}{\partial x} = \frac{A_{\psi y}}{A_{xy}} , \qquad \frac{\partial \psi}{\partial y} = \frac{A_{\psi x}}{A_{xy}} \tag{27}$$

with $A_{\psi x}$, $A_{\psi y}$, and A_{xy} also defined in Table 1.

TABLE 1

DEFINITION OF SYMBOLS FOR A TYPICAL AREA ELEMENT ijk

Symbols \ Indices	i	j	k
a	$x_j y_k - x_k y_j$	$x_k y_i - x_i y_k$	$x_i y_j - x_j y_i$
X	$x_k - x_j$	$x_i - x_k$	$x_j - x_i$
Y	$y_j - y_k$	$y_k - y_i$	$y_i - y_j$
ψ	$\psi_k - \psi_j$	$\psi_i - \psi_k$	$\psi_j - \psi_i$
$2A_{xy}$	$x_i Y_i + x_j Y_j + x_k Y_k$		
$2A_{\psi x}$	$\psi_i X_i + \psi_j X_j + \psi_k X_k$		
$2A_{\psi y}$	$\psi_i Y_i + \psi_j Y_j + \psi_k Y_k$		

In view of Equation (27), Equation (24) yields

$$I_{ijk} = \frac{A_{\psi x}^2}{A_{xy}} + \frac{A_{\psi y}^2}{A_{xy}} . \tag{28}$$

The evaluation of the line integral on the righthand side of Equation (25) using trapezoidal rule yields

$$\delta \Gamma_{ij} = \sqrt{2g} \ (\sqrt{x_i} + \sqrt{x_j}) \Delta s \delta Q - g(x_i \delta x_i + x_j \delta x_j) \Delta y$$

$$+ \ g(x_i \delta y_i + x_j \delta y_j) \Delta x \tag{29}$$

where for a typical line element ij on AB

$$\Delta y = y_j - y_i, \quad \Delta x = x_j - x_i, \quad \Delta s = \sqrt{[(\Delta x)^2 + (\Delta y)^2]} . \tag{30}$$

It should be noted that δy vanishes for all the nodes except the nodes on the downstream face BC. The constraint that the downstream face is a straight line $y = r_1 x + r_2$ requires that

$$\delta y_i = r_1 \delta x_i, \qquad i \text{ on BC.} \tag{31}$$

The discretized form of δU may be written, in general, in terms of the variations of the discrete unknowns x_m and Q as

$$\delta U = \sum_{m=1}^{n} f_m \delta x_m + f_{n+1} \delta Q . \tag{32}$$

Since

$$\delta I = \sum_{m=1}^{n} \left(\frac{\partial}{\partial x_m} \sum_{ijk} I_{ijk}\right) \delta x_m + \left(\frac{\partial}{\partial Q} \sum_{ijk} I_{ijk}\right) \delta Q \tag{33}$$

and $\sum_{ij} \delta \Gamma_{ij}$ may be written as

$$\sum_{ij} \delta \Gamma_{ij} = \sum_{m=1}^{n} q_m \delta x_m + q_{n+1} \delta Q \tag{34}$$

then the coefficients f_m and f_{n+1} in Equation (32) can be expressed as in Equations (35) and (36). The contributions to q_m and q_{n+1} by each line element on AB are determined later.

$$f_m = \frac{\partial}{\partial x_m} \sum_{ijk} I_{ijk} + q_m, \qquad m=1,2,\ldots,n, \tag{35}$$

$$f_{n+1} = \frac{\partial}{\partial Q} \sum_{ijk} I_{ijk} + q_{n+1}. \tag{36}$$

Thus, $\delta U = 0$ in discretized form reduces to the system of equations

$$f_m(x_1, x_2, \ldots, x_n, Q) = 0, \qquad m=1,2,\ldots,n+1. \tag{37}$$

To construct these equations explicitly, the partial derivatives of I_{ijk} appearing in Equations (35) and (36) must be evaluated and expressions for the contributions to q_m ($m=1,2,\ldots,n+1$) from a typical line element obtained. From the definitions of $A_{\psi x}$, $A_{\psi y}$ and A_{xy} given for a typical area element ijk in Table 1, the partial derivatives of I_{ijk} with respect to x_m can be expressed as

$$\frac{\partial}{\partial x_m} I_{ijk} = \begin{cases} 0, \; m \neq i \text{ or } j \text{ or } k, \\[2ex] -\dfrac{A_{\psi y}}{A_{xy}} \Psi_m - \dfrac{1}{2}\left[\left(\dfrac{A_{\psi x}}{A_{xy}}\right)^2 + \left(\dfrac{A_{\psi y}}{A_{xy}}\right)^2\right] Y_m, \\[2ex] \quad m=i \text{ or } j \text{ or } k \text{ and } y_m \neq r_1 x_m + r_2, \\[2ex] -\left(\dfrac{A_{\psi x}}{A_{xy}} - r_1 \dfrac{A_{\psi y}}{A_{xy}}\right) \Psi_m \\[2ex] \quad -\dfrac{1}{2}\left[\left(\dfrac{A_{\psi x}}{A_{xy}}\right)^2 + \left(\dfrac{A_{\psi y}}{A_{xy}}\right)^2\right](Y_m + r_1 X_m), \\[2ex] \quad m=i \text{ or } j \text{ or } k \text{ and } y_m = r_1 x_m + r_2. \end{cases} \tag{38}$$

The symbols Ψ, Y, and X are defined in Table 1. Using Equations (29) and (34), the contribution to q_m from a typical line element ij, q_m^{ij}, can be expressed as

$$q_m^{ij} = \begin{cases} 0, \; m \neq i \text{ or } j, \\[1ex] -gx_m \Delta y, \; m=i \text{ or } j \text{ and } m \neq B, \\[1ex] -gx_B \Delta y + gr_1 x_B \Delta x, \; m=j=B. \end{cases} \tag{39}$$

Noting that $\psi = Q$ on the free surface, the partial derivative of I_{ijk} with respect to Q can be expressed as

$$\frac{\partial}{\partial Q} I_{ijk} = \begin{cases} 0, \text{ none of the nodes of } ijk \text{ are on } AB, \\[2ex] \dfrac{A_{\psi x}}{A_{xy}} X_i + \dfrac{A_{\psi y}}{A_{xy}} Y_i, \text{ only node } i \text{ is on } AB, \\[2ex] -\left(\dfrac{A_{\psi x}}{A_{xy}} X_k + \dfrac{A_{\psi y}}{A_{xy}} Y_k\right), \text{ only nodes } i \text{ and } j \text{ are on } AB. \end{cases} \tag{40}$$

In view of Equation (29), the contribution to q_{n+1} from a line element ij on AB, q_{n+1}^{ij}, is given by

$$q_{n+1}^{ij} = \sqrt{2g} \; (\sqrt{x_i} + \sqrt{x_j}) \, \Delta s. \tag{41}$$

SOLUTION TECHNIQUE

In the system of $n+1$ algebraic nonlinear equations

$$f_i(x_1, x_2, \ldots, x_n, Q) = 0, \qquad i=1,2,\ldots,n+1 \tag{42}$$

let Q be replaced by an initial trial value Q^o. Then the resulting first n-equations in n unknowns $x_1, x_2, \ldots, x_n$ are

178

solved employing the Newton-Raphson method (Oden, 1967). The
iterative solution for the first n equations of the system

$$f_i(x_1, x_2, \ldots, x_n; Q^0) = 0, \qquad i=1,2,\ldots,n \tag{43}$$

can be obtained employing the recurrence relation

$$\underset{\sim o}{x}^{r+1} = \underset{\sim o}{x}^r - [J_r]_o^{-1} \underset{\sim}{f}(\underset{\sim o}{x}^r; Q^0), \qquad r=0,1,2,\ldots . \tag{44}$$

Here, subscript zero denotes that Q has the initial trial
value Q^0. The nxn Jacobian matrix at $(x_{\sim o}^r, Q^0)$ is given by

$$[J_r]_o = [\frac{\partial}{\partial x_j} f_i(\underset{\sim o}{x}^r; Q^0)] . \tag{45}$$

The initial trial location for Q^0 is denoted by $\underset{\sim o}{x}^o$. The iteration process described by Equation (44) is continued until

$$||\underset{\sim o}{x}^{r+1} - \underset{\sim o}{x}^r|| \le \varepsilon \tag{46}$$

for some prescribed accuracy ε.

The solution $\underset{\sim o}{x}^{r+1}$ and Q^0 does not satisfy the last
equation

$$f_{n+1}(x_1, x_2, \ldots, x_n, Q) = 0 \tag{47}$$

Depending on the residue $f_{n+1}(\underset{\sim o}{x}^{r+1}, Q^0)$, a new trial value Q^1
for Q is calculated. The solution $\underset{\sim o}{x}^{r+1}$ can be used as the
initial trial location $\underset{\sim 1}{x}^o$ for Q^1 for the next step in the
iteration process for Q. This iterative process in Q can be
repeated until Equation (47) is satisfied to some prescribed
accuracy.

COMPUTER PROGRAM

A computer program has been developed to construct Equation
(42) and solve the system of nonlinear Equations (43) for an
assumed total energy head on the crest, H_e (which defines the
stagnation level) or flow rate Q. Input to the program
consists of:

    (i)   Values of L_1 and L_2 to define upstream and down-
           stream boundaries.

    (ii) The mesh parameters $\alpha_1, \alpha_2, \ldots, \alpha_p$ and $\beta_1, \beta_2, \ldots, \beta_q$.

    (iii) x-Coordinates of the fixed nodes $(\alpha=0, \beta_i)$
           $i=1,2,\ldots,q$ on the reservoir bed and the spillway.

    (iv) A fixed value for the total energy head H_e and a
           corresponding initial trial value for Q or a fixed

value for the total flow rate Q and an initial trial
value for total energy head H_e.

(v) An initial set of trial values for the x-coordinates
     associated with the assumed set of streamlines ψ
     corresponding to the prescribed pair Q and H_e.
     These x-values must preserve the sequential ordering
     of streamlines; i.e., if $\psi_1 < \psi_2 < \psi_3$ then the trial
     values cannot be such as to put ψ_3 between ψ_1 and ψ_2.

The initially assumed streamlines are modified automatically by the iteration process defined in Equation (44) by
computing the new coefficient matrices for each step in the
iteration process to yield a solution for the assumed H_e or Q.
The adequacy of this solution is checked by computing the
residue of Equation (47). If this residue is too large the
value of H_e or Q is adjusted according to a prescribed maximum
step size ΔH_e or ΔQ and another solution is obtained. The
numerical results reveal that

$$f_{n+1}(\underset{\sim}{x}_m^{r+1};Q^m) \geq 0 \qquad\qquad (48)$$

and that the solution of the flow problem yields the minimum of
f_{n+1}. Therefore, the process of adjusting H_e or Q is continued
by an algorithm until the minimum of f_{n+1} is reached to a prescribed accuracy. This algorithm adjusts H_e^m or Q^m automatically to obtain H_e^{m+1} or Q^{m+1} within the prescribed maximum
step size ΔH_e or ΔQ. This adjustment is determined by a search
for the minimum for f_{n+1} considering the previous calculated
values of the residue in Equation (47), $f_{n+1}(\underset{\sim}{x}_o^{r+1};Q^o)$,
$f_{n+1}(\underset{\sim}{x}_1^{r+1};Q^1),\ldots,f_{n+1}(\underset{\sim}{x}_m^{r+1};Q^m)$.

APPLICATIONS

The proposed method is applied to the problem of flow over
Waterways Experiment Station (WES) standard spillway shapes
(Chow, 1959) with vertical upstream face. The shape of the
spillway for the design head H_d = 6 ft is illustrated in
Figure 3. Note that the $\bar{x}$-$\bar{y}$ axes are different from the axes
x-y used in the derivation. This change is made to facilitate
the comparison with the data given in (Chow, 1959). The discharge Q per unit length of the crest over a WES spillway can
be computed by the equation

$$Q = C H_e^{1.5} \qquad\qquad (49)$$

where H_e is the total energy head on the crest in ft, including velocity head in the approach channel, C is the discharge
coefficient. The height of the spillway (Figure 3) is denoted
by h. Two values of h are chosen for analysis. The first case
h/H_d = 4.13 corresponds to a high spillway; the second case is
a low spillway with h/H_d = 0.46. In both cases, the far

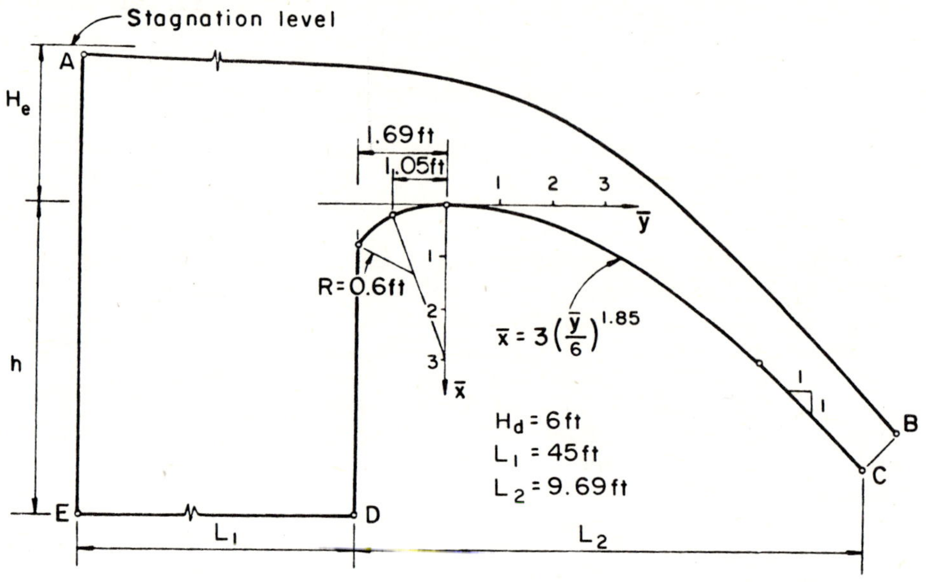

FIGURE 3   Geometry of Spillway Model CW 801

upstream boundary AE and the far downstream boundary CB are
defined by choosing L_1 = 45 ft and L_2 = 6.69 ft (Figure 3).

The upper nappe profiles for a high overflow spillway
were calculated for three values of the operating head (excluding velocity head) H = 8 ft, 6 ft, and 3 ft using the finite
element mesh defined by the parameters shown in Table 2.   The

TABLE 2

PARAMETERS DEFINING MESH

i	1	2	3	4	5	6	7
α_i	0	0.1	0.3	0.6	0.8	0.9	1
β_i	-1	-0.5	-0.25	-0.10	-0.05	-0.02	0

i	8	9	10	11	12	13	14	15
β_i	0.039	0.103	0.175	0.335	0.484	0.645	0.851	1

computed profiles are compared in Figure 4 with those measured
in model tests (Chow, 1959) using model spillway type CW 801
under the same operating heads. The agreement between the
calculated and experimental profiles is very good.

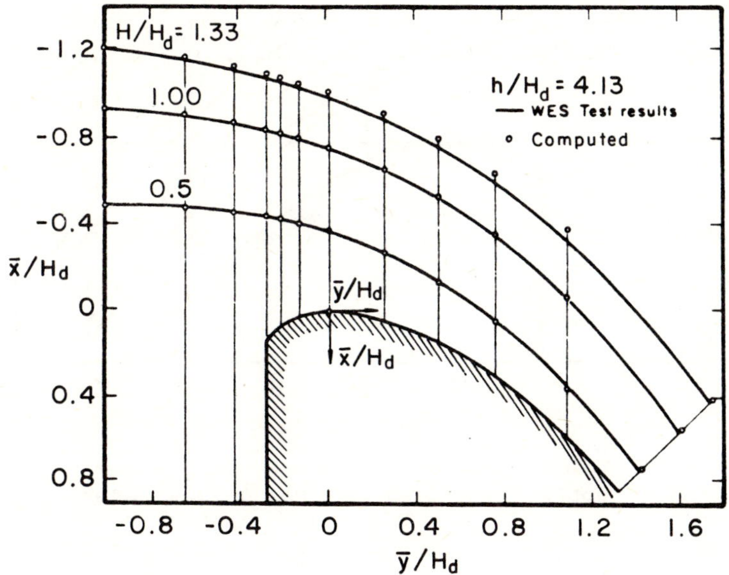

FIGURE 4   Comparison of Computed and Measured Upper
            Nappe Profiles for 3 Operating Heads

The upper nappe profiles were also computed for the high
overflow spillway for five different flow rates Q using the
same finite element mesh. The results are given in Figure 5
in which the corresponding computed stagnation levels are also
given. The results are compared with WES experimental data on
the given spillway (Chow, 1959) in Figure 6 for a high WES
spillway and in Figure 7 for a low WES spillway. There is
very close agreement between computed and experimental data.

These numerical results were obtained on an IBM 370/168
computer at the University of British Columbia. The total
processing time to obtain each upper nappe profile illustrated
in Figure 5 was 8-10 sec depending on the initial guess of the
energy head and the number of iterations required to determine
the final energy head. From 3 to 8 iterations were required
to obtain the streamlines associated with each fixed energy
head with a prescribed accuracy $\varepsilon = 10^{-4}$ (Equation 46). Using
the mesh defined in Table 2, each iteration took 0.22 sec of
central processing time.

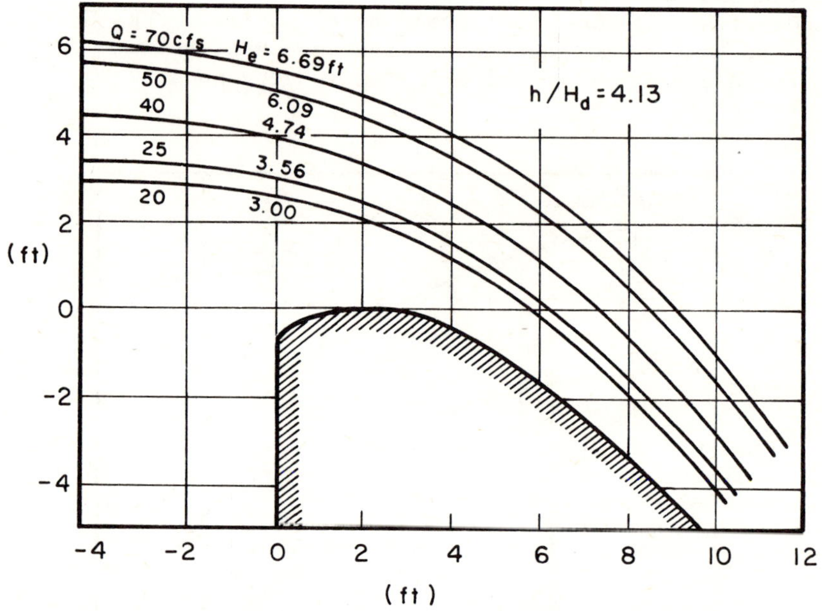

FIGURE 5   Upper Nappe Profiles for 5 Flow Rates

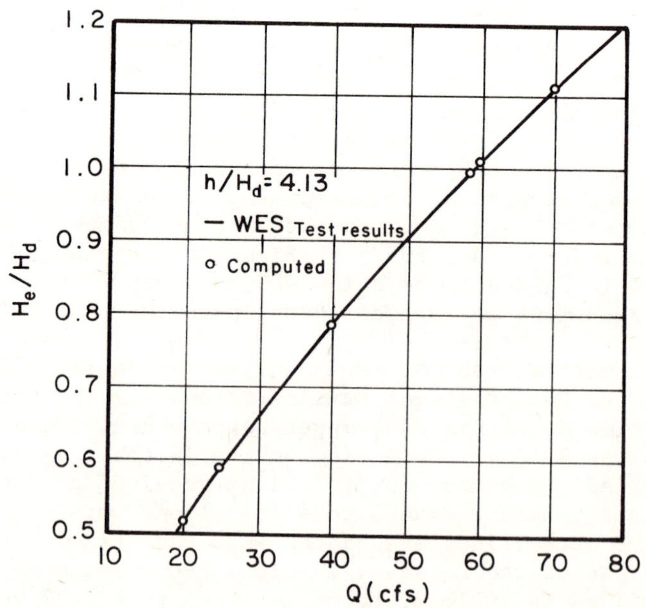

FIGURE 6   Comparison of Measured and Computed
Flow Rates for High WES Spillway

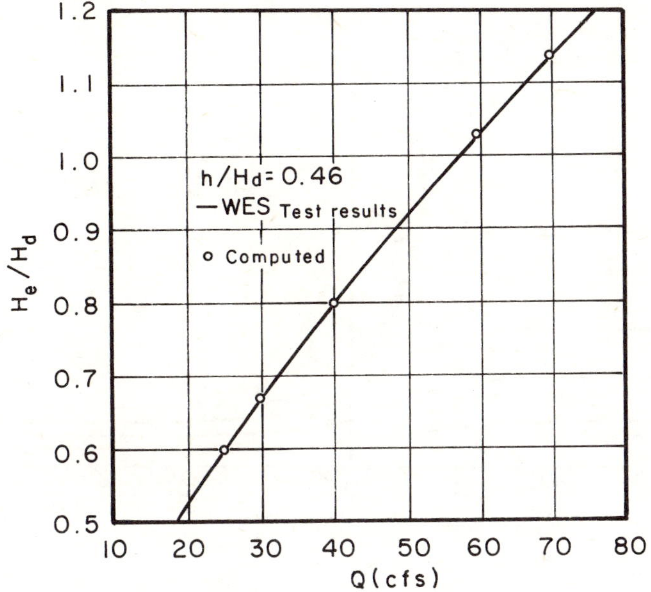

FIGURE 7   Comparison of Measured and Computed Flow
Rates for Low WES Spillway

EXTENSION TO GATED FLOW

The variational formulation embodied in Equations (13) and
(14) has recently been extended to gated flow. Typical problems in gated flow are illustrated in Figures 8 and 9 in which
flow in a channel and over a spillway are regulated by Tainter
gates. Gated flows for given stagnation levels are, at present, estimated on the basis of model and prototype data such
as that shown in Figures 10 and 11. The curves in these figures
are design curves taken from a hydraulic design manual developed by the Waterways Experiment Station of the U.S. Army,
Corps of Engineers. This manual is updated as required and
the reference WES (1981) is used to designate the manual as of
1981. The date does not imply that the referenced material
was updated in 1981.

The gated flow problems are characterized by two free
surfaces, one upstream and one downstream of the gate, on each
of which the two boundary conditions defined by Equations (2)
and (3) must be satisfied. The gate itself is a fixed boundary
segment on which one boundary condition is satisfied. Equation (14) is applied directly and the finite element discretization is carried out using the procedure outlined for free
flow over a spillway. The implications of the two free surfaces and the additional fixed boundary segment are taken into
account during the discretization procedure. As in the case

G

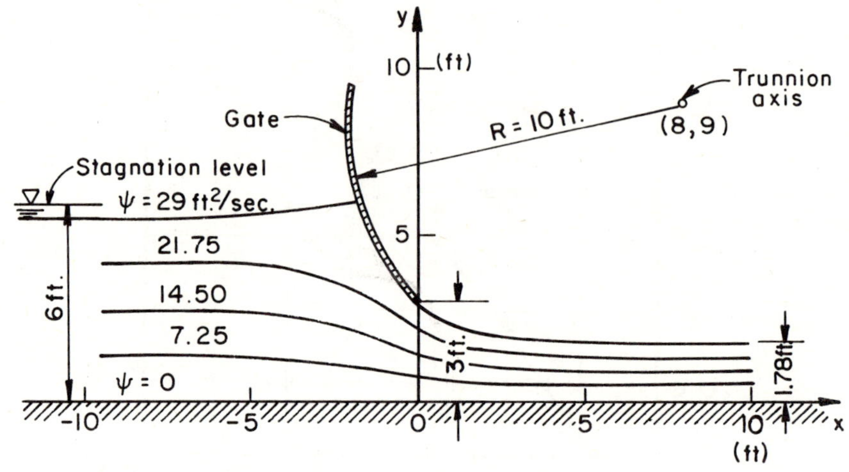

FIGURE 8   Flow in a Gated Channel

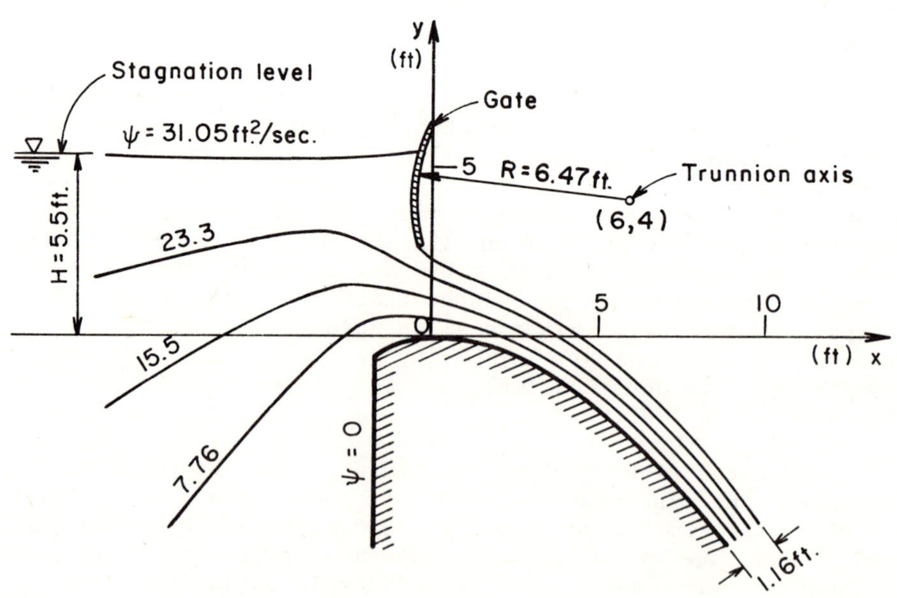

FIGURE 9   Flow over a Gated Spillway

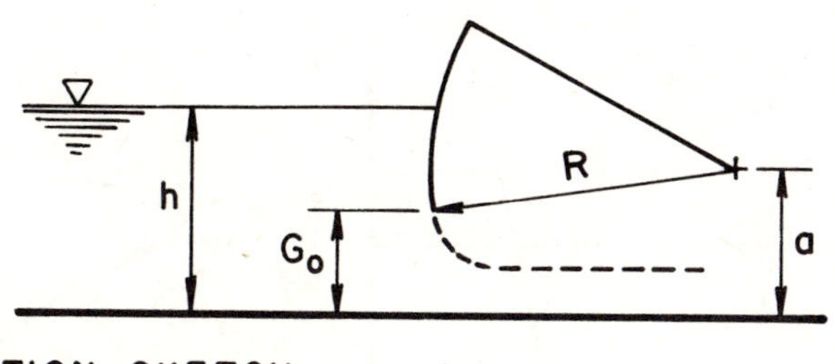

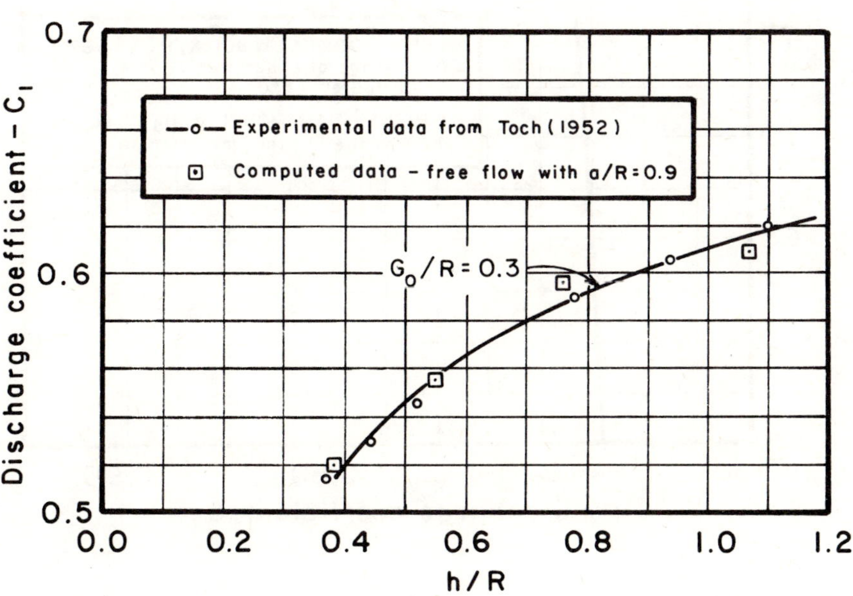

FIGURE 10   Comparison of Experimental and Computed Discharge Coefficients for Gated Channel Flow

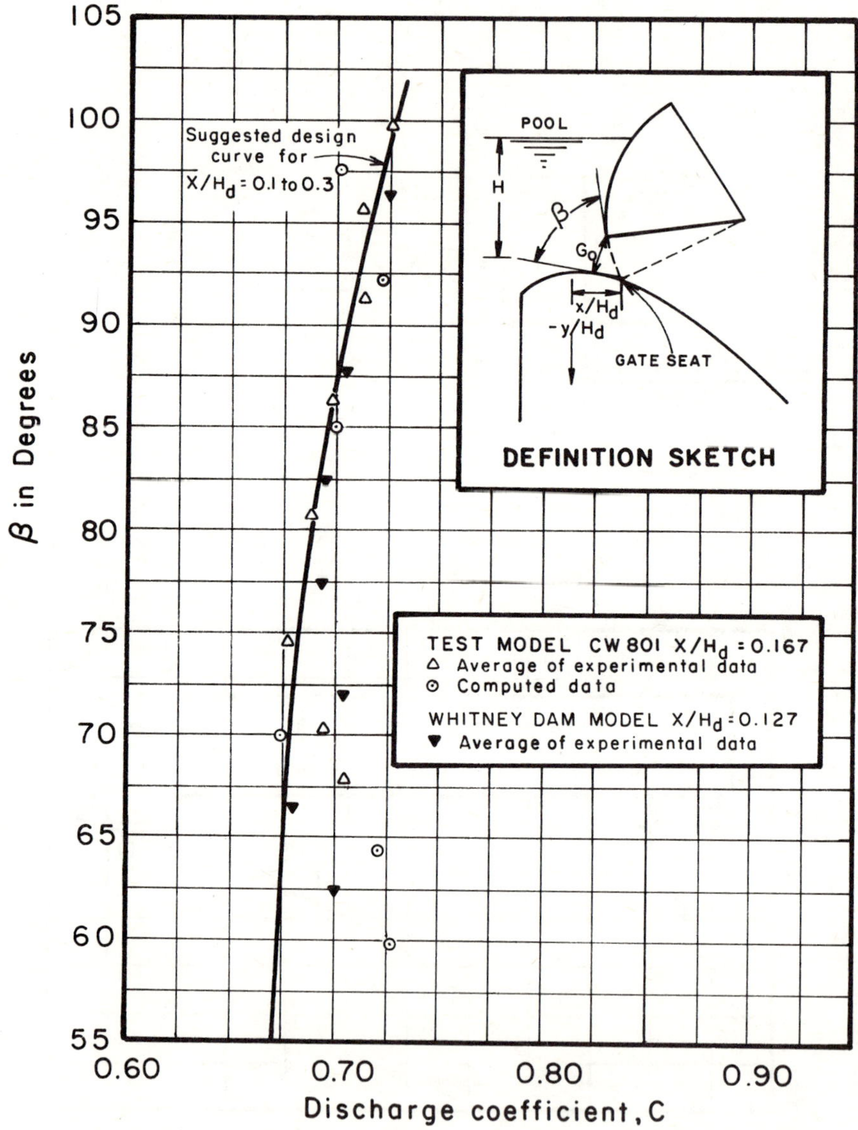

FIGURE 11   Comparison of Experimental and Computed Discharge Coefficients for Gated Spillway Flow

of free flow over a spillway, for each stagnation level a
unique solution was found by direct solution of the global
finite element equations.

Computed data points are shown in Figure 10 for channel
flow. The computed data agree very closely with the experimental data (Toch, 1952). The spillway used for the gated
spillway flow analyses is the standard Water Experiment
Station spillway model CW 801. This is the same model used
earlier for the analysis of free flow over a spillway. The
experimental data points in Figure 11 are each an average of
experimental values. Furthermore, there is a considerable
lumping of geometrical effects into one parameter, the angle
β, defined in Figure 11. Actual gate geometry is not given for
the specific data points. Comparison of computed and experimental data is, therefore, carried out solely on the basis of
equal angles β in the computational and experimental models.
The computed and experimental data agree very well. The
computed data seem to follow the proposed design curve very
closely over most of the range. The data defining the lower
end of the design curve for angles $\beta < 70^{\circ}$ are data from
prototype dams. Experimental and computed data for the model
CW 801 appear to fall to the right of the assumed design line
for smaller β angles.

The gated flow analyses were conducted at the University
of British Columbia Computing Centre on an Amdahl 470 V16.
Calculations were made to an accuracy of $\varepsilon = 10^{-4}$ (Equation
46). For the mesh used (150 unknowns) each iteration took
0.115 sec of central processing time.

CONCLUSION

A variational principle has been developed for free surface
gravity flow with free streamlines. The principle was derived
by allowing all admissible variations in the flow and the
region of flow. In particular, it incorporates the variations
in the stream function on the free surface as the free surface
is varied spatially. The variational principle is discretized
by final element procedures and the resulting global finite
element equations give unique stable solutions to a variety of
free surface problems of gravity flow. It is also applicable
to flow problems with more than one free surface such as gravity flow over a gated spillway.

At present, this variational procedure is the only one
which gives the unique solution to free surface gravity flows
directly. Other approaches require fixed domain flow solutions
for a series of trial free surfaces or free surface solutions
for a series of assumed flows from which the true solution is
selected by some ancillary criterion. These sequences of
ancillary solutions can usually be found only within limited
ranges of flow, otherwise they diverge.

From the practical point of view, a very efficient feature of the method is that the finite element mesh is not spatially fixed. The values of the stream function are used as one of the nodal co-ordinates. This allows the mesh to adapt freely, without any restructuring, to any variations in the region of flow. The method has been thoroughly tested by comparing computed data for free and gated flow over spillways and gated flow in channels with experimental data from the same models. The agreement is very good. Thus, there is confidence in the reliability and practical utility of the method.

ACKNOWLEDGEMENTS

Financial assistance under grant no. 1498 from the National Science and Engineering Research Council of Canada is gratefully acknowledged. The computations for gated flow were carried out by Mr. X.-K. Xin and Mr. N. Chontos.

APPENDIX I - NONLINEARITY

The determination of the free surface as a part of the solution in these problems involves the solution of an intrinsically nonlinear problem. This property of the problem can be demonstrated simply by the following example. Consider the boundary value problem in ordinary differential equations

$$\frac{d^2 y}{dx^2} - y = 0, \qquad 0 < x < L, \tag{A-1}$$

$$y(0) = 0, \quad y(L) = 2, \quad \frac{dy}{dx}(L) = 4. \tag{A-2}$$

Suppose the end point $x = L$ is not known. Since one boundary condition is prescribed on the fixed end point $x = 0$ and two boundary conditions on the unknown end point $x = L$, this problem has the same characteristics as a free surface problem. The general solution of the differential equation, Equation (A-1), can be written as

$$y = Ae^x + Be^{-x} \tag{A-3}$$

where A and B are arbitrary constants to be evaluated from the boundary conditions. The boundary conditions Equation (A-2) yield three equations

$$A + B = 0, \quad Ae^L + Be^{-L} = 2, \quad Ae^L - Be^{-L} = 4 \tag{A-4}$$

in three unknowns A, B and L. By eliminating A and B in Equation (A-4), a nonlinear equation

$$\cosh L = 0.5 \tag{A-5}$$

is obtained for the unknown L. Although Equation (A-1) is a
linear differential equation, solution for L involves the
solution of a nonlinear equation, Equation (A-5).

APPENDIX II - REFERENCES

Aitchison, J.M. (1979), "A Variable Finite Element Method for
the Calculation of Flow over a Weir", Report, RL-79-069,
Rutherford Laboratory, Chilton, Didcot, Oxon.

Aitchison, J.M. (1980), "A Finite Element Solution for Critical Flow over a Weir", Proceedings, 3rd Int. Conf. on Finite
Elements in Flow Problems, Banff, Alberta, June, Vol. II,
pp. 52-59.

Betts, P.L. (1976), "A Variational Principle in Terms of
Stream Function for Free Surface Flows and Its Application to
the Finite Element Method", University of Calgary, Mechanical
Engineering Report, No. 84.

Betts, P.L. (1978), "Computation of Stationary Water Waves
Downstream of a Two-dimensional Contraction", Proceedings,
2nd Int. Conf. on Finite Elements in Water Resources, Imperial
College, London, July, Edited by C.A. Brebbia, W.G. Gray and
G.F. Pinder, Pentech Press, London, Plymouth, pp. 2.177-2.193.

Chow, V.T. (1959), "Open-channel Hydraulics", McGraw-Hill,
New York, pp. 360-384.

Courant, R. and Hilbert, D. (1953), "Methods in Mathematical
Physics", Interscience, New York, pp. 260-264.

Finn, W.D. Liam (1967), "Finite Element Analysis of Seepage
through Dams", Journal of the Soil Mechanics and Foundation
Division, Proc. ASCE, SM6, pp. 41-48.

Finn, W.D. Liam and Varoḡlu, E. (1975), "Variable Domain
Finite Element Analyses of Free Surface Flow Problems",
Proceedings, 1st Int. Conf. on Finite Elements in Water
Resources, Princeton University, U.S.A., (First published,
Pentech Press, Plymouth, 1977), July, pp. 3.115-3.132.

Finn, W.D. Liam and Varoḡlu, E. (1976), "Numerical Solution of
Unconfined Seepage Problems", Proceedings, 2nd Int. Conf. on
Numerical Methods in Geomechanics, Virginia Polytechnic
Institute and State University, Blacksburg, Virginia, ASCE,
New York, June, pp. III-1465 - III-1474.

Ikegawa, M. and Washizu, K. (1973), "Finite Element Method
Applied to Analysis of Flow over a Spillway Crest", Int. J.
Num. Method. Engng. 6, pp. 179-189.

190

Luke, J.C. (1967), "A Variational Principle for a Fluid with a
Free Surface", J. Fluid Mech. 27, pp. 395-397.

Oden, J.T. (1967), "Finite Elements of Non-linear Continua",
McGraw-Hill, New York, pp. 279-283.

Southwell, R.V. and Vaisey, G. (1946), "Relaxation Methods
Applied to Engineering Problems XII. Fluid Motions Characterized by 'Free' Stream-lines", Phil. Trans. Roy. Soc., London,
240A, pp. 117-160.

Taylor, R.L. and Brown, C.B. (1967), "Darcy Flow Solutions
with a Free Surface", Journal of the Hydraulics Division,
Proc. ASCE, HY2, pp. 25-33.

Toch, A. (1952), "The Effect of Lip Angle upon Flow under a
Tainter Gate", Master's Thesis, State University of Iowa.

Tu, R.K. (1971), "Free Boundary Potential Flow using Finite
Elements", Ph.D. Thesis, University of Arizona, DA31/9B/5405.

Varoḡlu, E. and Finn, W.D. Liam (1977), "An Inverse Variable
Domain Finite Element Technique for Irrotational Flow Problems
with a Free Surface", Proceedings, Int. Symp. on Innovative
Numerical Analysis in Applied Engineering Science, Versailles,
France, May, pp. 4.21-4.25.

Varoḡlu, E. and Finn, W.D. Liam (1978), "Variable Domain
Finite Element Analysis of Free Surface Gravity Flow",
Computer and Fluids, Pergamon Press, Vol. 6, pp. 103-114.

WES (1981), "Hydraulic Design Criteria", U.S. Army Corps of
Engineers, Waterways Experiment Station, Vicksburg, Mississippi.

Zienkiewicz, O.C. (1971), "The Finite Element Method in
Engineering Science", McGraw-Hill, London.

CHAPTER 7

ITERATIVE PROCEDURE FOR SIMULTANEOUS CALCULATION OF A STRESS
TRANSPORT TURBULENCE MODEL IN TWO DIMENSIONAL ELIPTIC FIELDS

Dan Naot

Department of Nuclear Engineering, Technion, Haifa, Israel

INTRODUCTION

Although started with a delay of several years with respect to
the two-variables models (Launder and Spalding 1972), prediction methods based on the Reynolds Stress closures developed
to the stage that calculation of anisotropic fields is now
possible (Launder, Reece and Rodi 1975, Wolfshtein, Naot and
Lin 1975). The decision whether one resorts to these advanced
methods is mainly associated with the observation of high anisotropy or inhomogeniety. This is not discussed here restricting the philosophy behind the present work to the following:
Once we realize that the turbulent flow field of interest is
characterized by a degree of anisotropy and inhomogeniety so
high as to make the use of algebraic stress modeling inaccurate, still we have the choice of using the full stress transport turbulence models (Reynolds stress closures) in reasonable computing time and with no severe mathematical efforts.

The present chapter contains a description of a method suitable for the calculation of the three dimensional turbulent
flow using a stress transport turbulence model. The method is
assigned for eliptic nonhomogeneous fields where the diffusion
and the convection of turbulence in the diverse directions are
of equal great importance. The method however, is restricted
to cases where the mean flow parameters depend on two space
coordinates only. As examples we will refer to a few cases
classified under the two categories:

    (i)   Fully developed channel flow with stress induced
          secondary flow in the lateral plane (Naot, Shavit,
          Wolfshtein 1974).

    (ii) Axisymmetric flow with swirl or geometry induced
          recirculation in the rotated r,z plane (Naot and
          Launder 1975).

As the description of the relatively more simple first case can be derived by straightforward algebraic steps from the second one, complete description will be given only for the axisymmetric configuration together with reference to these steps.

On the basis of early experience with channel flow reconfirmed on performing the axisymmetric calculations it was found desirable to perform simultaneous solutions for the whole set of the ten model variables. This important point should be explained. Researchers who are familiar with the methods given by Patanker and Spalding (1972) can identify these as simultaneous solutions, commonly performed by a three diagonal matrix routine, where solutions for each individual variable are performed simultaneously for a whole line of grid nodes. These methods have an obvious advantage for small number of model variables and large number of grid nodes. Using an advanced turbulence model however, the number of variables becomes large (10 in the present case) and the advantage becomes less clear. If, in addition, these variables are strongly coupled as is the case with Reynolds stress model, one should also consider the possibility to choose the present phase of simultaneity. That is, to solve the equations at each grid node simultaneously for the whole set of variables.

Indeed, fast converging solutions can be obtained for the square duct and the rotating disk which represent two apparently different situations: one, where the lateral motion is but a weak secondary effect, and the other where it constitutes important percentage of the main flow. Both are common with respect to the fact that they are strongly influenced by the presence of solid walls which stabilize the calculations. Finally, two cases of free flows will be discussed. The first is the non-recirculating swirling free jet (Rose 1962, Chigier and Chervinsky 1965, and Pratte and Keffer 1969). The other is the well known phenomena of the recirculation bubble formed in the core of strongly swirling jet (Chigier and Gilbert 1968, Syred et al 1971 and Chigier 1971). Although fast converging solutions can be obtained for the first case it is difficult to establish fast convergence for the second case. Some methods to partially overcome this defect are also described.

THE MODEL EQUATIONS

Since Reynolds stress closures are well documented (Launder, Reece and Rodi 1975, Naot, Shavit and Wolfshtein 1975) the model here used is briefly specified. For the transformation to the cylindrical coordinate system used here the reader is referred to (Rodi 1970 , Priddin 1975 and Naot 1975).

The modeling of the Reynolds stresses equations is associated with three terms. First, the pressure redistribution term is divided into two parts:

$$P_{ij} = P_{ij} + P_{ij}^* \quad . \tag{1}$$

Here, the interaction forced by the presence of the mean velocity gradients is modeled by

$$P_{ij} = -\alpha \left(\pi_{ij} - 1/3 \, \pi_{11} \, \delta_{ij} \right) + \beta \left(D_{ij} - 1/3 \, D_{11} \, \delta_{ij} \right)$$
$$+\gamma \, 4/3 \, k \, \overline{\varepsilon_{ij}} \quad , \tag{2}$$

where:

$$D_{ij} = - \left(R_{im} \frac{\partial \overline{U}_m}{\partial x_j} + R_{jm} \frac{\partial \overline{U}_m}{\partial x_i} \right) \quad , \tag{3}$$

$$\pi_{ij} = - \left(R_{im} \frac{\partial \overline{U}_j}{\partial x_m} + R_{jm} \frac{\partial \overline{U}_i}{\partial x_m} \right) \quad , \tag{4}$$

$$4/3 \, k \, \overline{\varepsilon}_{ij} = 1/3 \, R_{11} \left(\frac{\partial \overline{U}_i}{\partial x_j} + \frac{\partial \overline{U}_j}{\partial x_i} \right) \quad , \tag{5}$$

and the numerical coefficients α, β, and γ are related to a single parameter c_2 given in Table 1:

$$\alpha = (c_2 + 8)/11 \quad , \tag{6a}$$

$$\beta = -(8c_2 - 2)/11 \quad , \tag{6b}$$

and

$$\gamma = -3(15c_2 - 1)/55 \quad . \tag{6c}$$

It is generally accepted to combine the modeling of the free interaction term together with the modeling of the dissipation term and use a single formula:

$$P_{ij}^* + L_{ij} = \frac{\varepsilon}{k} \left\{ R_{ij} \, c_1 + 2/3 \, k \, (1 - c_1) \, \delta_{ij} \right\} \quad , \tag{7}$$

where: ε - is the energy dissipation, k - is the energy of the turbulence, and c_1 - is a numerical constant given in Table 1.

Finally, the model for the turbulent diffusion taken from Donaldson (1969) is

$$T_{ij} = \frac{\partial}{\partial x_m} \left(\Gamma \frac{\partial}{\partial x_m} R_{ij} \right) \quad , \tag{8}$$

where the scalar turbulent diffusivity is

$$\Gamma = c_T \, k^2/\varepsilon \quad , \tag{9}$$

and c_T is a numerical constant given Table 1.

The modeling of the dissipation equation is associated with the following terms:

The production

$$\pi_\varepsilon = c_{\varepsilon 1} \, \frac{\pi_{11}}{2k} \, \varepsilon \quad , \tag{10}$$

The turbulent transport

$$T_\varepsilon = \frac{c_\varepsilon}{c_T} \frac{\partial}{\partial x_m} \left(\Gamma \frac{\partial}{\partial x_m} \varepsilon \right) \quad , \tag{11}$$

and the dissipation

$$L = - c_{\varepsilon 2} \, \frac{\varepsilon^2}{k} \quad . \tag{12}$$

This model equation contains three numerical constants c_ε, $c_{\varepsilon 1}$ given in Table 1. The modeling of the self generation by stretching the vortex tubes is not discussed here, instead it is assumed that it contributes to the coefficient $c_{\varepsilon 1}$.

Table 1. The Numerical Coefficients

The Coefficients of the Reynolds Stress Equations			The Coefficients of the Dissipation Equation		
c_1	c_2	c_T	c_ε	$c_{\varepsilon 1}$	$c_{\varepsilon 2}$
1.50	0.40	0.10	0.072	1.44	1.90

The two dimensional flow field

The present discussion is confined to flows which depend on
two space coordinates only. Due to this dimplification the
three dimensional mean velocity can be specified by means of
three variables only. Namely, the stream function Ψ
defined by

$$V_x = \frac{1}{r}\frac{\partial\Psi}{\partial r} \qquad \text{and} \qquad V_r = -\frac{1}{r}\frac{\partial\Psi}{\partial x} \quad , \tag{13}$$

the vorticity ω defined by

$$\omega = \left(\frac{\partial V_r}{\partial x} - \frac{\partial V_x}{\partial r}\right) \quad , \tag{14}$$

and the swirling velocity V_θ . This choice is also a significant simplification of the computational task as it automatically satisfies the continuity equation not dealt with any-
more. The set of the ten equations solved is composed of the
following groups:

   (i)   Three mean motion equations

   (ii) Six stress transport equations

   (iii) An equation for the energy dissipation rate

The mean motion is governed by the stream function equation
stemming directly from Eqs. (13) and (14)

$$\left(\frac{\partial}{\partial x}\left(\frac{1}{r}\frac{\partial}{\partial x}\right) + \frac{\partial}{\partial r}\left(\frac{1}{r}\frac{\partial}{\partial r}\right)\right)\Psi + \omega = 0 \quad , \tag{15}$$

the vorticity equation obtained from the mean lateral momentum
equations

$$r\left(rV_x\frac{\partial}{\partial x} + rV_r\frac{\partial}{\partial r}\right)\frac{\omega}{r} - \frac{\nu}{r}\left(\frac{\partial}{\partial x}\left(r^3\frac{\partial}{\partial x}\right) + \frac{\partial}{\partial r}\left(r^3\frac{\partial}{\partial r}\right)\right)\frac{\omega}{r}$$

$$+ \left\{ r\frac{\partial^2}{\partial x\partial r}(R_{rr} - R_{xx}) + \left(\frac{\partial}{\partial x}\left(r\frac{\partial}{\partial x}\right) - \frac{\partial}{\partial r}\left(r\frac{\partial}{\partial r}\right) + \frac{1}{r}\right)R_{rx} \right\}$$

$$- \frac{\partial}{\partial x}V_\theta^2 + \frac{\partial}{\partial x}(R_{rr} - R_{\theta\theta}) \quad = \quad 0 \quad , \tag{16}$$

and the tangential momentum equation

$$\left(rV_x \frac{\partial}{\partial x} + rV_r \frac{\partial}{\partial r} + V_r \right) V_\theta$$

$$- \nu \left(\frac{\partial}{\partial x} \left(r \frac{\partial}{\partial x} \right) + \frac{\partial}{\partial r} \left(r \frac{\partial}{\partial r} \right) - \frac{1}{r} \right) V_\theta$$

$$+ r \frac{\partial}{\partial x} R_{x\theta} + \left(r \frac{\partial}{\partial r} + 2 \right) R_{r\theta} = 0$$

$$(17)$$

For practical reasons it is convenient to express the solution for the Reynolds stresses in terms of the following variables: the energy $k = 1/2\, R_{11}$, the two differences $R_{rr} - R_{xx}$, $R_{rr} - R_{\theta\theta}$ and the off diagonal components R_{xr}, $R_{x\theta}$ and $R_{r\theta}$ The equations for these are:

$$\left(rV_x \frac{\partial}{\partial x} + rV_r \frac{\partial}{\partial r} \right) k$$

$$- \left(\frac{\partial}{\partial x} \left(r\Gamma \frac{\partial}{\partial x} \right) + \frac{\partial}{\partial r} \left(r\Gamma \frac{\partial}{\partial r} \right) \right) k + rDk$$

$$+ r \left\{ R_{rx} \left(\frac{\partial V_x}{\partial r} + \frac{\partial V_r}{\partial x} \right) + R_{xx} \frac{\partial V_x}{\partial x} + R_{rr} \frac{\partial V_r}{\partial r} + R_{\theta\theta} \frac{V_r}{r} \right.$$

$$\left. + R_{r\theta} \frac{\partial V_\theta}{\partial r} + R_{x\theta} \frac{\partial V_\theta}{\partial x} - \frac{V_\theta}{r} R_{r\theta} \right\} = 0 \quad ,$$

$$(18)$$

$$\left(rV_x \frac{\partial}{\partial x} + rV_r \frac{\partial}{\partial r} \right) (R_{rr} - R_{xx}) - 2V_\theta\, R_{r\theta}$$

$$- \left(\frac{\partial}{\partial x} \left(r\Gamma \frac{\partial}{\partial x} \right) + \frac{\partial}{\partial r} \left(r\Gamma \frac{\partial}{\partial r} \right) \right) (R_{rr} - R_{xx}) + \frac{2\Gamma}{r} (R_{rr} - R_{\theta\theta})$$

$$+ r\Lambda D\, (R_{rr} - R_{xx})$$

$$+ 2r \left\{ (1 - \alpha - \beta) \left(\frac{\partial V_r}{\partial x} - \frac{\partial V_x}{\partial r} \right) R_{xr} \right.$$

$$+ [(1-\alpha-\beta) R_{rr} - \frac{2}{3} \gamma k] \frac{\partial V_r}{\partial r} - [(1-\alpha+\beta) R_{xx} - \frac{2}{3} \gamma k] \frac{\partial V_x}{\partial x}$$

$$+ \beta (R_{r\theta} \frac{\partial V_\theta}{\partial r} - R_{x\theta} \frac{\partial V_\theta}{\partial x}) - (1 - \alpha) \frac{V_\theta}{r} R_{r\theta} \} = 0 ,$$

$$(19)$$

$$(rV_x \frac{\partial}{\partial x} + rV_r \frac{\partial}{\partial r}) (R_{rr} - R_{\theta\theta}) - 4 V_\theta R_{r\theta}$$

$$- \left(\frac{\partial}{\partial x} (r\Gamma \frac{\partial}{\partial x}) + \frac{\partial}{\partial r} (r\Gamma \frac{\partial}{\partial r}) - \frac{4\Gamma}{r} \right) (R_{rr} - R_{\theta\theta}) + r\Lambda D (R_{rr} - R_{\theta\theta})$$

$$+ 2r \{ R_{xr} \left(\beta \frac{\partial V_x}{\partial r} + (1 - \alpha) \frac{\partial V_r}{\partial x} \right)$$

$$+ [(1-\alpha+\beta) R_{rr} - \frac{2}{3} \gamma k] \frac{\partial V_r}{\partial r} - [(1-\alpha+\beta) R_{\theta\theta} - \frac{2}{3} \gamma k] \frac{V_r}{r}$$

$$- (1-\alpha-\beta) R_{r\theta} (\frac{\partial V_\theta}{\partial r} + \frac{V_\theta}{r}) - (1-\alpha) R_{x\theta} \frac{\partial V_\theta}{\partial x} \} = 0 ,$$

$$(20)$$

$$(rV_x \frac{\partial}{\partial x} + rV_r \frac{\partial}{\partial r}) R_{xr} - V_\theta R_{x\theta}$$

$$- \left(\frac{\partial}{\partial x} (r\Gamma \frac{\partial}{\partial x}) + \frac{\partial}{\partial r} (r\Gamma \frac{\partial}{\partial r}) - \frac{\Gamma}{r} \right) R_{xr} + r\Lambda D R_{xr}$$

$$+ r \{ [(1-\alpha) R_{rr} + \beta R_{xx} - \frac{2}{3} \gamma k] \frac{\partial V_x}{\partial r} + [(1-\alpha) R_{xx} + \beta R_{rr}$$

$$- \frac{2}{3} \gamma k] \frac{\partial V_r}{\partial x}$$

$$+ (1-\alpha+\beta) R_{xr} (\frac{\partial V_x}{\partial x} + \frac{\partial V_r}{\partial r})$$

$$+ R_{x\theta} [\beta \frac{\partial V_\theta}{\partial r} - (1-\alpha) \frac{V_\theta}{r}] + \beta R_{r\theta} \frac{\partial V_\theta}{\partial x} \} = 0 ,$$

$$(21)$$

$$\left(rV_x \frac{\partial}{\partial x} + rV_r \frac{\partial}{\partial r} \right) R_{x\theta} + V_\theta R_{xr}$$

$$- \left(\frac{\partial}{\partial x}\left(r\Gamma \frac{\partial}{\partial x} \right) + \frac{\partial}{\partial r}\left(r\Gamma \frac{\partial}{\partial r} \right) - \frac{\Gamma}{r} \right) R_{x\theta} + r\Lambda D\, R_{x\theta}$$

$$+ r\left\{ R_{r\theta}\left[(1-\alpha)\frac{\partial V_x}{\partial r} + \beta \frac{\partial V_r}{\partial x} \right] + (1-\alpha+\beta) R_{x\theta}\left(\frac{\partial V_x}{\partial x} + \frac{V_r}{r} \right) \right.$$

$$+ R_{xr}\left[(1-\alpha)\frac{\partial V_\theta}{\partial r} - \beta \frac{V_\theta}{r} \right] + \left[(1-\alpha) R_{xx} + \beta R_{\theta\theta} - \frac{2}{3}\gamma k \right]\frac{\partial V_\theta}{\partial x}$$

$$\left. \vphantom{\frac{\partial V_x}{\partial r}} \right\} = 0 \quad , \tag{22}$$

and

$$\left(rV_x \frac{\partial}{\partial x} + rV_r \frac{\partial}{\partial r} \right) R_{r\theta} + V_\theta (R_{rr} - R_{\theta\theta})$$

$$- \left(\frac{\partial}{\partial x}\left(r\Gamma \frac{\partial}{\partial x} \right) + \frac{\partial}{\partial r}\left(r\Gamma \frac{\partial}{\partial r} \right) - \frac{4\Gamma}{r} \right) R_{r\theta} + r\Lambda D\, R_{r\theta}$$

$$+ r\left\{ R_{x\theta}\left[\beta \frac{\partial V_x}{\partial r} + (1-\alpha)\frac{\partial V_r}{\partial x} \right] + (1-\alpha+\beta) R_{r\theta}\left(\frac{\partial V_r}{\partial r} + \frac{V_r}{r} \right) \right.$$

$$+ \left[(1-\alpha) R_{rr} + \beta R_{\theta\theta} - \frac{2}{3}\gamma k \right]\frac{\partial V_\theta}{\partial r} + (1-\alpha) R_{xr}\frac{\partial V_\theta}{\partial x}$$

$$\left. - \left[(1-\alpha) R_{\theta\theta} + \beta R_{rr} - \frac{2}{3}\gamma k \right]\frac{V_\theta}{r} \right\} = 0 \quad , \tag{23}$$

where D is an abbreviation for ε/k. Finally, the equation for the dissipation is:

$$\left(rV_x \frac{\partial}{\partial x} + rV_r \frac{\partial}{\partial r} \right) \varepsilon$$

$$- \frac{c_\varepsilon}{c_T}\left(\frac{\partial}{\partial x}\left(r\Gamma \frac{\partial}{\partial x} \right) + \frac{\partial}{\partial r}\left(r\Gamma \frac{\partial}{\partial r} \right) \right)\varepsilon + r D c_{\varepsilon 2}\, \varepsilon$$

$$+ \, Dr \, \{ \, R_{xr} \left(\frac{\partial V_x}{\partial r} + \frac{\partial V_r}{\partial x} \right) + R_{xx} \frac{\partial V_x}{\partial x} + R_{rr} \frac{\partial V_r}{\partial r} + R_{\theta\theta} \frac{V_r}{r}$$

$$+ \, R_{r\theta} \frac{\partial V_\theta}{\partial r} + R_{x\theta} \frac{\partial V_\theta}{\partial x} - R_{r\theta} \frac{V_\theta}{r} \; = \; 0 \quad .$$

$$(24)$$

Probably the most simple way to transform these model equations to the Cartesian coordinate system is by replacing the radius r by y + R and taking the limit R → ∞ properly. One can also perform this numerically by assuming that the fully developed channel flow where the streamwise coordinate Z replaces the θ coordinate, takes place in a spiral, the diameter of which is substantially large in comparison to the channel hydraulic diameter. Obviously one should not forget to add a constant pressure gradient term to Eq. (17). With these relatively simple steps the program can be modified to fully developed channel flows.

THE ITERATIVE-SIMULTANEOUS SOLUTION

Basically the iterative procedure is the Gauss-Seidel method. At every iteration the grid is swept and every grid node is visited. The equations are linearized by using for the non linear terms values of the variables calculated at the previous iteration. The linearized equations are then solved and the new values are immediately stored and used for the next calculation. As there are some ways to linearize the set of the equations the procedure is not unique. Here, further development of the proposal of Naot, Shavit and Wolfshtein (1974) suitable for elliptic fields is given forming a computing tool that is superior to the original one in reducing the computing time as was reported by Naot and Launder(1975).

<u>The coefficients matrix</u>
At each iteration for any grid node the set of the equations for the ten variables ϕ_i is linearized and algebraically transformed to

$$\phi_i \, C_{ij} \; = \; D_j$$

$$(25)$$

Where C_{ij} is the coefficients matrix calculated on basis of the former iteration, i = 1,2...10 is the variable index, and j = 1,2...10 is the equation index.
Then, the set of the linearized equations is solved by inverting the coefficients matrix.

In principle there are many ways to construct C_{ij}. For
example, using the methods given by Gosman et al.(1969) for
the convective and diffusive terms incorporating all the other
terms usually referred to as source terms in the D_j vector, a
diagonalized C_{ij} is obtained. This is the most simple way to
construct C_{ij}, leading to straightforward solutions. Unfortunately experience did not support this attractive method.
It became apparent that in certain zones of the elliptic field
where both the convective and the diffusive terms are small
some of the elements of the diagonalized C_{ij} become small,
thus causing instability. Another way to refer to this situation is to realize that the set of the equations loses its
second order operations and therefore degenerates. A typical
situation is the zone where local equilibrium turbulence pre-
vails. In that case the Reynolds stresses equations degenerate to first order equations relating the Reynolds stresses
to the first order derivatives of the mean velocities, thus
becoming a first order equation for the mean velocities
instead of a second order equations for the stresses. This
situation used to establish the algebraic stress modeling
turns to be a source for instability when the whole stress
transport equations arc used and deserves special treatment.
It is therefore necessary to use the source terms and form a
coefficient matrix which has non-zero off-diagonal elements,
and obtain a procedure that couples the calculation of the
Reynolds stresses with the calculation of the mean velocities.
To this end the former experience (Naot et al. 1974) where the
variables were solved in three groups is extended, and the
solution for the whole set of variables is performed simultaneously. On the other hand to make the task of the inversion
of the coefficient matrix simple, a compromise is suggested to
form non-zero elements only in the locations marked in
Table 2. These locations are deduced by the following
principles:

(i)   Every source term should be used to form part of
        the coefficients matrix

(ii)   Off-diagonal coefficients will be formed for all
        the Reynolds stresses which appear in the mean
        velocity equations

(iii)   Off-diagonal coefficients will be formed for the
        Lateral velocities (namely Ψ and V_θ) which appear
        in the stresses equations

Under these restrictions the matrix given in Table 2 couples
the solution of all the variables demonstrating relative
simplicity which makes its inversion economic.

Table 2. The Non-Zero Elements of the Coefficients Matrix

Variable Equation	V_θ	ψ	ω	$R_{x\theta}$	$R_{r\theta}$	R_{xr}	R_{rr} $-$ R_{xx}	R_{rr} $-$ $R_{\theta\theta}$	k	ε
V_θ	+	♣		+	+					
ψ		+	+							
ω	+	♣ *	+			+	+	+		
$R_{x\theta}$	+	+		+						
$R_{r\theta}$	+	+			+					
R_{xr}	+	+				+				
$R_{rr}-R_{xx}$	+	+					+	+		
$R_{rr}-R_{\theta\theta}$	+	+						+		
k	+	+							+	
ε	+	+								+

* With the application of the Multi-Point Circulation
Adjustment

♣ With the application of the Stream Function Dependent
Convection

Once a coefficients matrix is formed it is possible to solve
the whole set of the ten equations. Generally, this involves
an inversion of the matrix which is a wasteful procedure,
since many elements have null values. However, we may avoid
considering these by setting the off diagonal elements of the
upper right triangle of elements equal to zero using systematic subtraction of the suitable equations. The simplification is based on the special form of the matrix given Table 2,
where the off diagonal elements of the low left triangle of
elements have zero values except for the first and second
columns. This makes the procedure simple, and in fact it is
possible within eight subtractions of equation to reach the
stage where the only elements which have non-zero values are
the elements of the diagonal. The solution is then
straightforward.

Finite difference schemes

The various finite difference schemes used are now given. We begin with the upwind convection operation (Gosman et al 1969)

$$\left(rV_x \frac{\partial}{\partial x} + rV_r \frac{\partial}{\partial r} \right) \phi =$$

$$\frac{1}{\Delta V} \left[(a_E + a_W + a_N + a_S)\, \phi_P - a_E \phi_E - a_W \phi_W - a_N \phi_N - a_S \phi_S \right] \quad ,$$

$$(26)$$

where:

$$a_E = \tfrac{1}{2} \left(\; |\Psi_{NE} + \Psi_N - \Psi_{SE} - \Psi_S| \; - \Psi_{NE} - \Psi_N + \Psi_{SE} + \Psi_S \right) \quad ,$$

a_W, a_N, a_S are similar expressions, and

$$\Delta V = (X_E - X_W)(Y_N - Y_S) \quad .$$

The indices E, W, N, S, NE, NW, SE, SW denote the grid points in the vicinity of the central point P as is shown in Figure 1, and the y-coordinate is parallel to the radius.

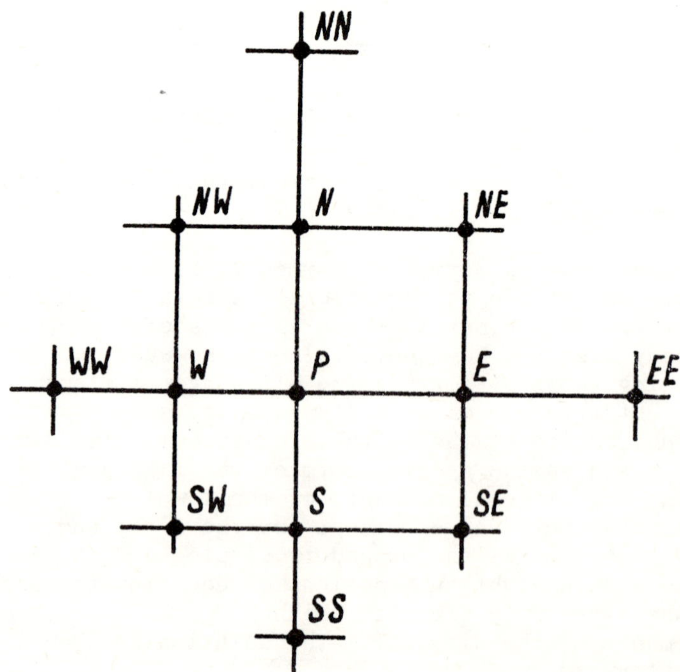

Figure 1.   Notation of the Grid Points in the Neighbourhood of the Central Point p.

The diffusion type operations are given by a conservative central point formula (Gosman et al 1969):

$$\left(\frac{\partial}{\partial x} \left(\alpha \frac{\partial}{\partial x} \right) + \frac{\partial}{\partial r} \left(\alpha \frac{\partial}{\partial r} \right) \right) \phi =$$

$$\frac{1}{\Delta v} \left[(c_E + c_W + c_N + c_S) \, \phi_P - c_E \phi_E - c_W \phi_W - c_N \phi_N - c_S \phi_S \right]. \tag{27}$$

Here:

$$c_E = \left(2(y_N - y_S)/(x_E - x_P) \right) \alpha_{\frac{1}{2}E} \quad ,$$

where $\alpha_{1/2E}$ denotes the value of α interpolated for the mid-point between the point P and the point E. The abbreviation α stands for:

$$\alpha = \begin{cases} 1/r & \text{for equation (15)} \\ r^3 & \text{for equation (16)} \\ r & \text{for equation (17)} \\ \Gamma r & \text{for equations (18) to (24)} \end{cases}$$

The treatment of the source terms deserves more attention since it is not so familiar. Mainly we should consider two aspects. (i) Some source terms are non-linear and are linearized by substituting values taken from the former iteration. (ii) Some source terms contain derivatives which have to be described by a non-central difference in order to keep the three rules stated formerly

We refer first to the sources of the vorticity equation. The first two source terms of Eq. (16) are given by:

$$\partial^2/\partial x \, \partial r \, (R_{rr} - R_{xx}) =$$

$$\tfrac{1}{2} \{ (R_{rr} - R_{xx})_{NE} - (R_{rr} - R_{xx})_E - (R_{rr} - R_{xx})_N +$$

$$(R_{rr} - R_{xx})_P \} / (y_N - y_P)(x_E - x_P)$$

$$+ \tfrac{1}{2} \{ (R_{rr} - R_{xx})_P - (R_{rr} - R_{xx})_S - (R_{rr} - R_{xx})_W +$$

$$(R_{rr} - R_{xx})_{SW} \} / (y_P - y_S)(x_P - x_W), \tag{28}$$

and an expression derived from Eq. (27) for the derivatives of the stress R_{xr}, thus forming coefficients for these two stresses.

The other two terms which express centrifugal forces, commonly referred to as the Vorticity Sources, are given by means of a backward difference forming coefficients for both V_θ and $(R_{rr} - R_{\theta\theta})$. The two source terms of the swirling velocity equation (17) are given by a forward difference formula for $(\partial/\partial x) R_{x\theta}$ and a backward difference formula for $(\partial/\partial r) R_{r\theta}$, resulting in two coefficients for the stresses $R_{x\theta}$ and $R_{r\theta}$ in the swirling velocity equation.

The sources of the Reynolds stresses equations contain multiplications of components of the Reynolds stresses by the first order derivatives of the mean velocities and also by the mean velocities themselves. These are used to form coefficients for the mean velocities after being linearized by substituting old values for the Reynolds stresses. The derivatives of the swirling velocity V_θ are given by backward difference schemes for $(\partial/\partial x) V_\theta$ and $(\partial/\partial r) V_\theta$, except for equation (23) for $R_{r\theta}$ where a forward difference scheme is used for $(\partial/\partial r) V_\theta$. In this way the non-central differences used to describe $(\partial/\partial x) V_\theta$ and $(\partial/\partial r) V_\theta$ were taken in the direction opposite to those used to describe $(\partial/\partial x) R_{x\theta}$ and $(\partial/\partial r) R_{r\theta}$, keeping the ellipticity of the numerical procedure in the case of local degeneracy.

The mean lateral velocities V_x and V_r, as well as their derivatives, were related to Ψ only. The mean velocity $V_r \propto \partial\Psi/\partial x$ was given by a backward difference. The derivatives $\partial V_x/\partial r$ and $\partial V_r/\partial x$ are related to $\frac{\partial}{\partial r}\left(\frac{1}{r}\frac{\partial\Psi}{\partial r}\right)$ and $\frac{\partial}{\partial x}\left(\frac{1}{r}\frac{\partial\Psi}{\partial x}\right)$ and are described by means of expressions of the type of Eq. (27). The other derivatives, $\partial V_x/\partial x$ and $\partial V_r/\partial r$ are related to $\partial^2\Psi/\partial x\,\partial r$ which is given by:

$$\frac{\partial^2\Psi}{\partial x\partial r} = \tfrac{1}{2}\{\Psi_{NE} - \Psi_E - \Psi_N + \Psi_P\}/(x_E - x_P)(y_N - y_P)$$

$$+ \tfrac{1}{2}\{\Psi_P - \Psi_S - \Psi_W + \Psi_{SW}/(x_P - x_W)(y_P - y_S) \quad \cdot(29)$$

Negative vorticity sources

In the presence of high internal negative vorticity sources the procedure was found to be unstable, and it is necessary to adjust it by incorporating two techniques: The method of the Multi-Point-Circulation-Adjustment, (Gosman & Spalding, 1970, and Crosley, 1975), as well as a procedure (Naot and Launder 1975) referred to as the Stream-Function-Dependent- Convection. In this method the perturbations of the convection, due to changes imposed on the stream function are estimated and taken into consideration in advance, thus accelerating the calculations slowed down by the introduction of the M.P.C.A. method.

Using the finite difference equation for the stream function one can calculate the vorticity at any point by means of the values of the stream function at that point and at the neighbouring points. Thus, it is possible to replace the values of the vorticity used to satisfy the vorticity equation by expressions which contain the values of the stream function:

$$\Psi_P,\ \Psi_N,\ \Psi_S,\ \Psi_E,\ \Psi_W,\ \Psi_{NE},\ \Psi_{NW},\ \Psi_{SE},\ \Psi_{SW},\ \Psi_{NN},\ \Psi_{SS},\ \Psi_{EE} \text{ and } \Psi_{WW}$$

at the field points near the central point p, as shown in Fig. 1.

The main idea of this scheme is to satisfy the stream function equation on as wide as possible a domain. In other words, while the equations for the other variables are satisfied exactly only when the point p is visited the stream function equation is satisfied on wider range. The direct outcome of this scheme is a positive element $C_{\Psi\omega}$, shown in Table 2, which contributes to the stabilization of the calculations. One should note that the coefficient $C_{\Psi\omega}$ obtained is not a result of the particular choice of the finite difference scheme used to express the convection. There are no limitations, and the M.P.C.A. method can also be applied to other schemes. (Atias, Wolfshtein and Israeli, 1975).

There is however, another way to relate the convection operation to the value of the stream function at the point P. Application of such method to the vorticity and the swirling velocity equations results in non-zero coefficients $C_{\Psi\omega}$ and $C_{\Psi V_\theta}$ and produce feedback mechanisms between the vorticity and the stream function and between the swirling velocity and the stream function, which were shown to stabilize the calculation. Let denote by $(\text{conv } \phi)_J$ the finite difference scheme used to calculate the convection of the variable ϕ on the basis of values taken from a single iteration J. Since the convection terms are non-linear we cannot use $(\text{conv } \phi)_J$ directly and resort to another form where some parts are calculated with values taken from the former iteration. In principle, we should use the linearized approximation

$$(\text{conv } \phi)_J = \underbrace{(\text{conv } \phi)_{J-1}}_{A}$$

$$\underbrace{+ \left(\frac{\partial}{\partial \phi_P} \text{conv } \phi\right)_{J-1} (\phi_J - \phi_{J-1})_P}_{B} \quad \underbrace{+ \left(\frac{\partial}{\partial \Psi_P} \text{conv } \phi\right)_{J-1} (\Psi_J - \Psi_{J-1})_P}_{C} \qquad (30)$$

plus similar terms for the neighbouring points.

Indeed, the existing schemes described by Atias, Wolfshtein
and Israeli (1975) can be regarded as the approximation

$$(\text{conv } \phi)_J \sim A + B \quad . \tag{31}$$

with the term C omitted as it cannot be explicitly evaluated.

It is suggested to use for the third part C an approximation
noting that with the progress of the calculation this part is
expected to vanish. To do so we estimate the variation of the
convection developing during calculations made between two
successive visits at a nodal point which are caused by a small
perturbation $\delta\Psi$ initially imposed at this point. The estimation is restricted to the simple case where the initial values
of the stream function form a solution, the variation of the
vorticity is negligible, and the grid is uniform. Knowing the
order by which the neighbouring points are visited an estimation of the way the perturbation $\delta\Psi$ propogates in the grid
between two successive visits at P can be estimated resulting
in:

$$\left(\frac{\partial}{\partial\Psi_p} \text{conv } \phi\right) \sim \frac{1}{\Delta V} \left\{0.125\left((\phi_E - \phi_P)a_E + (\phi_P - \phi_W)a_W\right)\right.$$

$$a_E = \begin{cases} 1 \text{ for negative } V_x \\ 0 \text{ for positive } V_x \end{cases}$$

and

$$a_W = 1 - a_E \quad , \tag{32}$$

Equation (32) is restricted to calculation where the y index

is advanced first and the x index second.

It was found that the application of the method to the description of the convection of both V_θ and ω/r accelerates the
calculations made with the Multi-Point-Circulation-Adjustment
but cannot replace it.

Boundary conditions
The use of a Reynolds stress closure and the application of the
simultaneous solution technique raises some problems associated
with the specification of the boundary conditions which can be
divided to three categories:

    (i) Special equations and solution procedures are needed
          near the axis of revolution.

(ii)   On line of symmetry some variables have zero values and the others have zero gradients.

(iii)   It is necessary to specify the Reynolds stresses near walls.

On the axis of revolution several variables have known values and should not be dealt any more. This applies to:

$$\Psi = \Psi_0 \qquad\qquad V_\theta = 0 \qquad\qquad R_{xr} = 0$$

$$R_{x\theta} = 0 \qquad\qquad R_{r\theta} = 0 \qquad\qquad R_{rr} - R_{\theta\theta} = 0 \tag{33}$$

The other variables ω, $R_{rr} - R_{xx}$, k and ε need special equations since the regular procedure does not apply to this limiting case.

To obtain these equations we assume for Ψ a polynomial expression

$$\Psi = \Psi_0 + \sum_{n=1}^{N} a_n (x) \, r^n \quad . \tag{34}$$

Limiting the degree of the polynomial to N = 3 we obtain both a condition for the vorticity

$$\left(\frac{\omega}{r}\right)_{r=0} = 0 \quad , \tag{35}$$

as well as the axial velocity used to calculate the convection terms

$$(V_x)_{r=0} = 2 \, \frac{\Psi_N - \Psi_0}{R_N^2} \quad , \tag{36}$$

Using these and considering the non vanishing mean velocity gradients we obtain finite difference equations which result in a diagonalized coefficient matrix and therefore can be solved separately.

On certain boundaries some variables are known, while the others have specified gradients. Usually this is the case of symmetry lines where some Reynolds stresses vanish and the others have zero gradients. To also use the regular procedure for this line the grid is extended in an additional line and the values of the variables for which the gradients are known are properly substituted for the additional nodes. Then the equations for the variables the values of which are known are replaced by the trivial equations: $\phi_j = \Phi_j$, where Φ_j are the known values. Practically the elements of the coefficients matrix are set equal zero for all $i \neq j$ and 1 for $i = j$ and the D_j element is set equal to Φ_j, thus making the application of the regular simultaneous solution procedure possible.

As the turbulence model does not apply close to solid walls, the technique of matching the solution with an approximate solution for the wall region is used. Since this practice is widely used we will refer only to points typical of the present turbulence model.

In a swirling flow the angle of inclination ϕ, defined by

$$tg\ \phi\ =\ V_\theta / V_r \tag{37}$$

is no longer small and therefore we specify the ten variables in the $y^+ \sim 70$ region by means of two local parameters: the inclination angle ϕ, and the shear velocity v^*. To do so we assume

$$V_r\ =\ U \cos \phi \quad \text{and} \quad V_\theta\ =\ U \sin \phi \tag{38}$$

where

$$U\ =\ v^*(2.5 \ln y^+ + 5.) \quad . \tag{39}$$

Integration, together with the assumption of $U^+ = y^+ < 10$, results for the stream function:

$$\Psi\ =\ \Psi_{wall} - [2.5\ y^+ (\ln y^+ + 1) - 32]\ vR \cos \phi \tag{40}$$

and differentiation results in:

$$\omega/r\ =\ 2.5\ \frac{v^*}{yR} \cos \phi \quad . \tag{41}$$

Assuming local equilibrium we obtain

$$k\ =\ (v^*)^2/A \tag{42a}$$

$$R_{rr} - R_{xx}\ =\ \frac{k}{\Lambda} [\ 1 - \alpha - 2\beta + (1 - \alpha) \cos 2\phi\ ] \tag{42b}$$

$$R_{rr} - R_{\theta\theta}\ =\ \frac{k}{\Lambda}\ 2(1 - \alpha) \cos 2\phi \tag{42c}$$

$$R_{r\theta} = \frac{k}{\Lambda}(1 - \alpha)\sin 2\phi \qquad (42d)$$

$$R_{xr} = -k\ A\ \cos\phi \qquad (42e)$$

$$R_{x\theta} = -k\ A\ \sin\phi \qquad (42f)$$

where

$$A = \sqrt{\frac{2}{3}}\left[\ 2\alpha - 4\alpha\beta - \beta^2 + 4\beta - \alpha^2 - 1 + \Lambda(1 + \beta - \alpha - \gamma)\ \right]/\Lambda \qquad (43)$$

Finally, the assumption of local equilibrium yields for the energy dissipation:

$$\epsilon = 2.5\ (v^*)^3/y \qquad (44)$$

Equations (40) to (41) are sufficient to provide full specification for the ten variables at the first grid line adjacent to the wall, in terms of two parameters. The matching procedure is practically established by deriving these two parameters from Eq. (40) and Eq. (41) assuming that these hold also at the second grid line.

Technical details
The results presented here were obtained by using a 22 x 22 non equally spaced grid. The differences between the grid nodes formed a geometrical series with a ratio smaller than 1.16.

Typical runs consume 0.75 second per iteration on a CDC 6600 computer to get impression on the quality of the calculations residues were calculated:

$$R_{es} = (\phi\ old - \phi\ new)/\phi_{ref} \qquad (45)$$

where ϕ_{ref} is a typical value. The calculations for the swirling jet, square duct, and the rotating disk converged rapidly. A 300 iterations run consumed 220 seconds. During that time the residues reached values of less than 10^{-7}. Unfortunately, this is not the situation when the recirculation bubble is concerned as the use of the Multi-Point-Circulation-Adjustment slows the calculations. The inclusion of the Stream-Function-Dependent-Convection method accelerates the calculation by reducing the residues by about half. Even with this practice it was necessary to perform long runs of 1200 iterations which consumed 900 seconds. During this time the residues reached the value of 10^{-5} only, which is an expensive calculation. Still, to get more balanced impression the reader should note that most engineering problems fall within the category of the 200 seconds runs, which is compatible with other methods. Probably the most significant advantage of the present method is the possibility to prepare all the numerical expressions for all the equations at each point at the same time, thus economizing and simplifying considerably the programming task.

TWO DIMENSIONAL ELIPTIC FLOW FIELDS

As stated in the introduction the method presented here is suitable for eliptic two dimensional flow fields. To get an impression of the variety of the fields contained in this category some special cases are briefly described here.

Case A: Stress induced secondary motion in fully developed flow in square channel.

Case B: Recirculation induced by a rotating disc in semi-shrouded annular domain.

Case C: Free swirling jet immerging from a rotating pipe into semi-infinite domain.

Case D: The recirculation bubble immerging from a vortex chamber into semi-infinite domain.

These represent a variety of physical and numerical conditions, as we will see by some common features listed here:

•	Intensive lateral motion	Case B & D
•	Weak lateral motion	Case A & C
•	Strong influence of solid walls proximity	Case A & B
•	Weak influence of solid walls	Case C & D
•	Lateral motion induced by unisotropy	Case A
•	Lateral motion induced by centrifugal forces	Case B & D
•	Mainly eliptic calculation	Case A & B
•	Mainly parabolic calculation	Case C
•	Eliptic and parabolic terms of similar importance	Case D

Unfortunately satisfactory experimental data does not exist for cases such as B & D for which the use of an advanced turbulence model is fully justified thus limiting the theoretical outcome of such a research for the present time.

<u>Stress induced secondary motion</u>

The fully developed turbulent flow in a square duct characterized by relatively weak lateral motion was calculated as
shown in Figure 2. Here the inequality between the normal
stresses, reasonably predicted by the present model, is the
main source for the lateral motion commonly termed as the
stress induced secondary currents. In this case the features
of the flow are dominated by the wall treatment which also
stabilize the calculation. Recently the physical interpretation and the accuracy of the turbulent model her presented
was argued by its own initiators. Therefore the results given
here are taken after Naot et al (1974) directly.

This case as well as the semi-shrouded disk problem were
designed to serve as examples for eliptic calculation fully
determined by the boundary conditions. The fact that the
walls are stationary in one case and moving in the other is
not principal and the main difference between these problems
is the motivating force for the lateral motion, i.e. the
dominant term of the vorticity equation, which is the difference between the normal stresses in the first and the
centrifugal forces in the second. Recently it was shown that
algebraic stress modeling based on quasi local equilibrium approximation yield reasonable results for the secondary currents in non-circular ducts. Still there are situations where
the turbulent transport is of importance and the present
approach is needed. Examples are ducts with discontinuity in
the surface roughness and the end region of thin fins, characterized by strong lateral gradients of the normal stresses
caused by the surface structure discontinuity. Another
example is open channel flow with wind induced turbulence and
shear stress which penetrate the channel depth by turbulent
transport from the open surface exposed to the strong wind.

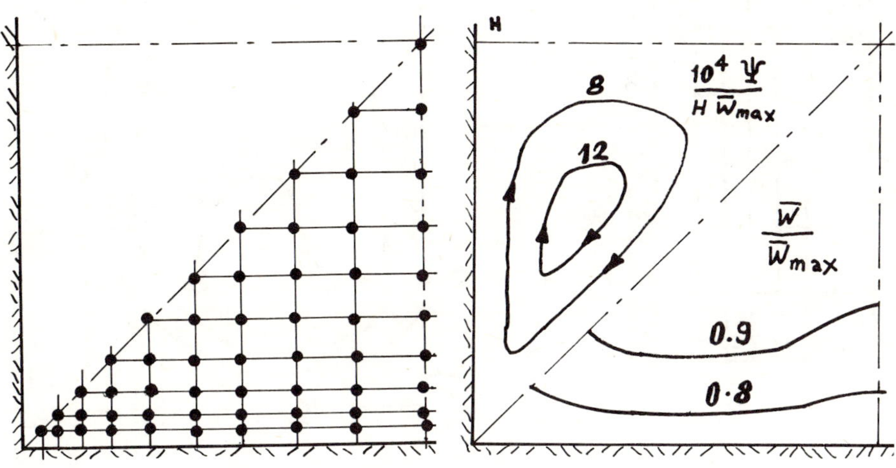

Figure 2. Secondary Currents in Square Duct Flow

<u>Recirculation induced by rotating disc</u>
The turbulent flow field formed between two semi-shrouded
discs, shown in Figure 3 was calculated. Here, one disc is
stationary and the other rotates. The fluid is inserted uniformly through a honeycomb and emerges also uniformly from the
periphery through a second honeycomb. This artificial semi-
shrouded configuration was chosen as to simplify the calculation. To obtain reasonable results the rotation velocity was
chosen high enough as to justify the use of the boundary
conditions. The swirling velocity gradients are strong and
relatively high recirculation is obtained due to the centrifugal forces. Still, due to the stabilizing effect of the
boundary condtion there was no need to use the M.P.C.A. or the
S.F.D.C. methods, and the calculation converged rapidly.

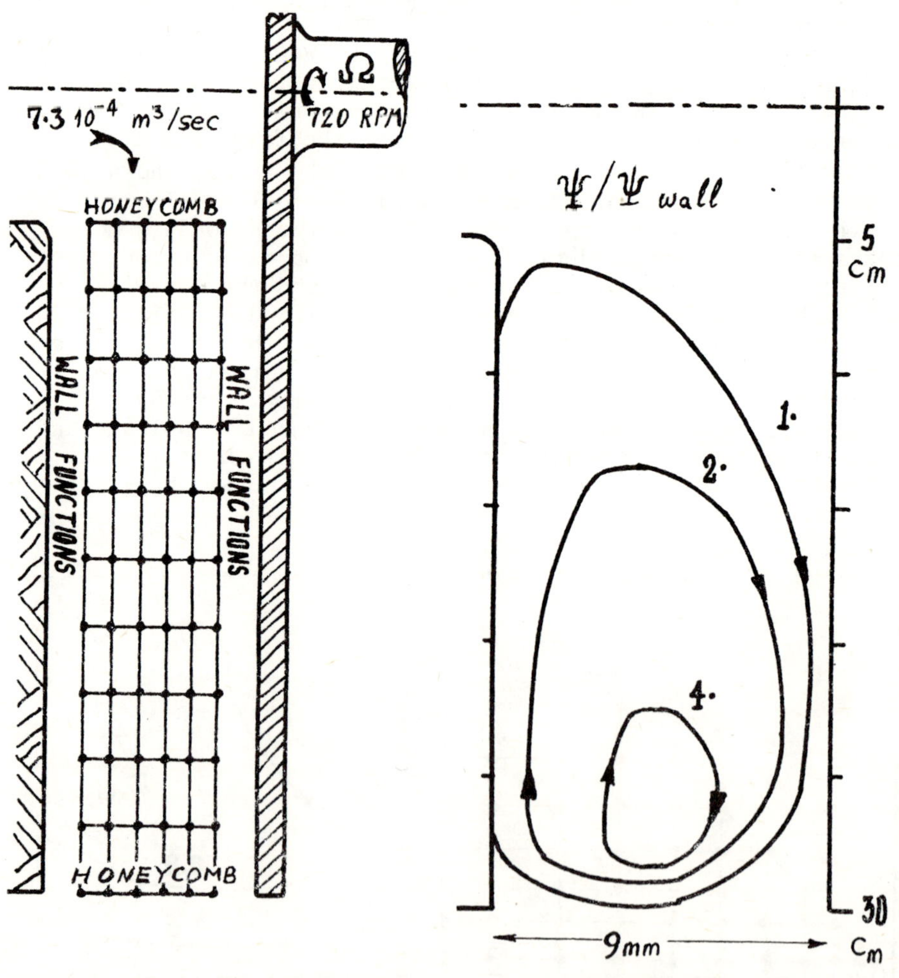

Figure 3.   Lateral Motion Induced by Semi-Shrouded Rotating
            Disk.

<u>The swirling free jet</u>

The results of a calculation made for the swirling round free
jet immerging from a rotating pipe into a semi-infinite still
media is shown in Figures 4 and 5. Comparison with the data
of Rose (1962), Chigier and Chervinsky (1966), and Pratte and
Keffer (1969) is also given, although as formerly stated,
there is evidence that the model is not sufficiently accurate.
The results are given as function of a nondimensional radius η
which measures the radius in terms of the half velocity width.
The calculation was performed in a cylindrical domain bounded
by three types of boundaries: stressless straight wall,
Laminar radial flow towards the jet axis, and matching with
proper asymptotic solutions.

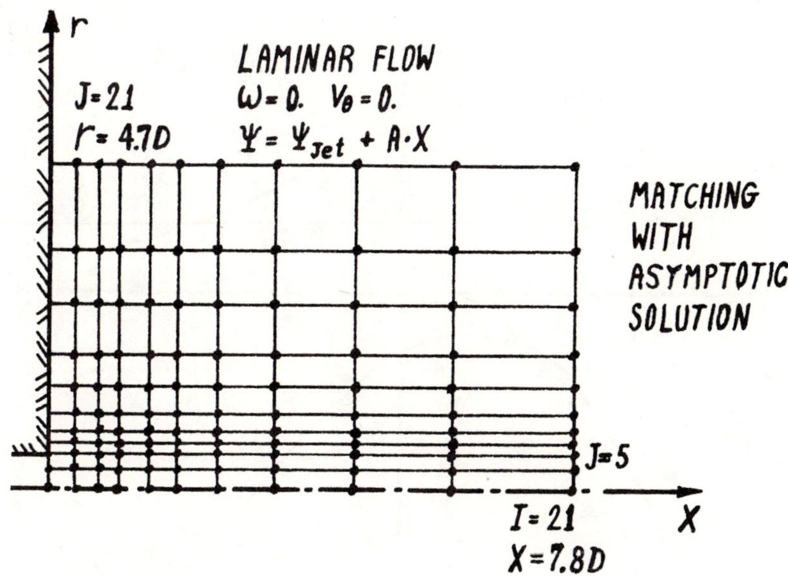

Figure 4. Swirling Free Jet in Semi-Infinite Media -
The Grid Used

A linear dependency between the stream function and the distance from the wall with numerical coefficient adjusted as to
obey an overall mass conservation was used for the first
boundary condition. The second condition was performed by
substituting the values of the variables in an additional grid
line using asymptotic proportions. Here momentum conservation
was directly applied. However, the downstream boundary conditions hardly effected the calculation which is almost
parabolic as far as recirculation bubble is not formed by
strong swirl.

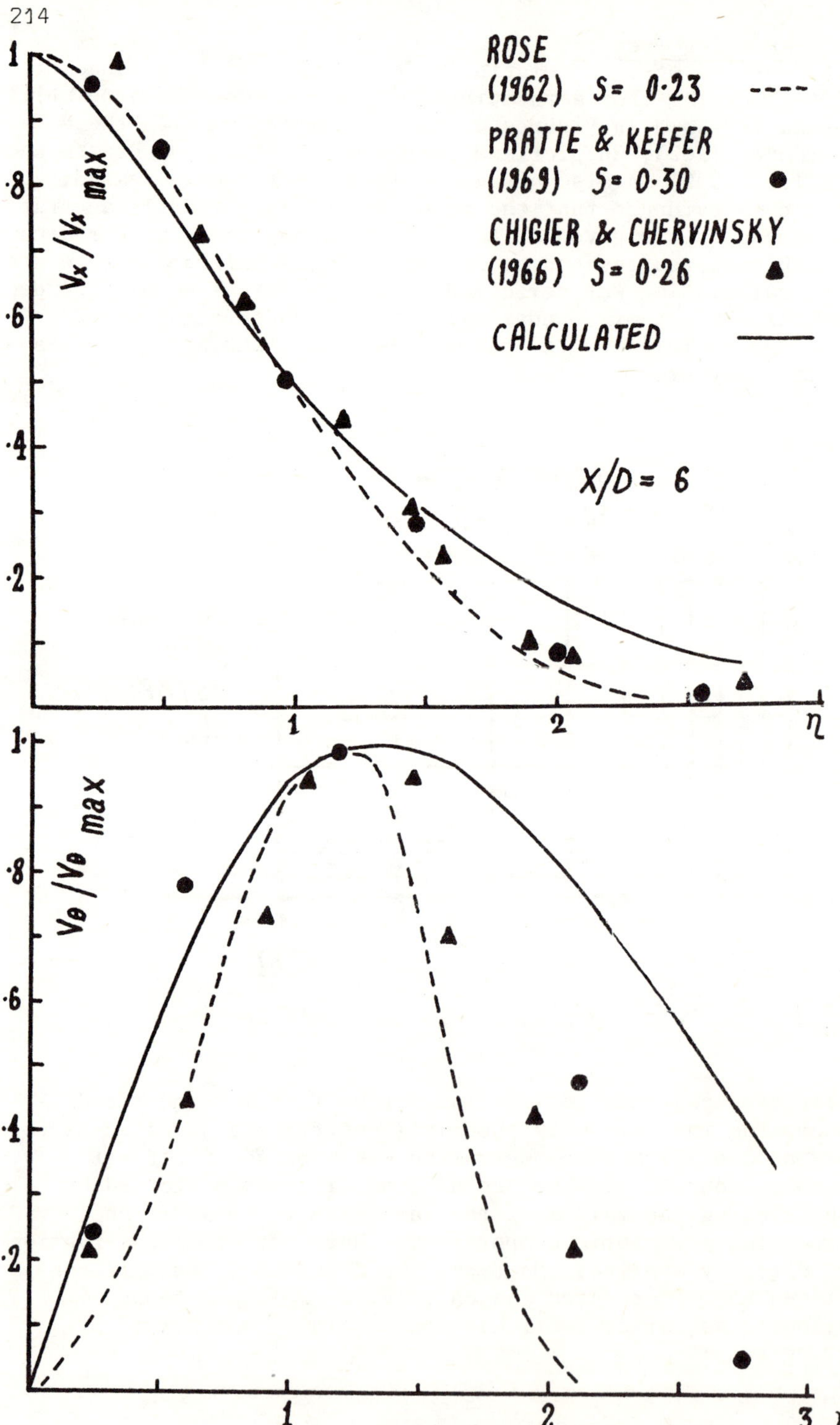

Figure 5.   Swirling Free Jet in Semi-Infinite Media -
Mean Velocity Profiles.

The recirculation bubble

Finally the numerical simulation of the recirculation zone
formed in the core of a highly swirling jet is shown. Here
the flow immerges from a vortex chamber into a semi-infinite
still media. Due to the centrifugal forces a backflow is
observed along the jet axis, and the maximum axial velocity is
shifted away from this axis. Profiles of the axial velocity
and the tangential velocity are given as well as the full set
of the Reynolds stresses. These should be regarded with
caution as it became evident that the model is not accurate
enough near the field axis. In addition, due to the large
computing time needed in this case, reduced by the use of the
M.P.C.A. and S.F.D.C. methods, it was necessary to limit the
length of the field and artificially specify downstream conditions. Yet, in the absence of measurements the specification of the Reynolds stresses has its own value.

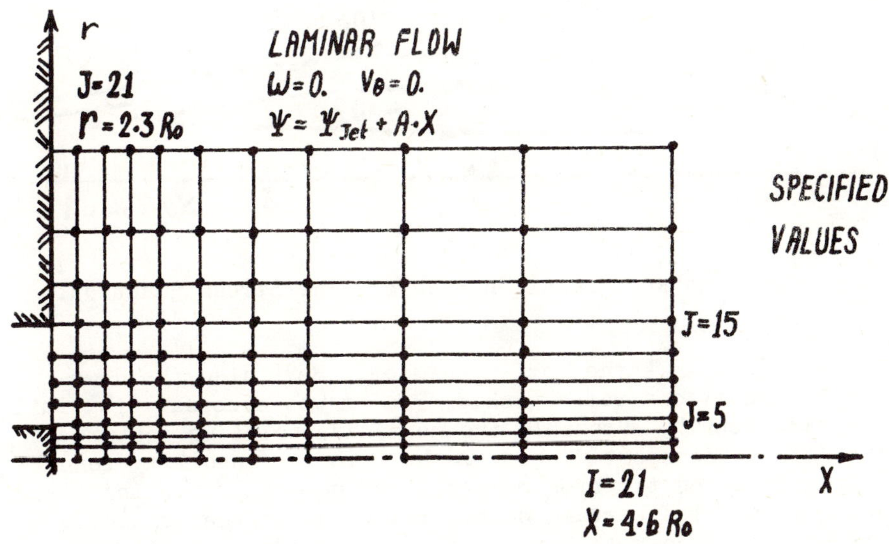

Figure 6: The Recirculation Bubble - The Grid Used

The case was designed to serve as an example of a calculation
where all the terms of the Reynolds Stresses equations are important and contribute to the calculation. Although in the
main stream enveloping the recirculation bubble the calculation
is parabolic in nature, the core region demonstrates eliptic
features as the turbulence convected upstream along the axis
has distinctively different features from that convected downstream directly from the flow origin. The turbulent diffusion
is very important everywhere, and it is doubted whether an
approach based on local equilibrium (Algebraic Modelling) will
describe this flow accurately. The value of the results is
limited by lack of satisfactory boundary conditions typical of
the use of a Reynolds stress closure. This mainly applies to

H

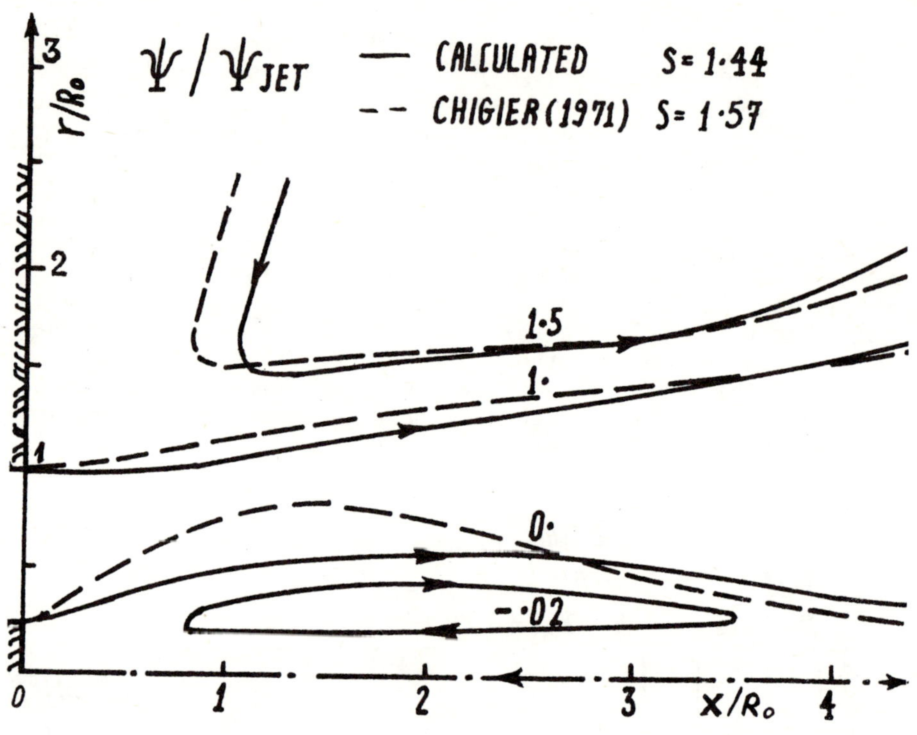

Figure 7.   The Recirculation Bubble - The Stream Function.

the flow origin - the vortex chamber. Here also the down-stream conditions raise problems as it was necessary to limit the grid length. Probably the most disappointing feature of the stress transport model used here is its inability to correctly describe the broadening of the non-recirculating free jet. This, plus certain doubts raised about the predicted $R_{x\theta}$ stress slowed down the progress of this model. Still, we may think about cases where these defects are of minor importance in comparison with the transport processes. Examples are - vortex chamber, turbulent axial air bearing, cyclon separator and others. These depend less on the structure of the incoming turbulence and their volume is bounded by rigid walls. With vanishing coefficient of the turbulent transport term the turbulence model here described reduces to the local equilibrium description used to form the algebraic models. As the numerical calculation converges also for this case, we obtain a tool that is superior to the algebraic models by the fact that no velocity derivative is ignored, and may be used to confirm the algebraic modeling usually associated with simplified equations.

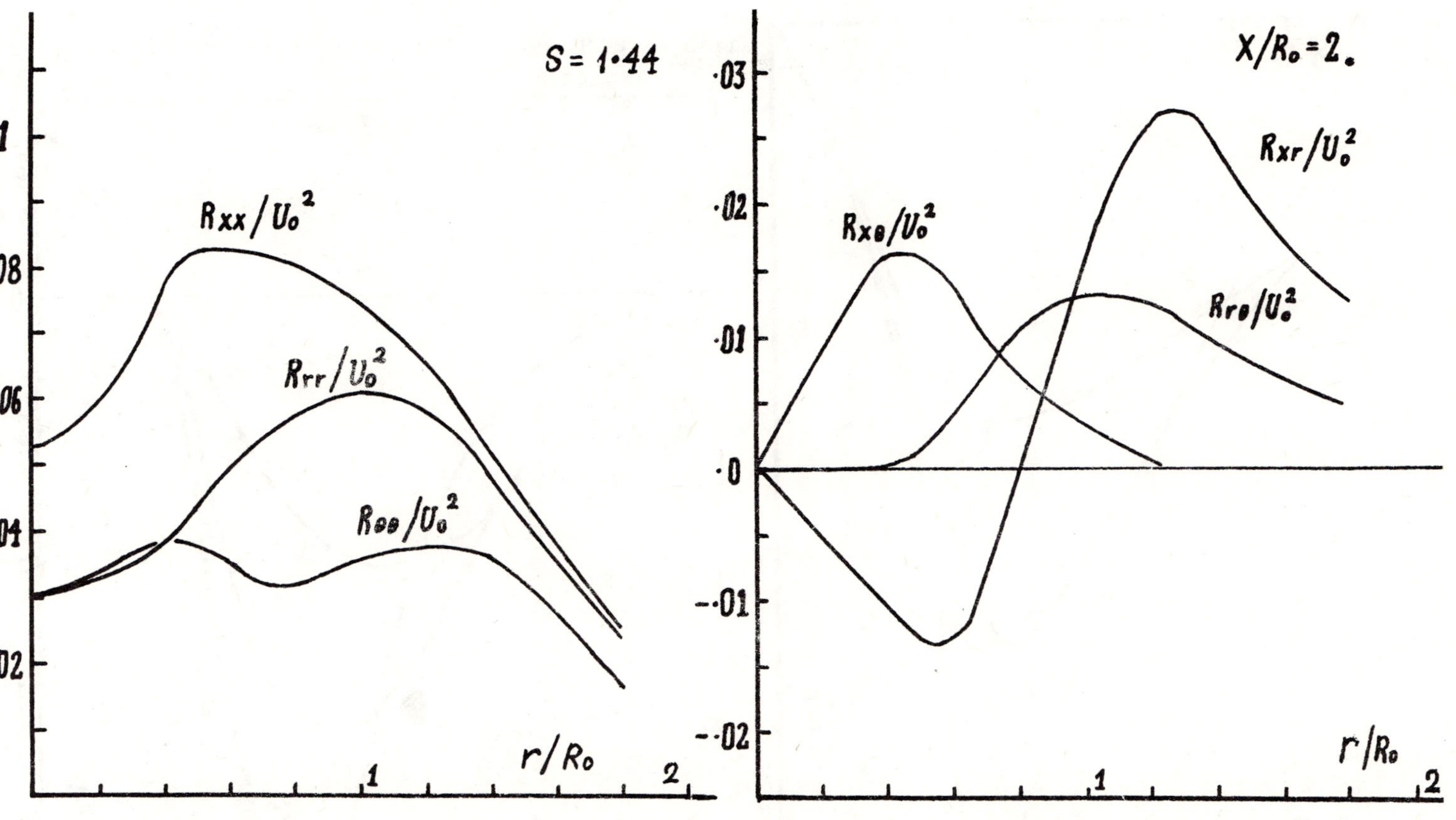

Figure 8.   The Recirculation Bubble - The Turbulent Stresses

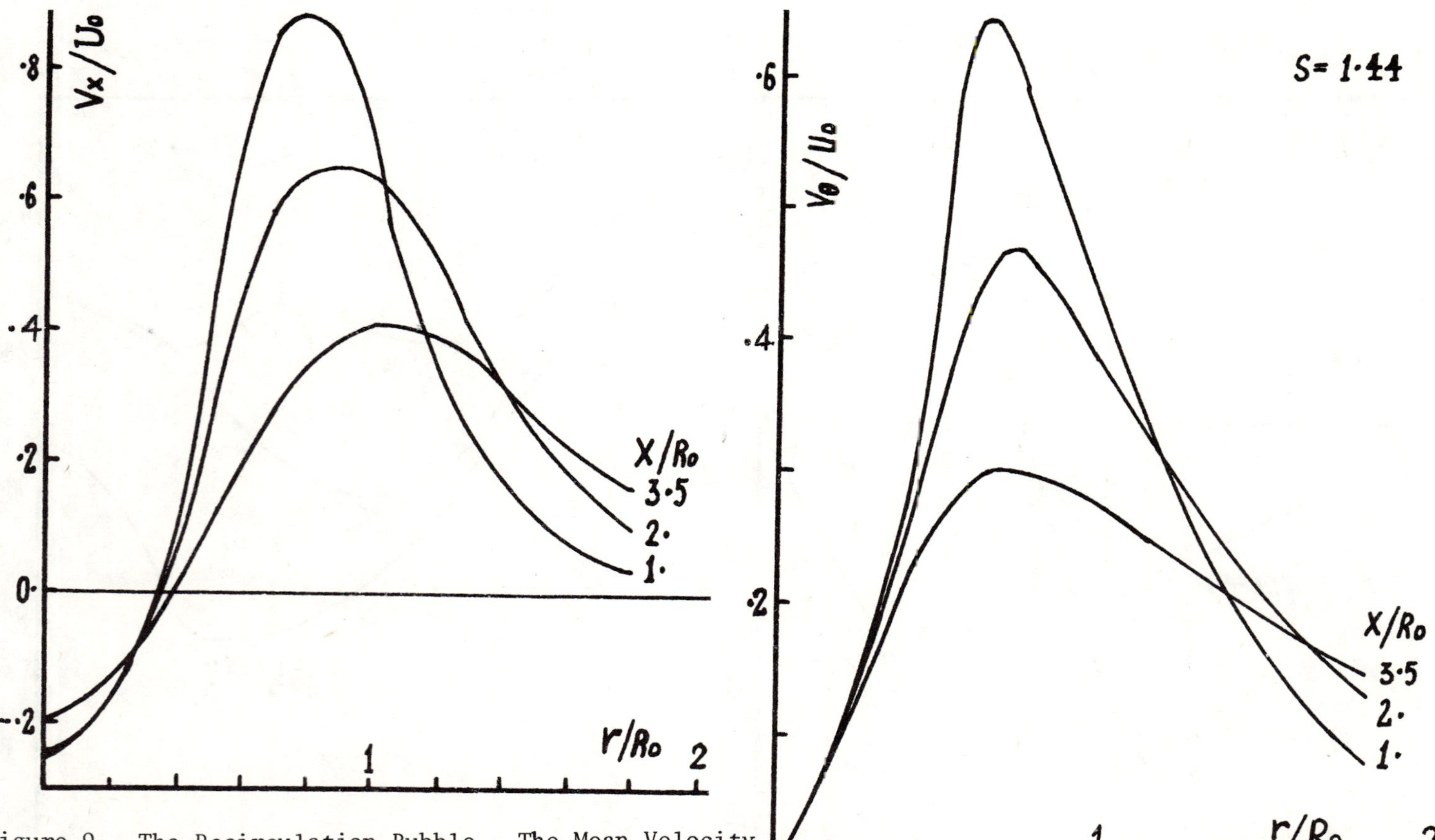

Figure 9. The Recirculation Bubble - The Mean Velocity

CONCLUDING REMARKS

The potentiality of a numerical procedure based on a simultaneous solution of the whole set of the ten equations linearized in a way that couples the solution for the mean
velocities with the solution for all the Reynolds stresses,
which is suitable for eliptic two dimensional fields was
described. The reasonable computing times and the relative
simplicity in programming call for further work along the
line. Although the considerations which lead to the use of
an advanced turbulence model are directly associated with the
nature of the flow field of interest the knowledge that the
use of this model is not accompanied by a substantial increase
in the computing time may help to take such a decision.

The application of the solution procedure results in fast
converging calculations as far as the negative vorticity
sources do not play an important part in the balance of the
vorticity. In the case when the centrifugal forces are
strong enough to form a recircualation zone the solution
becomes unstable and special techniques have to be used. The
two techniques used here, the M.P.C.A. and the S.F.D.C.,
resulted in a moderate convergency suggesting that the clue to
the stabilization of the calculation is in the existence of
positive $C_{\Psi\omega}$ and $C_{\Psi V_\theta}$ which form a feedback mechanism between
the variation of the stream function and the variation of the
convection of both the vorticity and the swirling velocity.
Finally, it is expected that when turbulence models, more
complicated than the Reynolds stress models, will become
available (e.g. multi-scales - Reynolds stresses models), the
simultaneity in the variables phase will become an essential
part of the calculation procedures.

REFERENCES

Atias, M., Wolfstein, M. and Israeli, M. (May 1975) A Study
of the Efficiency of Various Navier-Stokes Solvers. ICASE
Report No. 75-11, NASA's Langley Research Center.

Caretto, L.S., Gosman, A.D. and Spalding, D.B. (1971)
Removal of an Instability in a Free Convection Problem.
Imperial College, Mec. Eng., Report HTS/71/12.

Chigier, N.A. (1972) Gasdynamics of Swirling Flow in Combustion Systems. Astronautica Acta., 17, pp. 387-395,
Pergamon Press.

Chigier, N.A. and Chervinsky, A. (June 1967) Experimental
Investigation of Swirling Vortex Motion in Jets. Trans. of
the ASME, Jour. of Applied Mech., p. 443.

Chigier, N.A. and Chervinsky, A. (1966) Experimental and Theoretical Study of Turbulent Swirling Jets Issuing from a Round Orifice. Israel Jour. of Tech., 4, No. 1, p. 44.

Chigier, N.A. and Gilbert, J.L. (Jan. 10, 1968) Recirculation Eddies in the Wake of Flameholders. Symposium on Combustion in Marine Boilers, The Institute of Marine Eng., London.

Crossley, P.S. (1975) Recirculating Flows with Body Forces. Ph.D. Thesis, Imperial College, Mech. Eng., London.

Gosman, A.D., Pun, W.M., Runchal, A.K., Spalding, D.B. and Wolfshtein, M. (1969) Heat and Mass Transfer in Recirculating Flows. Academic Press, London and New York.

Gosman, A.D. and Spalding, D.B. (1970) Computation of Laminar Recirculating Flow between Shrouded Rotating Discs. Imperial College, Mech. Eng., Report IITS/70/8.

Hinze, J.O. (1959) Turbulence. McGraw-Hill Book Co., N. Y.

Koo Sin Lin, M.L. (1974) Turbulent Transport Properties in Swirling Two Dimensional Boundary Layers. Ph.D. Thesis, Imperial College, Mech. Eng., London.

Launder, B.E., Reece, G.J. and Rodi, W. (1975) Progress in the Development of a Reynolds Stress Turbulence Closure. J. Fluid Mech., 68, p. 537.

Launder, B.E. and Spalding, D.B. (1972) Mathematical Models of Turbulence. Academic Press.

Lumley, J.L. and Khajeh-Nouri, B. (1974) Computational Modeling of Turbulent Transport. Advances in Geophysics, 18A, p. 169.

Naot, D. (1976) On the Reynolds Stress Closure in Curvilinear Orthegonal Coordinates with Application to the Cylindrical System. Ben Gurion University of the Negev, Report ME-761(N).

Naot, D. and Launder, B.E. (1975) Numerical Calculation of the Turbulent Swirling and Recirculating Flows with a Reynolds Stress Closure. Ben Gurion University of the Negev, Report ME-22/75.

Naot, D., Shavit, A. and Wolfshtein, M. (1974) Numerical Calculation of Reynolds Stresses in a Square Duct with Secondary Flow. Warme und Stoff Ubertragung, 7, No. 3, p. 151.

Pratte, B.D. and Keffer, J.F. (1969) Swirling Turbulent Jet
Flows, Part I: The Single Swirling Jet. University of
Toronto, Report UT, Mech. Eng., TP 6901.

Roberts, L.W. (1971) The Prediction of Turbulent Swirling
Pipe Flow. Imperial College, Mech. Eng., Report EF/TN/A/37.

Rodi, W. (1970) Basic Equations for Turbulent Flow in
Cartesian and Cylindrical Coordinates. Imperial College,
Mech. Eng., Report BL/TN/A/36.

Rodi, W. (1970) On the Equation Governing the Rate of
Turbulence Energy Dissipation. Imperial College, Mech. Eng.,
Report TM/TN/A/14.

Rose, W.G. (Dec. 1962) A Swirling Round Turbulent Jet.
Trans. of the ASME, Jour. of Applied Mech., p. 615.

Syred, N., Chigier, N.A. and Beer, J.M. (1971) Flame Stabili-
zation in Recirculation Zone of Jets with Swirl. Thirteenth
Symposium on Combustion, The Combustion Institute, Pittsburgh,
Pennsylvania.

Tennekes, H. and Lumley, J.L. (1972) A First Course in
Turbulence. MIT Press, Cambridge, Massachusetts.

Wolfshein, M., Naot, D. and Lin, A. (1975) Models of Turbulence. In Current Topics in Thermal Science, Gutfinger, C.,
Editor, Scripta Book Co.

Acknowledgements

The author wishes to acknowledge Dr. Gosman of the Imperial
College, London, Prof. Launder of the University of California
at Davis, Dr. Leshziner of the S.F.B. 80 University of
Karlsruhe and Prof. Wolfshtein of the Technion, Israel, for
many helpful discussions. The program was developed in
governmental non-profitable universities and can be supplied
to institutes of similar character.

CHAPTER 8

COMPUTER ANALYSIS OF FLUID FLOW AND HEAT TRANSFER

Suhas V. Patankar

Department of Mechanical Engineering
University of Minnesota, Minneapolis, MN 55455, U.S.A.

INTRODUCTION

Purpose of the Chapter

Fluid flow and heat transfer phenomena abound in technological
and environmental processes. The recent drive for energy conservation, the search for newer energy sources, and the need
for the preservation of the natural environment require a
deeper understanding of the mechanisms of fluid flow and the
associated heat and mass transfer. Further, quantitative prediction techniques are needed to analyze existing and proposed
devices and configurations. With the prediction ability, de-
signs can be optimized, existing equipment can be operated
most efficiently and within safety limits, and the environmental impact can be estimated and controlled.

This chapter describes the details of a numerical prediction technique for fluid flow and heat transfer. The technique has been developed and improved over the past ten years.
It has also been applied to a very wide variety of physical
situations. This chapter provides a nearly self-contained
treatment of the various aspects of the technique.

There are many different methods for the solution of the
governing partial differential equations that arise in fluid
flow and heat transfer problems. Even after choosing a particular method, one must frequently exercise freedom of
choice in a multitude of major and minor details that finally
characterize the complete technique. Although the major
choices are usually guided by proven scientific findings,
many other details must be decided in a more-or-less ad hoc
manner influenced by considerations of convenience and personal
taste. The technique described in this chapter represents the
set of practices and procedures currently adopted by the present author and his co-workers. It will not be possible here
to document the complete reasons underlying these choices, nor
will an attempt be made to list the possible alternatives.
What is presented here is believed to be a good method, and it

has been very extensively tested; but there is no implication that other approaches are unworthy of attention.

A primary objective that has guided the choice of the method here is a very wide applicability of the method. It was decided that the method should be able to handle steady or unsteady situations, which may be one- , two - , or three-dimensional. The flow may be laminar or turbulent; the density may depend on the local temperature, concentration, etc.; the participating chemical species may react; and a variety of different boundary conditions may be present. These requirements disqualify, as a general prediction procedure, a number of elegant and computationally efficient techniques that are but limited to only a small class of flow situations.

Among the methods that would provide the desired generality, attention is given to computationally economical procedures. Often, elaborate computational procedures are devised, and rather impressive computations are performed, in research environments where almost unlimited computing resources are available. These methods, although they represent laudable scientific achievements, are impractical for regular use in engineering design. The method described in this chapter is the result of a constant search for techniques that give the required accuracy with only a modest amount of computer storage and time.

Although the basic purpose of this chapter is to describe a calculation method of great generality, some restrictions will be adopted here for the ease of presentation. Thus, attention will be given to only steady-state situations; no reference to time-dependent phenomena will be made. Further, the numerical scheme will be worked out for Cartesian co-ordinates, although the method can be employed, without any novelty, for any orthogonal coordinate system. Finally, figures and equations will often refer to <u>two-dimensional</u> situations; but their conceptual extension to three dimensions should not be very difficult.

<u>Examples of physical situations</u>
The development of the numerical technique can be well appreciated if its potential applications are clearly kept in mind. Therefore, a few practical situations are described here. The prediction procedure should be able to handle these.

<u>Buoyancy-driven flow in a cavity</u> Figure 1 shows a cavity with the opposite walls maintained at different temperatures. Under these thermal boundary conditions, a circulating flow as shown is produced by the action of the gravitational force. It is of interest to predict the distributions of temperature, pressure, velocity components, and associated quantities throughout the cavity. If the view in Figure 1 is considered as the cross section of a duct, prediction of the longitudinal flow, as influenced by the secondary flow, is

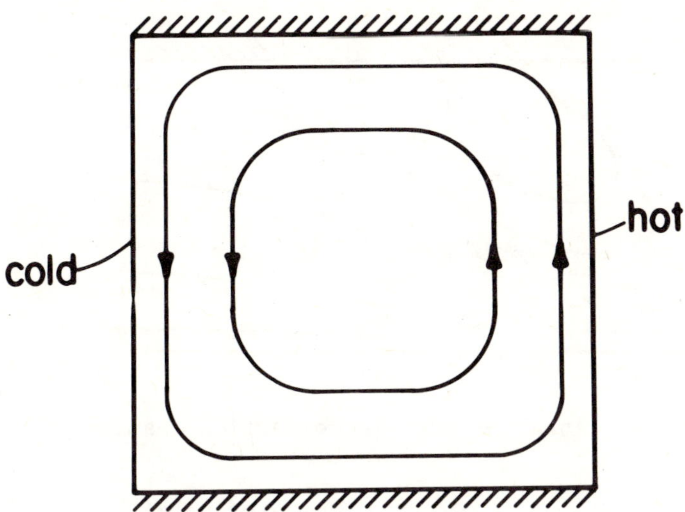

Figure 1   Circulating flow in a cavity

also of interest.

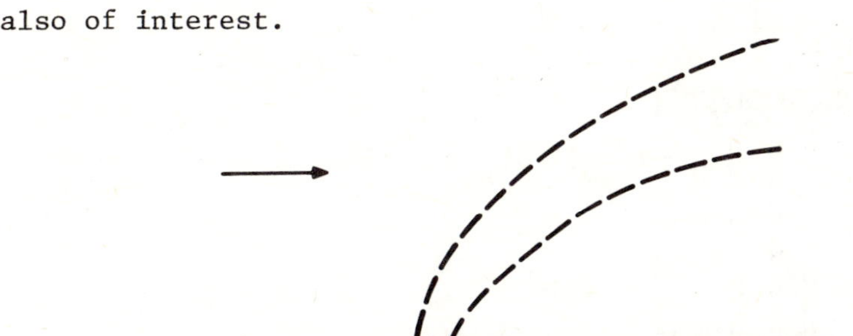

Figure 2   A jet deflected by the main stream

<u>Deflected-jet situation</u>   If a vertical jet from a circular orifice is deflected by a horizontal stream as shown in Figure 2, an interesting three-dimensional flow results.   Such flows are encountered in chimney plumes, film-cooling arrangements, V/STOL aircraft, combustion chambers, etc.   The numerical technique should solve for the flow field and the associated temperature and/or concentration field.

<u>Combustion chamber</u>   An axi-symmetric combustion chamber shown in Figure 3 involves the mixing and reaction between the fuel and air streams, a recirculation zone that keeps the flame ignited, and the outflow of the products of combustion.   In predicting this situation, one must be able to account for the simultaneous influence of fluid flow, heat and mass transfer, chemical reaction, turbulence, and perhaps radiation.

<u>Steam generator of a nuclear power plant</u>   A representative example of a somewhat different class of flows is sketched in

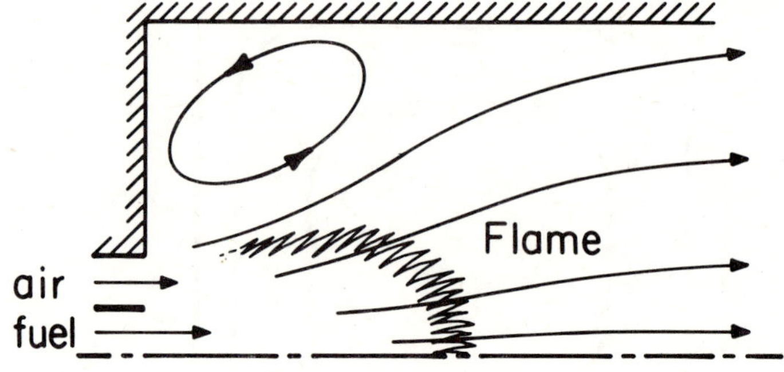

Figure 3   An axi-symmetric combustion chamber

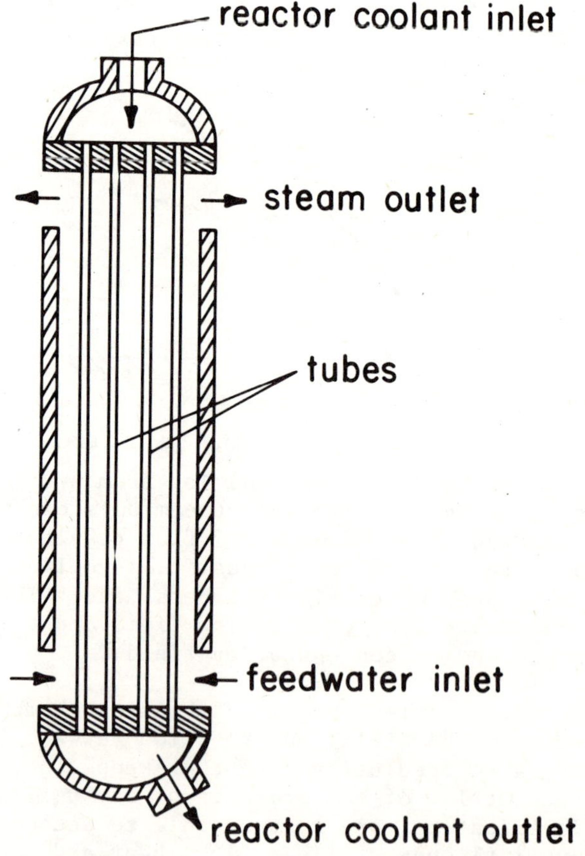

Figure 4   A once-through steam generator

Figure 4. It shows a once-through steam generator used in a
nuclear power plant. The hot coolant from the reactor flows
through the tubes as shown. Feedwater enters the shell at
the bottom and steam leaves radially near the top end. Other

examples of this class of flows are heat exchangers, conden-
sers, nuclear reactors, cooling towers, and packed beds. All
these are characterized by the presence of a large number of
solid objects, such as tubes, rods, slats, interspersed in the
region available for the flow. A "macroscopic" calculation
for such flows can be performed by disregarding the detailed
presence of the solid objects and replacing them by equivalent
resistance terms in the momentum and energy equations. This
approach, first formulated by Patankar and Spalding (1974),
has been successfully applied to steam generators in Patankar
and Spalding (1978), and Oberjohn, Rice, and Patankar (1980).

Role of the numerical method

Since a quantitative prediction for the practical situations
such as those described so far yields significant economic
benefits, good prediction methods are of great value. An experimental investigation is, of course, a very reliable way of
determining how a given configuration would perform. It is
only by actual experimentation that one can hope to learn
about a new physical process or phenomenon. However, for most
complex applications, an experimental study is either impossible or too expensive. Consequently, with limited guidance
from experiments, one turns to numerical techniques as a major
prediction method. If the equations governing the processes
are known with reasonable certainty, a numerical technique
provides a flexible and economical method for predicting the
details of practical situations.

The advantages offered by a numerical prediction procedure are many. The foremost advantage is the low cost. This
consideration becomes increasingly important as the size and
complexity of the physical situation grows. Second, numerical
predictions can be obtained at remarkable speed, thus enabling
rapid decisions in the design process. Unlike an experimental
investigation (which must be limited to a rather small number
of measurements), a numerical solution provides complete and
detailed information about the distributions of all the relevant variables. Since the physical limitations of handling
large sizes, high temperatures, toxic substances, etc. are not
experienced by a numerical computation, it is easy to simulate
all the actual conditions in a numerical experiment than in a
physical one. In a fundamental study, a numerical method enables the many desired idealizations (such as constant density,
two-dimensional flow, zero viscosity, etc.) to be set up easily
and exactly; they can be barely approximated even in very care-
ful experiments.

A preliminary idea about the task of a numerical method
can be obtained by reference to Figure 5, where a certain flow
situation is shown. A grid is drawn to cover the flow domain.
With a sufficiently fine grid, the complete distributions of
the relevant variables can be expressed in terms of their
values at the grid points. Thus, the task of a numerical
method is to evaluate velocity, pressure, temperature, con-

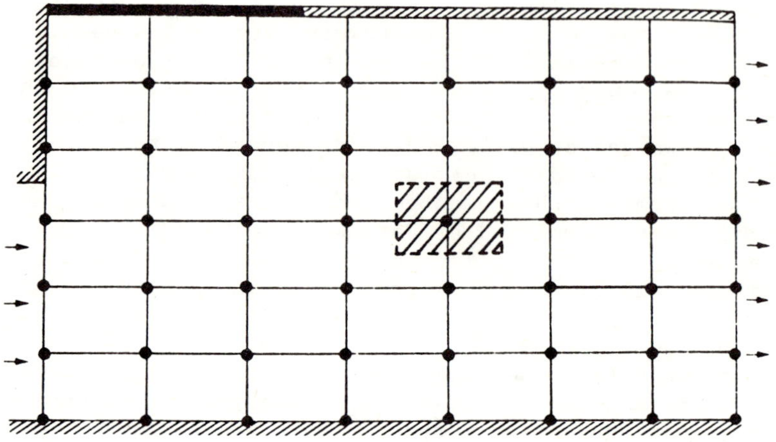

Figure 5   A calculation domain and the grid

centration, etc. at the chosen grid points.

From the differential equations governing the relevant variables, algebraic equations are derived for the grid-point values of the variables. The derivation of these discretization equations and a procedure for solving them constitute a numerical method.

The particular practice that is chosen here for the derivation of the discretization equations is known as the control-volume approach. The calculation domain is divided into subdomains or control volumes such that there is one control volume around a grid point; a typical control volume is shown as a shaded region in Figure 5. The differential equation is integrated over the control volume to yield the discretization equation. Thus, the discretization equation represents the same conservation principle over a finite region as the differential equation does over an infinitesimal region. This direct interpretation of the discretization equation makes the method easy to understand in physical terms; the coefficients in the equation can be identified, even when they appear in a computer program, as familiar quantities such as flow rates, conductances, areas, volumes, heat capacities, etc.

The control-volume approach can be regarded as a special case of the method of weighted residuals (Finlayson, 1972), in which the weighting function is chosen to be unity over a control volume and zero everywhere else. Although the main reasons for choosing the control-volume formulation are its simplicity and easy physical interpretation, the formulation has also been shown by Ramadhyani and Patankar (1980), for a limited set of test problems, to be more accurate than the Galerkin method, which is a more popular weighted-residual technique.

Outline of the Chapter

The necessary mathematical formulation will be given in the
next section. Then the complete details of the numerical
method will be developed. Towards the end of the chapter,
illustrative results will be presented from some recent applications of the method.

MATHEMATICAL FORMULATION

Governing differential equations

The physical situations involving fluid flow and heat and mass
transfer are governed by the conservation principles of mass,
momentum, energy, chemical species, etc. These principles can
be expressed in terms of partial differential equations. A
close examination of these equations reveals that they possess
a common form. This recognition provides a major convenience
for numerical formulation. All that is necessary is to derive
a numerical procedure for solving the differential equation of
the common form.

If the dependent variable is denoted by ϕ , the corresponding <u>steady-state</u> form of the general differential equation is

$$\frac{\partial}{\partial x_i} (\rho u_i \phi) = \frac{\partial}{\partial x_i} (\Gamma \frac{\partial \phi}{\partial x_i}) + S \tag{1}$$

where the Cartesian-tensor notation has been used. The symbol
u_i denotes the velocity component in the x_i direction, while ρ
stands for the fluid density. The term on the left-hand side
of Equation (1) is known as the convection term, representing
the flux of ϕ convected by the mass flow rate ρu_i. The diffusion coefficient Γ and the source term S will shortly be discussed in detail. The diffusion term is included in Equation
(1) because it is almost always present, and often plays an
important role, in any particular manifestation of ϕ. The
source term is primarily meant for the mechanisms that cause
to the generation (or destruction) of ϕ. But it can also be
used as a general "dumping" ground; whatever cannot be conveniently expressed through the convection and diffusion terms
can always be lumped into the source term. Because this flexibility is available, the assumption of the general equation
does not restrict the processes or the types of the dependent
variable that can be accommodated in the calculation procedure.

When the flow is turbulent, time-averaged forms of the
governing equation are used. Then the quantities in Equation
(1) should be interpreted as the time-mean values of the relevant variables. The additional terms arising from the time-
averaging operation are expressed via appropriate redefinition
of Γ and S. For example, Γ may stand for the turbulent viscosity or the eddy diffusivity.

Research into mathematical models of turbulence has shown that often it is desirable to compute, in addition to the time-mean properties of the flow variables, the time-averaged properties of the fluctuating motion itself. Thus, additional differential equations may have to be solved for turbulence parameters, such as the kinetic energy of turbulence and the length scale of turbulent eddies. Fortunately, these differential equations also possess the general form of Equation (1), and the general solution technique continues to be applicable.

The dependent variable ϕ can stand for a variety of physical quantities. It can represent velocity components, enthalpy (or temperature), mass fractions of chemical species, turbulence parameters, radiation fluxes, and so on. Each meaning of ϕ is accompanied by appropriate expressions for Γ and S and appropriate boundary conditions. Once a portion of the computer program is written for the solution of Equation (1), it can be repeatedly used with different meanings of ϕ, the "DO loop" being executed the required number of times.

Complete expressions for Γ and S for different dependent variables will not be given here. These are available from standard textbooks. When one is faced with a new situation, it is assumed that at least a tentative mathematical model has been written down before beginning the computational task. However, some general comments about Γ and S will now be made.

Diffusion coefficient

The diffusion coefficient Γ is a general representation for the fluid properties such as viscosity or conductivity, which in conjunction with the gradient of the appropriate variable lead to the diffusion flux such as a viscous stress or a heat flux. For turbulent flow, many mathematical models replace the laminar (or molecular) values of Γ by the corresponding turbulent (or effective) properties, for example, the turbulent viscosity or eddy diffusivity. If the diffusion flux for a given ϕ is not driven by the gradient of ϕ, then Γ may be set equal to zero, and the diffusion flux expressed as a part of S. In general, the diffusion coefficient Γ may be nonuniform; it may depend on position, as in a composite material, or on quantities such as the local velocity, temperature, density, etc.

Source term

Internal heat generation in a fluid, production or destruction of a chemical species in a chemical reaction, the body force acting on a fluid element are common examples of the source term S appearing in the equations for relevant variables. In a turbulence model, the equations for the turbulence parameters may contain production (and/or dissipation) terms. The source term S in a momentum equation for the velocity component u_j includes the pressure gradient $(-\partial p/\partial x_j)$, the body forces in the i direction, and a contribution from the viscous stress terms that cannot be accommodated in the diffusion term.

For the type of application shown in Figure 4, the source term takes on an added significance. Since the detailed presence of the tubes is ignored, their influence is incorporated into the momentum equations in the form of a source term representing the distributed resistance of the tubes. Similarly, in the energy equation for the shell fluid, a source term accounts for the heat transfer from the tube fluid.

Continuity equation

Although the behavior of any dependent variable ϕ is governed by Equation (1), the mass flow rate ρu_i appearing in it must satisfy an additional equation. This is the continuity equation for the flow field, which can be written for steady flows as

$$\frac{\partial}{\partial x_i} (\rho u_i) = 0 \tag{2}$$

What is the additional degree of freedom that can be invoked to satisfy this extra equation? It is the pressure field, which enters the source terms of the momentum equations and which must adjust so that the resulting velocity field satisfies the continuity equation. This coupling between the velocity and pressure fields will be used in constructing the calculation procedure for fluid flow.

Boundary conditions

A calculation domain such as the one shown in Figure 5 has usually three kinds of boundaries. They are no-flow, inflow, and outflow boundaries. At a no-flow boundary, an impermeable wall coincides with the boundary location. At such a boundary, either the value of the dependent variable ϕ is known or a relationship about the diffusion flux at the boundary is known. Thus, if ϕ stands for temperature, one may know the wall temperature, the wall heat flux, or an external convective heat transfer coefficient. At an inflow boundary, usually the value of ϕ in the incoming stream is known. At an outflow boundary, on the other hand, neither the value nor the flux of ϕ is known. However, if the numerical method is properly formulated, no information is indeed needed at an outflow boundary. This feature will later be explained with reference to the particular formulation.

One-way and two-way coordinates

Partial differential equations are often classified as parabolic, elliptic, and hyperbolic. There is a corresponding computational classification, which is based on physical concepts and which has direct computational implications. This classification is based on the ideas of one-way and two-way coordinates.

<u>Definitions</u> Normally the conditions at a given location in a coordinate are influenced by the happenings on <u>either</u> side of the location. Such a coordinate is called two-way. In a one-way coordinate, the conditions at one point are influenced by the happenings on only <u>one side</u> of that point.

Time is a natural one-way coordinate, since the conditions at a given instant of time are dependent on the happenings <u>before</u>, and not after, that instant. Even in steady flows, a near one-way behavior can be observed in a <u>space</u> coordinate if there exists a unidirectional significant flow in that coordinate. Under these conditions, the flow properties at a given location are affected only by the <u>upstream</u> events, and not by the downstream ones. The mathematical term "parabolic" indicates the presence of at least one one-way coordinate. When no one-way coordinate is present, the situation is called elliptic. (The hyperbolic problems do not fit neatly into the computational classification. They are to be handled as elliptic problems for computational purposes.)

<u>Computational implications</u> The one-way nature of the time coordinate is routinely taken advantage of in all unsteady computations. Rather than storing and handling the values of ϕ at all the future instants of time, one "marches" from a set of known ϕ values at one instant to the ϕ values at the <u>next</u> instant of time. A similar marching procedure is available in steady flows, where one space coordinate has a one-way behavior everywhere. Examples of such situations are boundary layers and duct flows. A two-dimensional boundary layer can be computed by solving for the ϕ distribution over one cross-stream line at a time and advancing the line in the mainstream direction. A three-dimensional duct flow requires consideration of only one cross-stream plane at a time. Thus, the identification of a one-way coordinate limits the attention to only one or two locations in that coordinate. The computer storage requirements then are those of a problem of lower dimensionality, and less computer time is required since the ϕ distributions are decoupled from all the downstream conditions. In this manner, the recognition of a one-way coordinate leads to significant saving of computer storage and computer time.

Two-dimensional boundary-layer situations are fully discussed by Patankar and Spalding (1970) and Spalding (1977). A procedure for three-dimensional parabolic flows has appeared in Patankar and Spalding (1972). In this chapter, direct reference to parabolic situations will be made only occasionally; however, it is interesting to note that the formulation described here would imply a one-way behavior under appropriate flow conditions.

NUMERICAL TECHNIQUE

<u>Control volume definition</u>
When a grid such as the one shown in Figure 5 is drawn to cover the desired calculation domain, control volumes must be

constructed around the grid points. This can be done in many
ways. Although a particular definition is employed here, other
definitions can also be used in conjunction with the calculation method described here. Figure 6 shows the construction of
the control volumes. Indeed, it is convenient to decide the
control-volume boundaries first and then place the grid points.

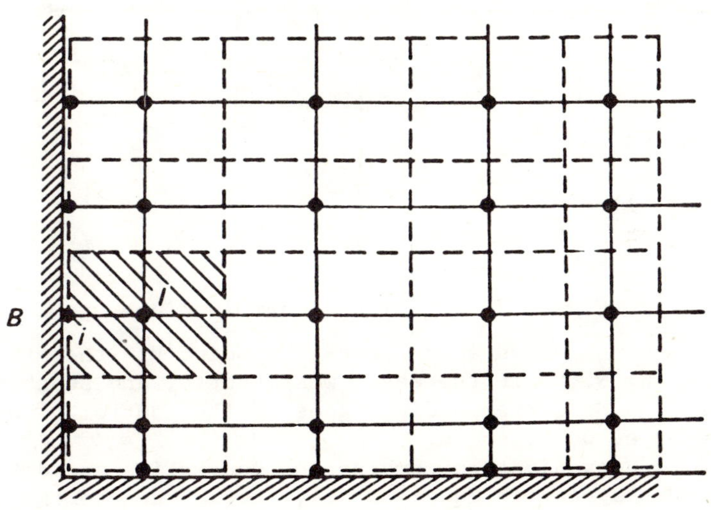

Figure 6 Control volumes and the associated grid

The calculation domain is thus subdivided into control volumes;
the dashed lines in Figure 6 denote the control-volume boundaries. Then, grid points are placed at the geometrical centers
of the control volumes. The solid lines in Figure 6 are the
grid lines, and the dots denote the grid points. For the two-
dimensional situation shown, a given grid point communicates
with four neighboring grid points through the four faces of the
control volume. The situation with a near-boundary control
volume is somewhat different; such a control volume is shown
shaded in Figure 6. Here, one face of the control volume coincides with the boundary of the calculation domain, and a
boundary grid point is placed at the center of the control-
volume face. (Another way of looking at this situation is to
imagine a control volume of infinitesimal thickness located at
the boundary.) This arrangement makes it rather easy to treat
different boundary conditions; the shaded control volume can
easily accept a given value of ϕ at the boundary or a given
flux through the boundary surface.

<u>Conservation equation for the control volume</u>
The general differential equation (1) can now be integrated
over the control volume shown in Figure 7. Only a two-dimensional representation is shown for convenience. The control

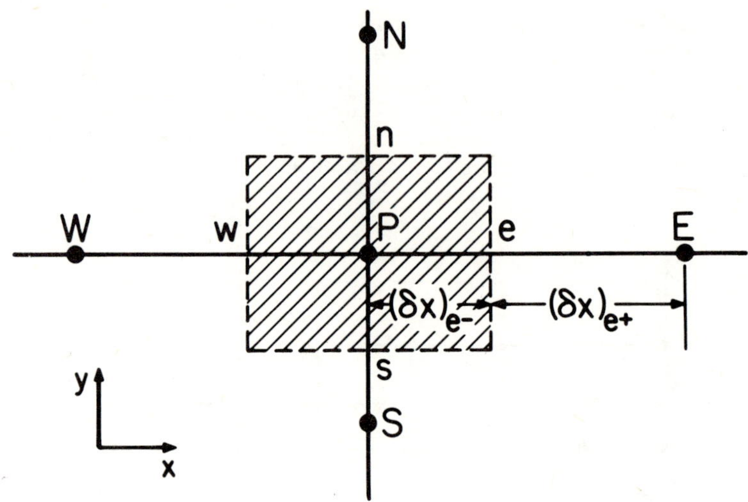

Figure 7   A typical control volume

volume is constructed around the grid point P; the other grid
points E, W, N, S are the east, west, north, and south neighbors of P. The corresponding faces of the control volume are
denoted by e, w, n, and s.

It is convenient to combine the convection and diffusion
fluxes that appear in Equation (1). Let J_i denote the total
(i.e., convection plus diffusion) flux in the i direction.
Then,

$$J_i = \rho u_i \phi - \Gamma \frac{\partial \phi}{\partial x_i} \tag{3}$$

The resulting differential equation is

$$\frac{\partial J_i}{\partial x_i} = S \tag{4}$$

The integration of this equation over the control volume shown
in Figure 7 leads to

$$J_w A_w - J_e A_e + J_s A_s - J_n A_n + \overline{S}\Delta V = 0 \tag{5}$$

where the J's represent appropriate total fluxes at the control-volume faces, the A's are the areas of the faces, $\overline{S}$ is
the average source term over the control volume, and ΔV is the
volume of the control volume. Although for the Cartesian coordinates, the areas A_w and A_e are equal, separate symbols are
used for generality.

The next two subsections deal with the total fluxes and
the source term respectively.

Total fluxes

Consider the region between grid points P and E in Figure 7. If a one-dimensional convection-diffusion problem is solved between the points P and E, an exponential profile of ϕ is obtained as an exact solution. This is given by Spalding (1972) and by Raithby and Torrance (1974). The exponential solution leads to

$$J_e = F_e \{ \phi_P + (\phi_P - \phi_E)/(\exp P_e - 1) \} \tag{6}$$

where the Peclet number P_e is given by

$$P_e = F_e/D_e \tag{7}$$

Here F_e is the flow rate $(\rho u)_e A_e$ and D_e is the diffusion conductance. If the diffusion coefficient Γ is regarded as uniform over each control volume, the appropriate expression for D_e is

$$D_e = A_e \{ (\delta x)_{e-}/\Gamma_P + (\delta x)_{e+}/\Gamma_E \}^{-1} \tag{8}$$

where the distances $(\delta x)_{e-}$ and $(\delta x)_{e+}$ are shown in Figure 7. The rationale for and the advantages of defining D_e in this manner are discussed in Patankar (1978).

Source term

If the source term $\bar{S}$ in Equation (5) is known, no difficulty arises. But often the source term depends on the variable ϕ itself. In order that the resulting discretization equation remains (at least nominally) linear, the source term S is expressed as a linear function of ϕ_P. Thus, the linearization formula can be written as

$$\bar{S} = S_C + S_P \phi_P \tag{9}$$

where S_P is the coefficient of ϕ_P, and S_C is the part of $\bar{S}$ that does not explicitly depend on ϕ_P.

Detailed discussion on different ways of linearizing $\bar{S}$ is given by Patankar (1980). Here two main points will be noted. First, the source-term linearization is often a very crucial operation; it is responsible for computational success in many complex situations. Second, it is desirable that the linearization results in a negative S_P; a positive S_P can cause divergence.

Final discretization equation

Substitution of the relations derived in the last two subsections into Equation (5) now leads to the final discretization equation. It can be written as

$$a_P\phi_P = a_E\phi_E + a_W\phi_W + a_N\phi_N + a_S\phi_S + b \tag{10}$$

where

$$a_E = \frac{F_e}{\exp(P_e) - 1} \tag{11}$$

$$a_W = \frac{F_w \exp(P_w)}{\exp(P_w) - 1} \tag{12}$$

$$a_N = \frac{F_n}{\exp(P_n) - 1} \tag{13}$$

$$a_S = \frac{F_s \exp(P_s)}{\exp(P_s) - 1} \tag{14}$$

$$b = S_C \, \Delta V \tag{15}$$

$$a_P = a_E + a_W + a_N + a_S - S_P \, \Delta V \tag{16}$$

Here F_e, F_w, F_n, and F_s are the mass flow rates through the respective control-volume faces. The Peclet numbers P_e, P_w, P_n, and P_s are defined along the lines of Equation (7), and the corresponding diffusion conductances according to Equation (8). The expression for a_P as given by Equation (16) results from the assumption that the flow rates F_e, F_w, F_n, and F_s satisfy the continuity equation for the control volume.

Because exponentials are expensive to compute, approximations have been worked out for the coefficient expressions in Equations (11)-(14). These include the hybrid scheme of Spalding (1972) and the power-law scheme of Patankar (1980). Those details, however, are not needed for the further development in this chapter.

At this stage, it is useful to write Equation (10) in a generalized form

$$a_P\phi_P = \Sigma a_{nb}\phi_{nb} + b \tag{17}$$

where the subscript nb denotes the neighbor grid points of P; the summation is to be taken over all the neighbors. For the two-dimensional situation for which Equation (10) was derived, there are four neighbors. Six neighbors must be considered for a three-dimensional problem.

Boundary conditions

With reference to Figure 6, it can be seen that, since there is a control volume around each internal grid point, there will be a corresponding discretization equation such as (10). In this system of equations, the values of ϕ at the boundary points will appear. If these values (such as ϕ_B) are known, the solution of the equations is straight-forward. If instead some information about the boundary flux is given, it is necessary to write an equation similar to Equation (6) for the boundary flux in terms of ϕ_B and ϕ_I. By the use of this equation, the unknown value ϕ_B can be eliminated from the control volume equation for the grid point I.

The outflow boundary condition requires special mention. Usually no information is available at the locations where the fluid _leaves_ the calculation domain. The difficulty is resolved by a closer examination of the total flux formulation. The coefficient expressions given by Equations (11)-(14) are such that, when the magnitude of the Peclet number is large (i.e., greater than about 5), the coefficient of the _downstream_ neighbor is almost zero. Thus, the value ϕ_B at an outflow boundary influences ϕ_I only through a vanishing coefficient. No knowledge of ϕ_B is, therefore, needed! (It is true that this reasoning is based on the premise that the Peclet number is large; but usually it is not a serious approximation to _assume_ that the value of Γ at the out-flow boundary is small and hence the Peclet number large.)

Nonlinearity

From the discussion about the boundary conditions, it can be concluded that Equation (10) represents a set of enough equations for the evaluation of the grid-point values of ϕ. If these equations were truly linear, a straight-forward solution would yield the final answer. However, it must be recognized at this stage that these equations are only nominally linear. The coefficients in Equation (10) may themselves depend on the values of ϕ. Further, since ϕ can stand for a number of physical quantities, the coefficients for one meaning of ϕ may be influenced by some of the other ϕ's. Thus, when ϕ stands for temperature, its discretization coefficients depend on velocity, turbulence parameters, concentration, and so on.

Because of these interlinkages and nonlinearities, the final solution is to be obtained by iteration. At any given stage, the discretization coefficients can be calculated from the current estimates of all the ϕ values. The solution of equations such as (10) then gives an improved estimate. When, after many repetitions of this process, all the ϕ values

238

cease to change, the final converged solution is reached.
Thus, the solution to a set of nonlinear and interlinked equations can be obtained via many intermediate solutions of nominally linear and decoupled algebraic equations.

It does not always follow that successive iterations
would lead to a converged solution. At times, the values of
ϕ oscillate or drift away from what can be considered a reasonable solution. Such divergence of the iteration process
must be avoided.

Various ingredients of the discretization equations have
been developed so as to minimize the chances of divergence.
Further, it is desirable to slow down the changes in the coefficients from iteration to iteration. This can be accomplished by restricting the changes in the ϕ values via a technique
called underrelaxation, which is discussed next.

<u>Underrelaxation</u>
Equation (17) can be rewritten as

$$\phi_P = \phi_P^* + [(\Sigma a_{nb}\phi_{nb} + b)/a_P - \phi_P^*] \qquad (18)$$

where ϕ_P^* stands for the value of ϕ_P from the previous itera-
ation. The contents of the square brackets in Equation (18)
can be interpreted as the change in ϕ_P in the current iteration. To reduce this change, an underrelaxation factor α
(lying between o and 1) can be introduced so that

$$\phi_P = \phi_P^* + \alpha \, [(\Sigma a_{nb}\phi_{nb} + b)/a_P - \phi_P^*] \qquad (19)$$

It should be noted that, when convergence is reached (i.e.,
$\phi_P = \phi_P^*$), the satisfaction of Equation (19) also guarantees
the satisfaction of Equation (17). Further rewriting of
Equation (19) leads to

$$(a_P/\alpha)\phi_P = \Sigma a_{nb}\phi_{nb} + b + (1-\alpha)a_P\phi_P^*/\alpha \qquad (20)$$

The use of Equation (20) with an appropriate value of α introduces the desired underrelaxation into the system. If α
is close to zero, the values of ϕ change very slowly. When
α is unity, no underrelaxation is effected. If α is chosen
to be greater than unity, the values of ϕ change more rapidly
than implied by Equation (17); this behavior is known as
overrelaxation.

<u>Solution of the algebraic equations</u>
The solution procedure described so far has assumed that a
set of (nominally) linear algebraic equations can be solved

by a convenient method. If the discretization equation such
as (10) is written for a <u>one-dimensional</u> situation, it leads
to a system with tri-diagonal matrix of coefficients. There
is a particularly efficient algorithm to solve this system;
it is known as the Thomas algorithm or the tri-diagonal-
matrix algorithm (TDMA). It is commonly used in computations
of one-dimensional unsteady heat conduction and of two-dimensional boundary layers.

However, such a convenient direct method for solving the
discretization equations for two- or three-dimensional problems does not exist. The direct methods that are available
require excessive storage and computer time. Further, since
the set of nominally linear equations must be solved every
iteration, the expense of direct solution seems unacceptable.
Therefore, an iterative method for solving the algebraic equations is employed.

The iterative method uses the TDMA as its basic unit. In
a multi-dimensional problem, the equations for all the ϕ
values along <u>one grid line</u> are considered simultaneously.
These equations, of course, contain the ϕ values also along
the neighboring grid lines. If these neighbor-line values of
ϕ are substituted from the best available estimate, then the
equations along the chosen line have the same <u>appearance</u> as
the one-dimensional equations and can be solved simultaneously
by the TDMA. When all the lines are visited in this manner,
one iteration of the line-by-line technique is complete. The
process can be repeated by choosing lines in different directions.

The line-by-line solution procedure can also be supplemented by some form of the additive-correction method described
by Settari and Aziz (1973).

Calculation of fluid flow

The general differential equation (1) has been cast into the
discretization form with the understanding that the flow field
ρu_i as well as Γ and S are known. The density ρ is algebraically related to the other variables such as concentration and
temperature. The velocity components u_i are governed by the

momentum equations, which are special cases of the general
differential equation (1). Thus, the fields of u_i are to be

obtained simply by interpreting the general dependent variable
ϕ as the velocity u_i. In this sense, a calculation procedure

for the flow field has already been described. That the discretization coefficients depend on the flow field is simply a
case of nonlinearity, which can be handled by iteration. Is
there any reason for giving special attention to the flow
field calculation?

<u>Main difficulty</u> A closer look at the governing equations re-
veals that it is not strictly true that the velocity components u_i can be obtained by solving the momentum equations.

240

These equations contain the pressure gradient $(-\partial p/\partial x_i)$ as an important source term, which is not expressible in terms of u_i or other ϕ's. What indirectly determines the pressure field is the continuity equation, which was stated as Equation (2) earlier. This indirect specification is not convenient as a computational procedure; a direct method of determining the pressure field must be found.

There are some other related difficulties in handling the momentum and continuity equations. These have been discussed in detail elsewhere (Patankar, 1980), and the discussion will not be included here. What is shown there is that, if the velocity components and the pressure are calculated for the same grid points, some physically unrealistic fields arise as solutions. A remedy for this ailment is the staggered grid, which is described next.

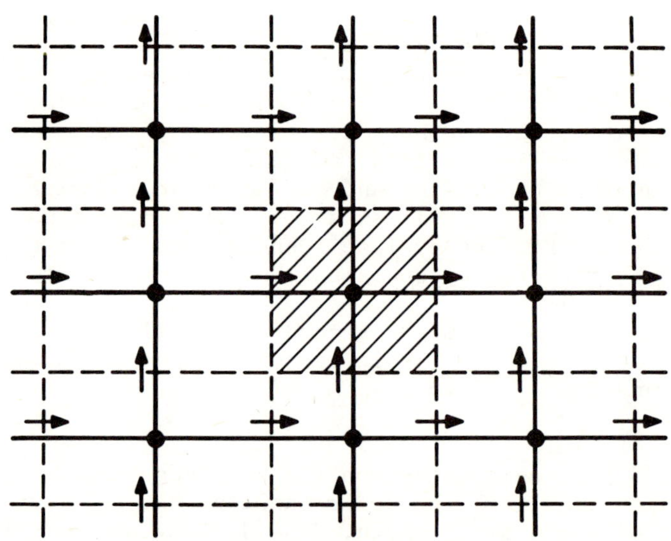

Figure 8   Staggered grid for velocity components

<u>Staggered grid</u>   Figure 8 shows a portion of a two-dimensional grid. The locations for which the velocity components are calculated are shown by short arrows. In the staggered grid, the velocity components are computed for the points lying on the control-volume faces. All other variables including pressure are calculated at the grid points shown by dots. A consequence of this arrangement is that the normal velocity components are directly available at the control-volume faces, where they are needed for the calculation of mass flow rates. Further, the pressure difference between two grid points can be used to "drive" the velocity component located between them.

<u>SIMPLER procedure</u> A procedure for the calculation of the
flow field was worked out by Patankar and Spalding (1972); it
was named SIMPLE (Semi-Implicit Method for Pressure-Linked
Equations). A more recent variant of the procedure called
SIMPLER (SIMPLE--Revised) has been described in Patankar
(1980). It is this improved procedure that will be described
here.

The staggered locations for the velocity components determine corresponding control volumes such that their faces
normal to the direction of the component now pass through the
grid points. Thus, with reference to Figure 7, two faces of
the control volume around velocity u_e will pass through the
grid points P and E. The corresponding momentum equation can
be written as

$$a_e u_e = \Sigma a_{nb} u_{nb} + b + A_e (p_P - p_E) \qquad (21)$$

where the term b includes the source terms other than the
pressure gradient. Equation (21) can be rewritten as

$$u_e = \hat{u}_e + D_e (p_P - p_E) \qquad (22)$$

where

$$\hat{u}_e = (\Sigma a_{nb} u_{nb} + b)/a_e \qquad (23)$$

and

$$D_e = A_e / a_e \qquad (24)$$

It should be noted that $\hat{u}_e$ does not contain the pressure term
but is defined in terms of the neighbor velocities u_{nb}.

It is possible to solve the momentum equations for a
<u>given</u> pressure field. Let u^* denote the velocity field based
on an estimated pressure field p^*. (In general, velocities
like u^* will not satisfy the continuity equation.) This
implies

$$a_e u_e^* = \Sigma a_{nb} u_{nb}^* + b + A_e (p_P^* - p_E^*) \qquad (25)$$

Further, a pressure correction p' and a velocity correction
u' can be defined as

$$p = p^* + p' \qquad (26)$$

242

$$u = u^* + u' \tag{27}$$

Equations (21) and (24)-(27) can be manipulated to yield a velocity-correction formula. This formula can be simplified if the term $\Sigma a_{nb} u'_{nb}$ is considered negligible. The result is

$$u_e = u_e^* + D_e (p'_P - p'_E) \tag{28}$$

The similarity between Equations (22) and (28) should be noted. If Equation (28) is used to calculate the pressure correction p' (so that the corrected velocity satisfies the continuity equation), the resulting values of p' are likely to be in error (because of the approximation made in arriving at Equation (28)), but the implied velocity corrections will still be reasonable. This feature is used in constructing the SIMPLER procedure.

The continuity equation for the control volume shown in Figure 7 can be written as

$$(\rho u A)_w - (\rho u A)_e + (\rho v A)_s - (\rho v A)_n = o \tag{29}$$

where u and v denote the velocity components in the x and y directions respectively. If u_e, u_w, v_n, and v_s are expressed via equations like (22), a discretization equation for pressure can be obtained and cast into the form

$$a_P p_P = a_E p_E + a_W p_W + a_N p_N + a_S p_S + b \tag{30}$$

where

$$a_E = (\rho DA)_e \tag{31}$$

$$a_W = (\rho DA)_w \tag{32}$$

$$a_N = (\rho DA)_n \tag{33}$$

$$a_S = (\rho DA)_s \tag{34}$$

$$a_P = a_E + a_W + a_N + a_S \tag{35}$$

$$b = (\rho \hat{u} A)_w - (\rho \hat{u} A)_e + (\rho \hat{v} A)_s - (\rho \hat{v} A)_n \tag{36}$$

A similar equation can be derived for the pressure correction p' if, instead of Equation (22), Equation (28) is used for substitution for u_e. Thus

$$a_P' p_P' = a_E' p_E' + a_W' p_W' + a_N' p_N' + a_S' p_S' + b \qquad (37)$$

where the coefficients a_E, a_W, a_N, a_S, and a_P are given by Equations (31)-(35), and b is calculated from

$$b = (\rho u^* A)_w - (\rho u^* A)_e + (\rho v^* A)_s - (\rho v^* A)_n \qquad (38)$$

It can be seen that the expressions for b given by Eqns. (36) and (38) closely resemble the continuity equation (29). The term b can be regarded as the "mass source" that is implicit in the velocity fields $\hat{u}$, $\hat{v}$ or u^*, v^*.

As in the rest of this chapter, the equations have been written here for a two-dimensional situation; but the extension to three dimensions is very straight-forward.

By now, the required ingredients for the calculation of fluid flow have been assembled; the procedure can best be described through the actual steps in the computation process.

Computation sequence
(i)   Start with guesses for the velocity field and related quantities.
(ii)   Calculate the coefficients in the momentum equations and hence obtain the values of $\hat{u}$, $\hat{v}$, $\hat{w}$ from equations like (23).
(iii)   Hence compute b from Equation (36) and solve Equation (30) to obtain the pressure field.
(iv)   Regarding this pressure as the estimated field p^*, solve the momentum equations to yield u^*, v^*, w^*.
(v)   Hence compute b from Equation (38) and solve Equation (37) for p'.
(vi)   Use the p' values to correct the velocity field via equations such as (28). Do not correct the pressure itself.
(vii)   Solve for other ϕ's such as temperature, turbulence parameters, etc. that influence the velocity field.
(viii) With the new values of all variables as improved guesses, return to step (ii) and repeat until convergence.

SOME RECENT APPLICATIONS

The calculation method described in this chapter has been extensively used in a wide variety of applications. A list of most of these applications is given in Chapter 9 of Patankar (1980), where a number of investigations are described in some

244

detail. Here, three different computational analyses are
presented to illustrate the variety and complexity of problems
that are now within the reach of the method.

<u>Mixed convection in a shrouded fin array</u>
Acharya and Patankar (1980) have reported the numerical solutions for the situation shown in Figure 9. The axial flow in

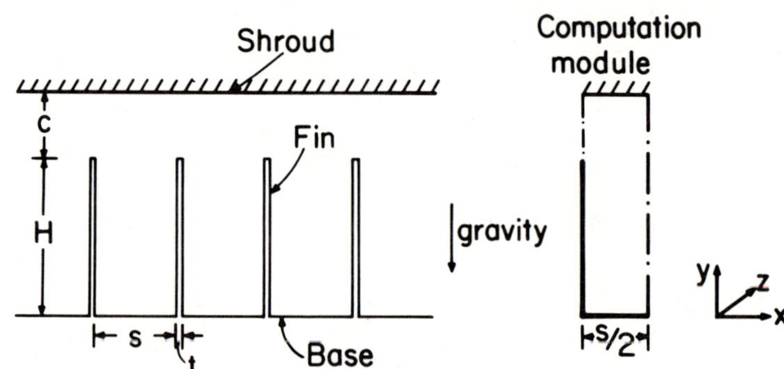

Figure 9   Shrouded fin array

the shrouded fin array is influenced by the presence of a
buoyancy-induced secondary flow. Two types of heating conditions are considered. In one, the fins and the base are at a
temperature higher than that of the fluid. In the other, the
fins and the base are colder than the fluid.

     Typical streamline patterns in the cross section for the
hot fin case are shown in Figure 10 for three values of the
clearance parameter. The quantity Ψ_m is the maximum value of
the dimensionless stream function, which is a measure of the
recirculating mass flow rate. The up-flow caused by buoyancy
near the hot fin can be clearly seen. The magnitude of the
circulating flow increases as the restricting shroud is moved
further away. For the cold fin case shown in Figure 11, the
flow direction has reversed. Also, a multiple-eddy structure
is beginning to form in the clearance region. For this kind
of heating condition, there is a tendency for the hot fluid
to accumulate near the top, thus reducing the strength of
buoyancy near the fin; the value of Ψ_m is, therefore, seen to
decrease with increasing clearance.

     The two heating conditions give markedly different heat
transfer coefficients. In Figure 12, the ratio of the actual
Nusselt number to the zero-buoyancy Nusselt number is plotted
as a function of the buoyancy parameter Gr^+. Whereas the hot
fin case exhibits a significant increase in the heat transfer
coefficient, the effect is non-existent for the cold fin cases
with clearance. In these cases, as mentioned earlier, the
presence of nearly stagnant hot fluid in the clearance space
reduces the temperature gradients near the fin. More complete
analysis and results are given in Acharya and Patankar (1980).

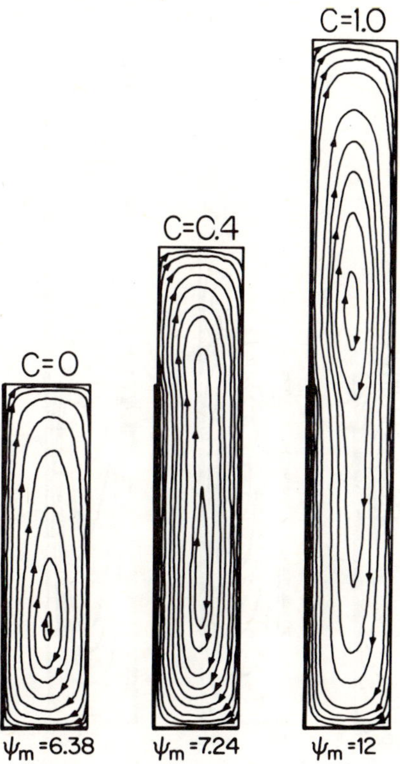

Figure 10 Streamline patterns for the hot fin case

Turbulent flow in a rotating duct

The flow in a centrifugal impeller passage can be idealized
as a flow in rotating duct. In such a situation, the Coriolis
forces acting in the cross-sectional direction produce significant secondary flows. Further, there is evidence that even
the turbulence structure is influenced by the Coriolis forces.
Howard, Patankar, and Bordynuik (1980) have presented a computational analysis for the flow in rotating ducts. They used
the k-ε turbulence model, in which differential equations are
solved for the kinetic energy of turbulence k and for its
dissipation rate ε. The standard k-ε model was modified to
include the effect of the Coriolis force. Many comparisons
with experimental data were made; a typical example is shown
in Figure 13. The figure shows the predicted profiles of the
mainstream velocity in the duct for various rotation speeds.
The predictions are compared with experimental data of Wagner
and Velkoff (1972). The agreement can be regarded as satisfactory; the skewness of the profile especially at higher rotation speeds is predicted quite well.

Thermal-hydraulic analysis of a once-through steam generator

A steam generator of a nuclear power plant has been shown in
Figure 4. In such a device, a calculation of the fluid flow

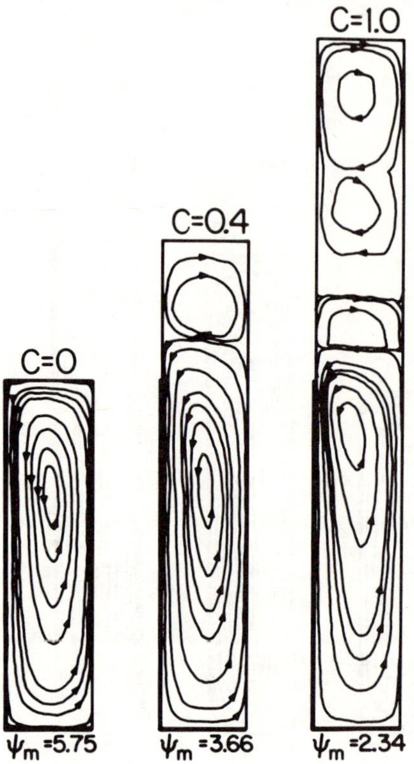

Figure 11 Streamline patterns for the cold fin case

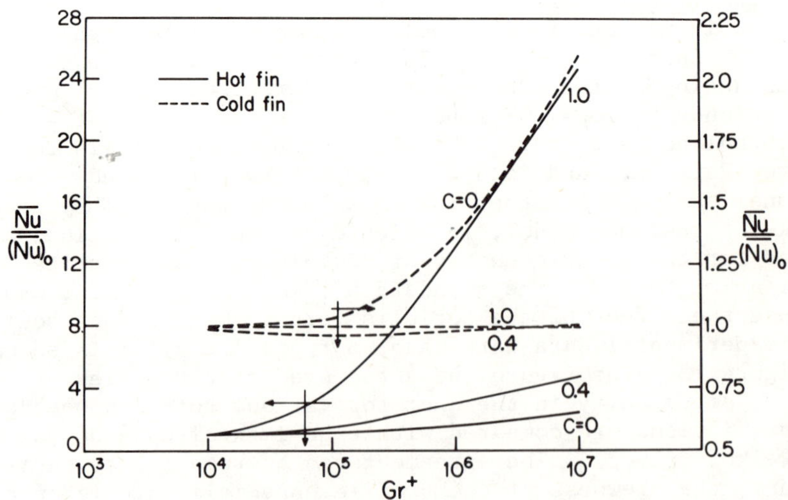

Figure 12 Effect of buoyancy on the Nusselt number

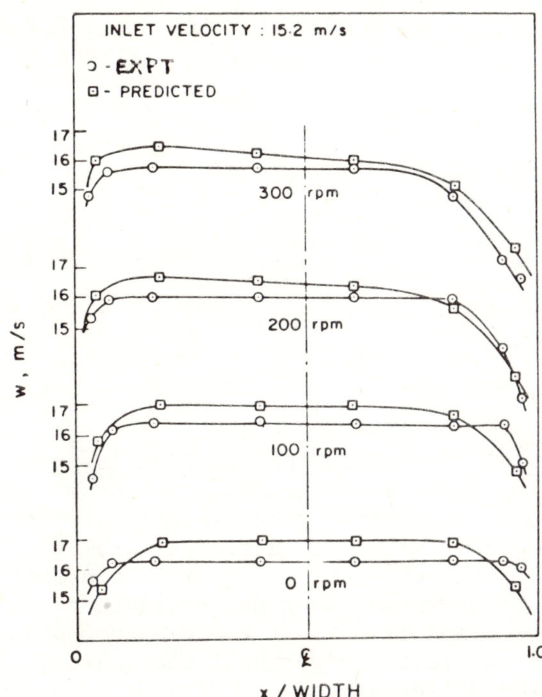

Figure 13   Mainstream velocity profiles

and heat transfer is conveniently possible if the presence of
the tubes is expressed as distributed resistance in the governing equations. Although only a few tubes are shown in
the schematic representation in Figure 6, the actual steam
generator contains hundreds of tubes. The shell-side flow
then is much like the flow through a porous material. The
flow resistance has, however, different formulas for flow
along the tubes and for flow normal to the tubes. The resistance formulas and the heat transfer coefficients used must
account for the fact that the shell-side flow is two-phase.

The thermal-hydraulic analysis of a steam generator involves the calculation of the three velocity components and
pressure on the shell side and of the the enthalpy distributions of the shell fluid and the tube fluid. The flow field
inside the tubes is assumed to be known. One such computational investigation has been reported by Oberjohn, Rice, and
Patankar (1980). They used a computer program called THEDA,
which incorporated the calculation scheme described in this
chapter. The empirical correlations used for the flow resist-
ances and heat transfer coefficients are given in the paper by
Oberjohn et al (1980). Some typical results will be summarized here.

For a given set of operating conditions, the predicted
performance of the steam generator is compared in Table 1
with the measured values. The agreement can be seen to be
very satisfactory.

I

Table 1   Overall performance of the steam generator

	Predicted	Measured
Steam Exit Temperature	310°C	312°C
Length to 100% Quality	8.5 m	9.1 m
Pressure Drop		
To Full Length	63.4 KPa	62.5 KPa
To Bleed Port	37.8 KPa	36.1 KPa

A more crucial test of the calculation is provided by the comparison of some local measurements. The steam generator includes an untubed inspection lane, in which the flow and temperature conditions are quite different from the rest of the tube bundle. Figure 14 shows the predicted shell-fluid temperature and the corresponding measurements along a horizontal diameter. The left half of the diameter passes through the untubed lane, while the right half goes through the tube bundle. For three different load conditions, the agreement

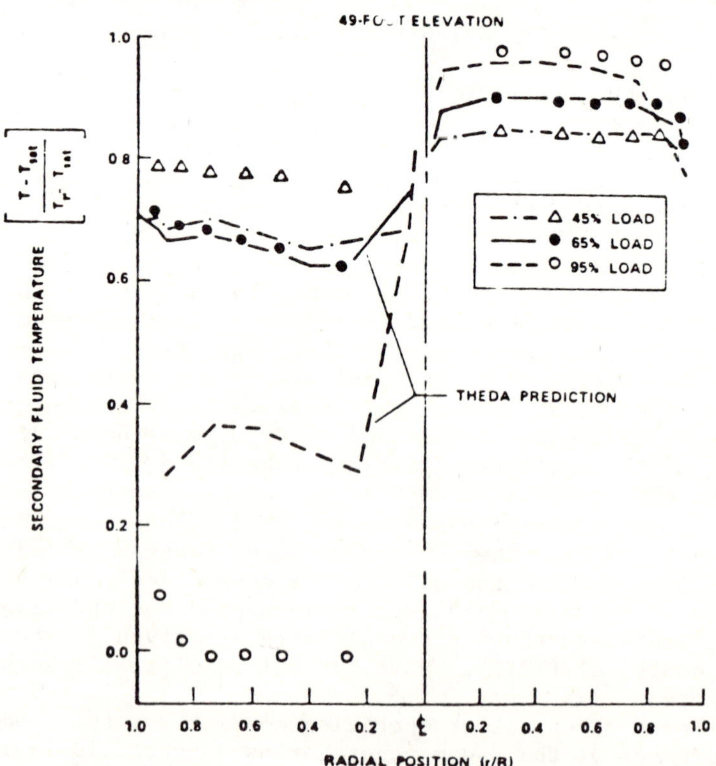

Figure 14   Radial temperature distribution

of the predictions with measurements is quite good in the
right half. The lack of close agreement on the left side can
at least partly be explained by measurement errors. (The
thermocouples tend to read saturation temperatures in super-
heated stream.)

The axial variation of the shell-fluid temperature is
shown in Figure 15. Here the computations agree well with
experimental data.

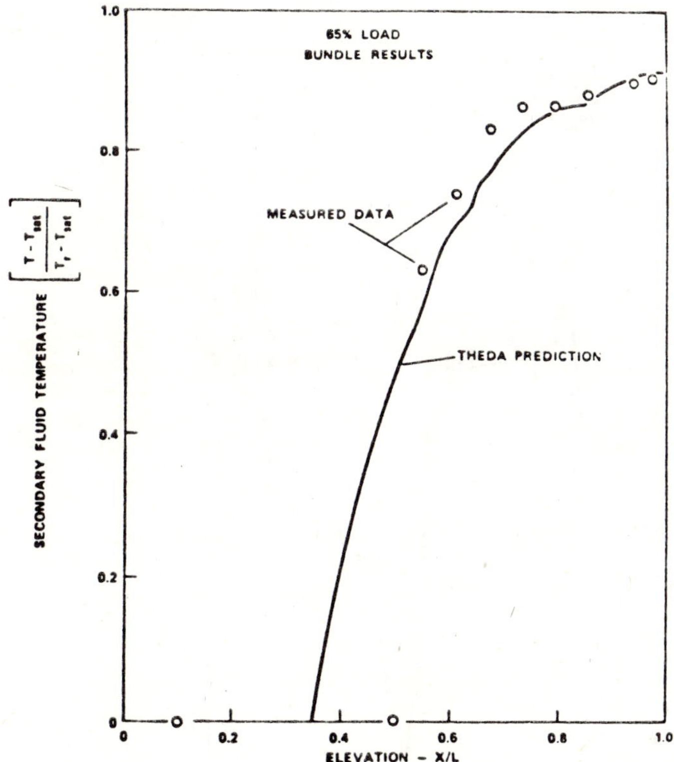

Figure 15   Axial temperature distribution

Further validation of the prediction procedure is made
with reference to the data from an isothermal air model, which
simulated the untubed lane in a tube bundle. The predictions
and data shown in Figure 16 represent the axial variation of
the axial velocity through the untubed lane. The sharp variations in the velocity are caused by the tube support plates,
whose locations are marked near the bottom of the figure.
The dashed curve and the data shown by circular dots refer to
the case in which a flow blockage was installed at the middle
support plate. The computed results in Figure 16 are in very
good agreement with the measurements.

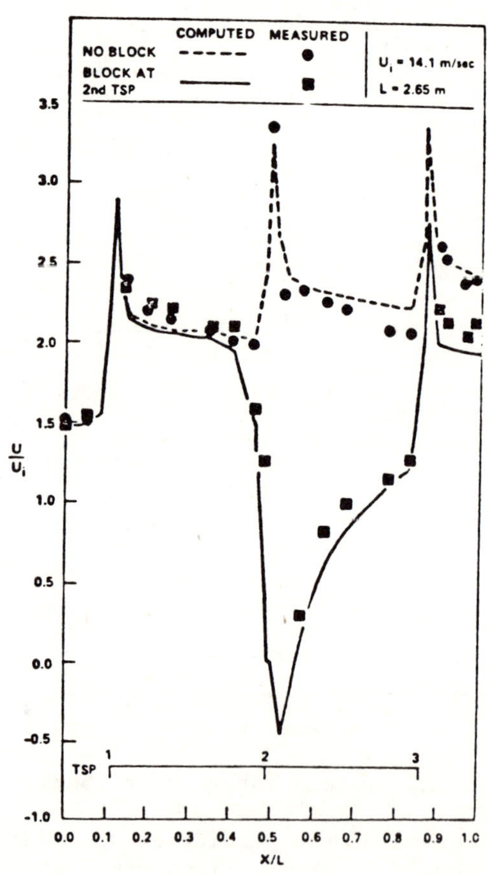

Figure 16   Axial velocity distribution
in the untubed lane

CONCLUDING REMARKS

In this chapter, a general calculation method for fluid flow, heat and mass transfer, and related processes has been described. The illustrative applications give a direct experience of the kind of solutions that can be obtained. Moreover, they enable one to envisage the many future applications that are within the reach of the method. The actual possibilities are largely limited by only the imagination of the user.

REFERENCES

Acharya, S. and Patankar, S. V. (1980) Laminar mixed convection in a shrouded fin array. Submitted to J. Heat Transfer

Finlayson, B. A. (1972) _The Method of Weighted Residuals and Variational Principles_, Academic, New York.

Howard, J.H.G., Patankar, S. V., and Bordynuik, R. M. (1980) Flow prediction in rotating ducts using Coriolis-modified turbulence models. ASME Paper 80-WA/FE-1. To be published in J. Fluids Eng., Dec. 1980.

Oberjohn, W. J., Rice, J. G., and Patankar, S. V. (1980) THEDA: A multi-dimensional steam generator analysis. Presented at the 19th National Heat Transfer Conference, Orlando, Florida, 1980.

Patankar, S. V. (1978) A numerical method for conduction in composite materials, flow in irregular geometries and conjugate heat transfer. Proc. 6th Int. Heat Transfer Conf., Toronto, $\underline{3}$, 297.

Patankar, S. V. (1980) _Numerical Heat Transfer and Fluid Flow_, McGraw-Hill--Hemisphere, New York.

Patankar, S. V. and Spalding, D. B. (1970) _Heat and Mass Transfer in Boundary Layers_, Second Edition, Intertext Books, London.

Patankar, S. V. and Spalding, D. B. (1972) A calculation procedure for heat, mass and momentum transfer in three-dimensional parabolic flows. Int. J. Heat Mass Transfer, $\underline{15}$, 1787.

Patankar, S. V. and Spalding, D. B. (1976) A calculation procedure for the transient and steady-state behavior of shell-and-tube heat exchangers. Chapter 7 in _Heat Exchangers: Design and Theory Sourcebook_, Hemisphere, New York.

Patankar, S. V. and Spalding, D. B. (1978) Computer analysis of the three-dimensional flow and heat transfer in a steam generator. Forsch. Ingenieurwes., $\underline{44}$, 47.

252

Raithby, G. D. and Torrance, K. E. (1974) Upstream-weighted differencing schemes and their application to elliptic problems involving fluid flow. Computers and Fluids, 2, 191.

Ramadhyani, S. and Patankar, S. V. (1980) Solution of the Poisson equation: comparison of the Galerkin and control-volume methods, Int. J. Num. Methods Eng., 15, 1395.

Settari, A. and Aziz, K. (1973) A generalization of the additive correction methods for the iterative solution of matrix equations, SIAM J. Numer. Anal., 10, 506.

Spalding, D. B. (1972) A novel finite-difference formulation for differential expressions involving both first and second derivatives, Int. J. Num. Methods Eng., 4, 551.

Spalding, D. B. (1977) GENMIX-A General Computer Program for Two-Dimensional Parabolic Phenomena. Pergamon, New York.

Wagner, R. E. and Velkoff, H. R. (1972) Measurements of secondary flows in a rotating duct, J. Eng. Power, 94, 261.

CHAPTER 9

APPROACHES TO THE FINITE ELEMENT SOLUTION OF TWO-DIMENSIONAL TURBULENT FLOWS

B.E. Larock, Civil Engineering Department, University of California, Davis, CA 95616, U.S.A.

D.R. Schamber, Civil Engineering Department, University of Utah, Salt Lake City, UT 84112, U.S.A.

INTRODUCTION

The solution of many engineering problems requires a knowledge of turbulent fluid flow within a domain of prescribed shape. This need has spurred the development of numerous mathematical models of turbulent flow in the past decade or so; reviews of these modeling efforts (e.g. Launder and Spalding, 1974; Reynolds, 1976; Rodi, 1980) describe the several major approaches and differing levels of complexity of these models. In all but the simplest instances investigators have had to use numerical approximation techniques, usually finite difference methods, to obtain solutions. While attention was primarily concentrated on simple jet and wake flows, solution difficulties due to geometric domain complexity were largely absent, and effective finite difference solution techniques, such as the Patankar-Spalding (1970) method, were developed. Since most practical engineering applications involve complex geometric flow domains, however, interest in solving the turbulent flow models in these regions has always been substantial.

Currently the analyst must choose between two avenues of approach when seeking to solve a turbulent flow problem in a nonregular domain. One may either employ finite difference methods, which have relatively proven solution techniques but treat irregular domains only approximately or with additional effort and difficulty (Oliver (1980) discusses this problem), or employ finite element techniques, which are completely flexible in treating nonuniform flow domains but for which there is little experience to date in solving the turbulent flow equations. As Oliver notes, most investigators choose to sacrifice some geometric flexibility to implement familiar finite difference methods. The present writers have chosen to explore the finite element alternative and began (Schamber and Larock, 1980a) by developing and testing some finite element solution procedures for several one-dimensional turbulent channel flows. The results seemed sufficiently promising to warrant an examination of these methods for two-dimensional flow, which will be described in this article.

The next section will present one finite element formulation for two-dimensional turbulent flow which incorporates the relatively thoroughly tested kinetic energy-dissipation, or k-ε, turbulence model. Subsequent sections will describe one viable solution scheme and its application to determination of the flow pattern in a circular sedimentation basin. Some critical comments on the present solution scheme will then precede an assessment of alternatives, tried and untried, to the current scheme.

A TWO-DIMENSIONAL FINITE ELEMENT MODEL FOR TURBULENT FLOW

In the present approach the flow is described by the conventional time-averaged mean-flow equations expressing conservation of mass and momentum, supplemented by the k-ε turbulence closure model. The Galerkin weighted residual procedure then produces a discretized equation set for subsequent solution. A fundamentally different alternative to this approach is the Large Eddy Simulation (LES) method which spatially filters the governing conservation equations and then solves the resulting inherently transient equation set by finite elements; this approach is described by Findikakis and Street (1979) and will be commented on in the section on alternatives at the end of the chapter.

Mean Flow
The familiar Reynolds approach decomposes the instantaneous value of each variable, e.g. velocity $\tilde{u}_i$, into a time-averaged mean value U_i and an instantaneous deviation or fluctuation u_i from the mean. Similarly, for the pressure $\tilde{p} = P + p$. Applying this device to equations expressing conservation of mass and linear momentum yields the governing equations for the time-invariant turbulent mean flow of an incompressible fluid, which are

$$\frac{\partial U_i}{\partial x_i} = 0 \tag{1}$$

$$U_j \frac{\partial U_i}{\partial x_j} = -\frac{1}{\rho}\frac{\partial P}{\partial x_i} + g_i - \frac{\partial}{\partial x_j}(\overline{u_i u_j}) \tag{2}$$

The time average of the product of fluctuating velocities is $\overline{u_i u_j}$. For two-dimensional problems the subscripts range over the values 1,2. The Einstein convention of summing over a repeated subscript is used. Also x_i = cartesian coordinate component, g_i = gravitational acceleration component, and ρ = constant fluid density. The laminar stress terms which normally appear in Equation 2 have been deleted since these entities are insignificant in comparison with the turbulent stress terms $\overline{u_i u_j}$ in fully turbulent flows. Equation 2 is historically called the Reynolds equation, and the terms $\overline{u_i u_j}$ are called Reynolds stresses when written as $-\rho\overline{u_i u_j}$, which then have the dimensions of stress.

The k-ε Turbulence Model

In this model the turbulent stress terms $-\overline{u_i u_j}$ are assumed to be related to the mean-flow rate of strain as follows:

$$-\overline{u_i u_j} = \nu_t \left(\frac{\partial U_i}{\partial x_j} + \frac{\partial U_j}{\partial x_i} \right) - \frac{2}{3} \delta_{ij} k \tag{3}$$

The proportionality factor between stress and rate of deformation is called the kinematic eddy viscosity ν_t; it may be further expressed as

$$\nu_t = c_\mu \frac{k^2}{\varepsilon} \tag{4}$$

The form of Equation 4 is supported by dimensional analysis and Kolmogorov's (1942) assertion that turbulence could be characterized by two scalar variables. Here δ_{ij} = Kronecker delta, c_μ = 0.09 is a constant, $k = \overline{u_i u_i}/2$ is the turbulence kinetic energy, and ε = dissipation rate of the turbulence energy.

Both k and ε are determined from individual model transport equations, which for steady flow are

$$U_j \frac{\partial k}{\partial x_j} = P_r - \varepsilon + \frac{\partial}{\partial x_j} \left[\frac{c_\mu}{\sigma_k} \frac{k^2}{\varepsilon} \frac{\partial k}{\partial x_j} \right] \tag{5}$$

and

$$U_j \frac{\partial \varepsilon}{\partial x_j} = c_{\varepsilon 1} \frac{P_r \varepsilon}{k} - c_{\varepsilon 2} \frac{\varepsilon^2}{k} + \frac{\partial}{\partial x_j} \left[\frac{c_\mu}{\sigma_\varepsilon} \frac{k^2}{\varepsilon} \frac{\partial \varepsilon}{\partial x_j} \right] \tag{6}$$

In these equations the scalar production P_r is

$$P_r = -\overline{u_i u_j} \frac{\partial U_i}{\partial x_j} \tag{7}$$

which operates primarily at the large scales of motion and supplies mean-flow energy to the turbulence. At the small-scale end of the turbulence spectrum where processes are presumably isotropic, the dissipation ε, which formally is

$$\varepsilon = \nu \; \overline{\frac{\partial u_i}{\partial x_j}} \; \overline{\frac{\partial u_i}{\partial x_j}} \qquad (8)$$

extracts energy from the turbulence and converts this energy to heat. In Equations 5 and 6 the left-side terms represent convective transport of k and ε respectively; the right-most term in each equation models the diffusive transport processes; the two terms immediately following the equals sign are source and sink terms which supply and remove energy. The four additional constants which appear in Equations 5 and 6 have the following values: $c_{\varepsilon_1} = 1.45$, $c_{\varepsilon_2} = 1.90$, $\sigma_k = 1.0$ and $\sigma_\varepsilon = 1.3$. These constants are inferred from experimental data and computer optimization (Schamber and Larock, 1980b).

A relatively complete discussion of the derivation of, and modeling assumptions implicit in, Equations 3-8 can readily be found in the literature (see e.g. Launder and Spalding (1974) and Schamber (1979)). At this point it must be noted that this turbulence model is valid only in fully turbulent flow, that is, flow sufficiently far from solid boundaries. Low Reynolds number turbulence models do exist, however, which allow the computational domain to coincide with wall boundaries (cf. Chieng and Launder, 1980).

Galerkin Finite Element Formulation
This finite element methodology is in principle applicable to any well-posed equation set, self-adjoint or not. An approximate functional representation for each dependent variable is chosen. These approximations are then substituted into the governing equations in place of the original variables. This replacement produces a set of equation errors or residuals. These residuals are then each multiplied by a member of a linearly independent set of weighting functions, integrated over the solution domain, and set to zero. The Galerkin approach is distinguished from other weighting methods by employing the same approximation functions in both the functional representation and the weighting functions. At least for symmetric, positive definite operators characterizing elliptic equation sets, this approach is known to produce the best solution from the chosen family of approximate solutions (Oden and Reddy, 1976). Using this approach, one can select a nonuniform graded computational mesh, seek solutions in irregular flow domains and avoid difficulties in applying boundary conditions (cf. Heubner, 1975, for additional background).

The finite element solution is obtained in a global domain Ω having boundary Γ; the domain is subdivided into finite elements e covering discrete subdomains Ω_e enclosed by boundaries Γ_e. Within an element the pressure is approximated by

$$P \simeq \hat{P} = P_i M_i \qquad (9)$$

in which M_i is the bilinear interpolation function. All other variables, for example U, are approximated as

$$U \simeq \hat{U} = U_i N_i \tag{10}$$

wherein N_i is the 8-node biquadratic interpolation function. The above choices for the approximation of the flow variables (i.e. pressure approximated one order lower than velocity) have been known to be the correct practical choices for a number of years. Recently Sani et al. (1981) have presented theoretical and numerical results which confirm this notion for the mean-flow variables.

Application of the Galerkin method with the appropriate weighting functions to Equations 1, 2, 5 and 6 at the element level produces the equation residuals

$$f_P = \int_{\Omega_e} M_i \left[\frac{\partial}{\partial r} (r^m U) + \frac{\partial}{\partial z} (r^m W) \right] d\Omega \tag{11}$$

$$f_U = \int_{\Omega_e} N_i \left[r^m (U \frac{\partial U}{\partial r} + W \frac{\partial U}{\partial z}) - m(P + \overline{vv}) \right] d\Omega$$

$$- \int_{\Omega_e} r^m \left[(P + \overline{uu}) \frac{\partial N_i}{\partial r} + \overline{uw} \frac{\partial N_i}{\partial z} \right] d\Omega$$

$$+ \int_{\Gamma_e} r^m N_i \left[(P + \overline{uu}) \ell_r + \overline{uw} \ell_z \right] d\Gamma \tag{12}$$

$$f_W = \int_{\Omega_e} r^m N_i \left[U \frac{\partial W}{\partial r} + W \frac{\partial W}{\partial z} \right] d\Omega$$

$$- \int_{\Omega_e} r^m \left[\overline{uw} \frac{\partial N_i}{\partial r} + (P + \overline{ww}) \frac{\partial N_i}{\partial z} \right] d\Omega$$

$$+ \int_{\Gamma_e} r^m N_i \left[\overline{uw} \ell_r + (P + \overline{ww}) \ell_z \right] d\Gamma \tag{13}$$

$$f_k = \int_{\Omega_e} r^m N_i \left[U \frac{\partial k}{\partial r} + W \frac{\partial k}{\partial z} - P_r + \varepsilon \right] d\Omega$$

$$+ \int_{\Omega_e} r^m \frac{\nu_t}{\sigma_k} \left[\frac{\partial k}{\partial r} \frac{\partial N_i}{\partial r} + \frac{\partial k}{\partial z} \frac{\partial N_i}{\partial z} \right] d\Omega$$

$$- \int_{\Gamma_e} r^m N_i \frac{\nu_t}{\sigma_k} \left[\frac{\partial k}{\partial r} \ell_r + \frac{\partial k}{\partial z} \ell_z \right] d\Gamma \tag{14}$$

$$f_\varepsilon = \int_{\Omega_e} r^m N_i \left[U \frac{\partial \varepsilon}{\partial r} + W \frac{\partial \varepsilon}{\partial z} - c_{\varepsilon 1} \frac{\varepsilon}{k} P_r + c_{\varepsilon 2} \frac{\varepsilon^2}{k} \right] d\Omega$$

$$+ \int_{\Omega_e} r^m \frac{\nu_t}{\sigma_\varepsilon} \left[\frac{\partial \varepsilon}{\partial r} \frac{\partial N_i}{\partial r} + \frac{\partial \varepsilon}{\partial z} \frac{\partial N_i}{\partial z} \right] d\Omega$$

$$- \int_{\Gamma_e} r^m N_i \frac{\nu_t}{\sigma_\varepsilon} \left[\frac{\partial \varepsilon}{\partial r} \ell_r + \frac{\partial \varepsilon}{\partial z} \ell_z \right] d\Gamma \tag{15}$$

From Equation 11 onward all variables are nondimensionalized on a characteristic velocity U_o, length h_o and density ρ. For $m = 1$ these equations describe axisymmetric flow in the r-z plane, and for $m = 0$ they describe planar flow in the r-z cartesian plane. The terms ℓ_r and ℓ_z are direction cosines to the outward unit normal vector to the element boundary Γ_e. The stress and production terms are defined with reference to Equations 3 and 7:

$$\overline{uu} = \frac{2}{3} k - 2\nu_t \frac{\partial U}{\partial r} \tag{16}$$

$$\overline{vv} = \frac{2}{3} k - 2m\nu_t \frac{U}{r} \tag{17}$$

$$\overline{ww} = \frac{2}{3} k - 2\nu_t \frac{\partial W}{\partial z} \tag{18}$$

$$\overline{uw} = -\nu_t \left[\frac{\partial U}{\partial z} + \frac{\partial W}{\partial r} \right] \tag{19}$$

and

$$P_r = \nu_t \left\{ 2 \left[\frac{\partial U}{\partial r}\right]^2 + \left[\frac{\partial U}{\partial z} + \frac{\partial W}{\partial r}\right]^2 + 2 \left[\frac{\partial W}{\partial z}\right]^2 + 2n \left[\frac{U}{r}\right]^2 \right\} \qquad (20)$$

For computational convenience the pressure has been redefined in these equations. Letting P' be the true nondimensional pressure, one can then define the nondimensional deviation P from hydrostatic pressure via the relation

$$P' = P + (h-z)/F_o^{\,2} \qquad (21)$$

where $F_o^{\,2} = U_o^{\,2}/(gh_o)$ is the Froude number and h = elevation of a free surface. Since the hydrostatic pressure term balances the body force term in Equation 13, neither appears. The second derivative terms in the transport equations were integrated once by parts, allowing C^o-continuous elements to be used. Finally, the pressure term was also integrated by parts to facilitate the application of normal-stress boundary conditions. The line integral contributions are only evaluated when the element boundary coincides with the global boundary Γ on which a derivative boundary condition is specified.

Once the element contributions are formed, they must be appropriately assembled to give a complete set of system equations for the flow problem. At some stage in the numerical algorithm one must apply the boundary conditions which make a problem solvable and unique. The final equation set is then solved. These steps can interact. Before the current solution scheme is described, pertinent boundary conditions are discussed.

<u>Boundary Conditions</u>
Each discrete variable in the system of global equations can be associated with its own equation. Logically one pairs the nodal values of U_i and W_i with the corresponding Galerkin representations of the component momentum equations, Equations 12 and 13. Likewise, nodal values k_i and ϵ_i are associated with Equations 14 and 15, respectively. Nodal values of pressure P_i must therefore be associated with Equation 11, the incompressibility constraint. Hence conservation of mass is enforced only at corner nodes, whereas the momentum and turbulence model equations are enforced at all nodes defining an element. Since the application of boundary conditions reduces the global number of unknown nodal variables, a corresponding deletion of equations from the global equation set is required.

Because the present turbulence model is only valid in a fully turbulent regime, the finite element mesh should not extend all the way to the boundary walls. At the edge of the mesh the normal component of the mean velocity is zero, and the tangential component is allowed to slip in accordance with the law of the wall

$$V/u_\tau = \kappa^{-1}\ln(Ey^+) \tag{22}$$

where V = velocity parallel to the wall, κ = von Karman constant (≈ 0.4), u_τ = friction velocity, E = constant $\simeq 0.9$ for smooth wall, and $y^+ = yu_\tau R_o$. Here y = normal distance from wall to mesh and $R_o = U_o h_o/\nu$. The distance y between mesh and wall should place the mesh boundary in the turbulent inertial subrange and also where the local shear stress is nearly constant; this occurs approximately when $30 \leq y^+ \leq 400$ is satisfied (Gosman and Ideriah, 1976). Turbulence production and dissipation almost balance near a wall; this condition leads to specified boundary values

$$k = u_\tau^2/c_\mu^{1/2} \quad , \qquad \varepsilon = \frac{c_\mu^{3/4} k^{3/2}}{\kappa \, \delta} \tag{23}$$

for k and ε, in which δ = distance between mesh and wall. Once V is found from mean-flow calculations, u_τ can be computed iteratively from Equation 22, and remaining boundary values can be specified explicitly.

A free surface is a constant-pressure boundary which no mass flux crosses. The free surface location is usually unknown a priori and, strictly speaking, must be determined as part of the overall solution. However, application of boundary conditions at an unknown location is usually difficult and consequently is to be avoided whenever possible. If the slope of the free surface is locally small, then it suffices to replace the free surface with a solid boundary (a "rigid lid") at a fixed horizontal position. The pressure at this boundary is not specified but computed as part of the solution. This pressure variation along the free surface then approximates the driving force in the momentum equation which is originally caused by the sloping actual free surface. The velocity normal to the boundary, the turbulent shear stress along the boundary and normal derivatives of k and ε are all set to zero.

Profiles of U, W, k and ε are specified along portions of the boundary Γ where flow enters the computational domain. Recourse to experimental data sets is advised in selecting appropriate profiles. In the example to follow the domain outflow was modeled as a point sink whose strength matched the inflow discharge. Local boundary values of U and W were directly chosen from the point sink model. Although boundary values of k and ε are undoubtedly influenced by prior flow events, no appropriate data are available. In this case the outer normal derivatives of k and ε were set to zero so that exit values would be computed directly by the computational algorithm; only if the flow were fully developed in the outflow region would this procedure be strictly correct, however.

From the pairing of unknowns and equations one can see that the direct specification of the pressure at boundary nodes will locally violate the incompressibility constraint; this specification should therefore be avoided. However, in the case of a contained flow one must nevertheless specify a pressure datum. This is achieved by setting P = constant at one interior node and deleting the corresponding nodal continuity equation. Additional examples of velocity boundary conditions and their relation to the pressure field can be found in Sani et al. (1981).

SOLVING THE MODEL EQUATIONS

A brief inspection of Equations 11-15 reveals the highly nonlinear and thoroughly coupled nature of this equation set. Furthermore, a typical finite element discretization with about 500 node points will have approximately 2000 unknowns and equations. The number of nonzero coefficients in the system coefficient matrix would be approximately 250,000. The element coefficient matrix, with five variables per node on an eight-node element, would contain 1600 entries. The equation nonlinearities and the near-impossibility of achieving a suitably accurate simultaneous initialization for all variables, in combination with the large size of the equation set, caused the writers to employ a partitioned Newton scheme for the solution of these equations.

The basic Newton scheme solves an equation set by writing

$$\mathbf{J}^{\ell} \, \delta\mathbf{x}^{\ell} \; = \; - \, \mathbf{f}^{\ell} \tag{24}$$

in which ℓ = iteration counter, $\mathbf{f}$ = vector of equation residuals, $\mathbf{J}$ = Jacobian of equation matrix, and $\delta\mathbf{x}$ = set of incremental corrections to be made to the vector of unknowns $\mathbf{x}$. Upon solution of Equation 24, the new estimate of the solution is

$$\mathbf{x}^{\ell+1} \; = \; \mathbf{x}^{\ell} + \alpha^{\ell} \, \delta\mathbf{x}^{\ell} \tag{25}$$

where $\alpha^{\ell} \leq 1$ is a relaxation parameter. Equation 24 is assembled from element contributions and solved directly by a Gaussian elimination procedure.

Presently flow problems are solved by partitioning the global equation system into a mean flow (UWP) equation set and a turbulence model (k-ε) equation set and solving the two sets alternately rather than simultaneously. This method is easier to initialize and reduces core storage requirements substantially, but the method requires more computational cycles to obtain a converged solution. The remainder of this section will describe the algorithm further.

Mean Flow Equations

To begin, this approach requires that initial distributions of U, W, P and ν_t be specified. The assumptions of uniform parallel flow and a constant value for ν_t usually provide an acceptable starting point for the numerical solution procedure. The mean flow equations are then solved by Newton's method. Generally the mean flow field solution converges in 5 to 7 iterations. Several underrelaxation schedules (i.e. sequences of α^ℓ) have been successful in obtaining a converged solution. One such schedule is

$$
\begin{aligned}
\alpha^\ell &= \ell/4 \quad , \quad \ell < 4 \\
\alpha^\ell &= 1 \quad , \quad \ell \geq 4
\end{aligned}
\tag{26}
$$

Within each iteration the coefficient matrix J_e and right-hand side f_e were formed for each element with the elements of $\delta \mathbf{x}_e$ being

$$
\delta \mathbf{x}_{ej} = \left[\delta U_j, \ \delta W_j, \ \delta P_j \right]^T
\tag{27}
$$

The contributions for each element were then assembled into the global mean flow equation matrix, Equation 24. The element Jacobian has the form

$$
J_e =
\begin{bmatrix}
\dfrac{\partial f_U}{\partial U_j} & \dfrac{\partial f_U}{\partial W_j} & \dfrac{\partial f_U}{\partial P_j} \\[2ex]
\dfrac{\partial f_W}{\partial U_j} & \dfrac{\partial f_W}{\partial W_j} & \dfrac{\partial f_W}{\partial P_j} \\[2ex]
\dfrac{\partial f_P}{\partial U_j} & \dfrac{\partial f_P}{\partial W_j} & 0
\end{bmatrix}
\tag{28}
$$

Full expressions for the individual entries in this matrix are listed in Appendix I; the evaluation of each coefficient expression is effected by using 3x3 Gaussian integration. Expressions for f_e are given in Equations 11-13. The derivatives formed with respect to the nodal pressures are of course evaluated only at the corner nodes, while the remaining derivatives are evaluated at all nodes. The matrix J_e has 24x24 = 576 entries. This represents a reduction in storage at the element level of approximately 2.8 in comparison with an algorithm which considers all five variables simultaneously.

Solution of the global mean-flow equation set (Equation 24) provides values of velocity and pressure at each node for a specified ν_t field.

Turbulence Equations

Newton's method is also used to solve for the global distribution of k and ε. Initial values for the k and ε fields are needed to begin the process; constant values or simple linear variations are usually employed. Values for U and W are obtained from the previous mean flow solution. The nonlinearity of the source and sink terms and the diffusion terms in Equations 14 and 15 require a modified Newton approach during the first iterations. The derivatives of the diffusion terms are formed with ν_t assumed constant. The derivative of the production term in the turbulence kinetic energy equation is also formed holding ν_t constant. In the dissipation equation the derivatives of the sink term $c_{\varepsilon_2}\varepsilon^2/k$ are formed holding the ratio ε/k constant. With these adjustments the modified element coefficient matrix J_e for k and ε is

$$
J_e = \begin{bmatrix} \dfrac{\partial f_k}{\partial k_j} & \dfrac{\partial f_k}{\partial \varepsilon_j} \\[2em] \dfrac{\partial f_\varepsilon}{\partial k_j} & \dfrac{\partial f_\varepsilon}{\partial \varepsilon_j} \end{bmatrix}
\tag{29}
$$

(See Appendix I for complete coefficient expressions.) In this case the coefficient matrix, Equation 29, has 16x16 = 256 entries. Several iterations are performed with $\alpha^\ell < 1$ using the modified Newton scheme to update the k and ε fields. The modified scheme and underrelaxation of the solution are required to prevent k and ε from becoming negative. Computational experience has shown that negative values of k and ε quickly destroy the iteration process and cause the solution to diverge. Only a few modified iterations are required to create k and ε fields that are partially compatible with the current UWP values.

The Iteration Process

Iterations with the modified k-ε coefficient matrix are not continued until the k and ε solutions fully converge, because the current UWP solution is only approximate. Control is now passed back to the mean flow equation solver with updated k and ε fields and hence new values of ν_t from Equation 4. Several iterations are performed with $\alpha^\ell = 1$ to align the UWP values with the updated distribution of ν_t. The solution scheme continues by passing from one equation set to the other, iterating within each set as required, until there is no significant change in any of the variables. Once the solution is close to the converged solution ("close" being determined by computational experience), the full Newton method may be implemented for the k-ε scheme. The Jacobian matrix, Equation 29, is then computed without use of the modified coefficients; Appendix I also lists these coefficient expressions. Thus the alternating or "flip-flop" method cycles between the two sets of equations, iterating within each set, until a fully converged solution is obtained.

APPLICATION

Sedimentation Basin

Much of the original impetus for the writers to study the numerical simulation of turbulent flow came from the writers' interest in predicting flow patterns in sedimentation basins. These units are an integral part of most water and wastewater treatment systems, but the measurement of velocity patterns is notoriously difficult because of the low speed of the flow and the common presence of turbidity. Modeling at sufficiently reduced scale will result in the occurrence of laminar, rather than turbulent, flow. For these reasons numerical modeling is relatively attractive. Furthermore, simulation of flow in these basins adequately demonstrates several features of the current methodology.

The velocity field is computed for a circular basin of 25 ft. radius (7.6 m) and inlet and exit depths of 10 ft. (3.0 m) and 8 ft. (2.4 m) respectively. A relatively high loading rate of 1500 gpd/ft^2 (0.07 cm/sec) was selected (1) to simulate conditions in an overloaded primary clarifier which may represent the lower limit of clarifier performance, and (2) to ensure that the flow over most of the domain was indeed turbulent.

The finite element computational mesh employed in the simulation is shown in Figure 1.

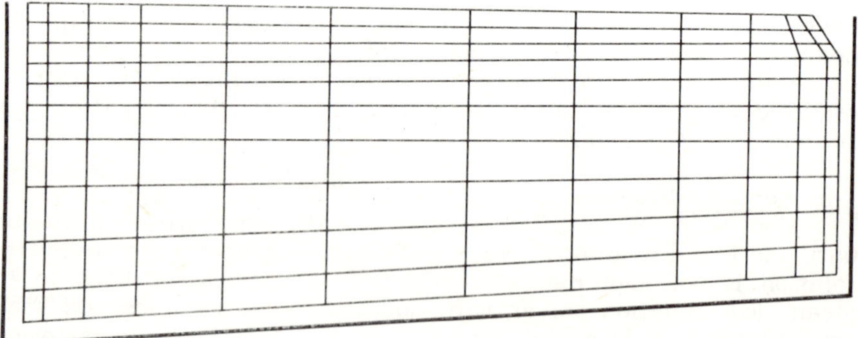

Figure 1. Finite element mesh for circular sedimentation basin.

The elements are graded to give a relatively dense mesh near the inlet and outlet where velocity gradients are substantial. Selection of this mesh was guided partly by the results of preliminary runs in which computational difficulties were encountered. This experience demonstrates one of the attributes of using the standard Galerkin finite element method; as noted by Gresho and Lee (1979), in many instances the appearance of aberrant behavior in the results of a working Galerkin code is a clear indication of the need for additional local mesh refinement.

For this example the jet inlet condition was applied over the entire left edge of the domain. The majority of the inflow discharge is concentrated, however, in the upper third of the inlet section. Experiments (Larsen, 1977) show that simple inlet structures produce jet-like velocity profiles; inlet velocity and turbulence boundary data were therefore modeled as half a plane jet. Outflow occurs at the top right and was modeled as a point sink. The free surface along the top of the domain was modeled as a rigid lid. The remainder of the domain periphery is composed of solid walls near which the law of the wall is applied. The computational mesh does not extend to the walls, keeping the solution domain in the fully turbulent flow region.

The two-dimensional simulation is expected to replicate full-scale basin flow under certain conditions. Montens (1954) has shown by direct measurement on full-scale circular basins that the flow is nearly axisymmetric in the radial-vertical plane in the absence of surface wind.

Computational Programme
The numbers of elements and nodes in this example are 120 and 405 respectively. This resulted in 835 UWP equations and 694 k-ε equations. The storage requirements for the system matrix for each set of equations are approximately 108,000 and 67,000. The computations required approximately 6 minutes of CPU time on a CDC 7600 computer.

Results
A vector plot of the computed velocity field is presented in Figure 2. The velocities decrease rather quickly with distance from the inlet owing to the rapid increase in cylindrical cross-sectional area. A reattachment point is located at $z/h_0 \simeq 0.74$ along the righthand vertical wall. The flow separates at this point and moves upward toward the exit and downward toward the tank floor; this bifurcation produces a significant return flow across the lower portion of the basin. Relatively rapid fluid accelerations occur as the fluid approaches the exit. It is encouraging to note that flow patterns measured by Clements and Khattab (1968) on scale-model circular basins with surface inlets are similar to the simulated pattern depicted in Figure 2; unfortunately the results in their paper are sufficiently qualitative in nature that duplication by numerical computation is not possible.

Figure 3 is a contour plot of the nondimensional kinematic eddy viscosity ν_t. Recall that $\nu_t = c_\mu k^2/\varepsilon = \nu_t^*/(U_o h_o)$, where ν_t^* is the dimensional kinematic eddy viscosity. Evidently one would encounter extreme difficulty in attempting to specify, even qualitatively, such a distribution of ν_t a priori. Similar nonuniformities are easily

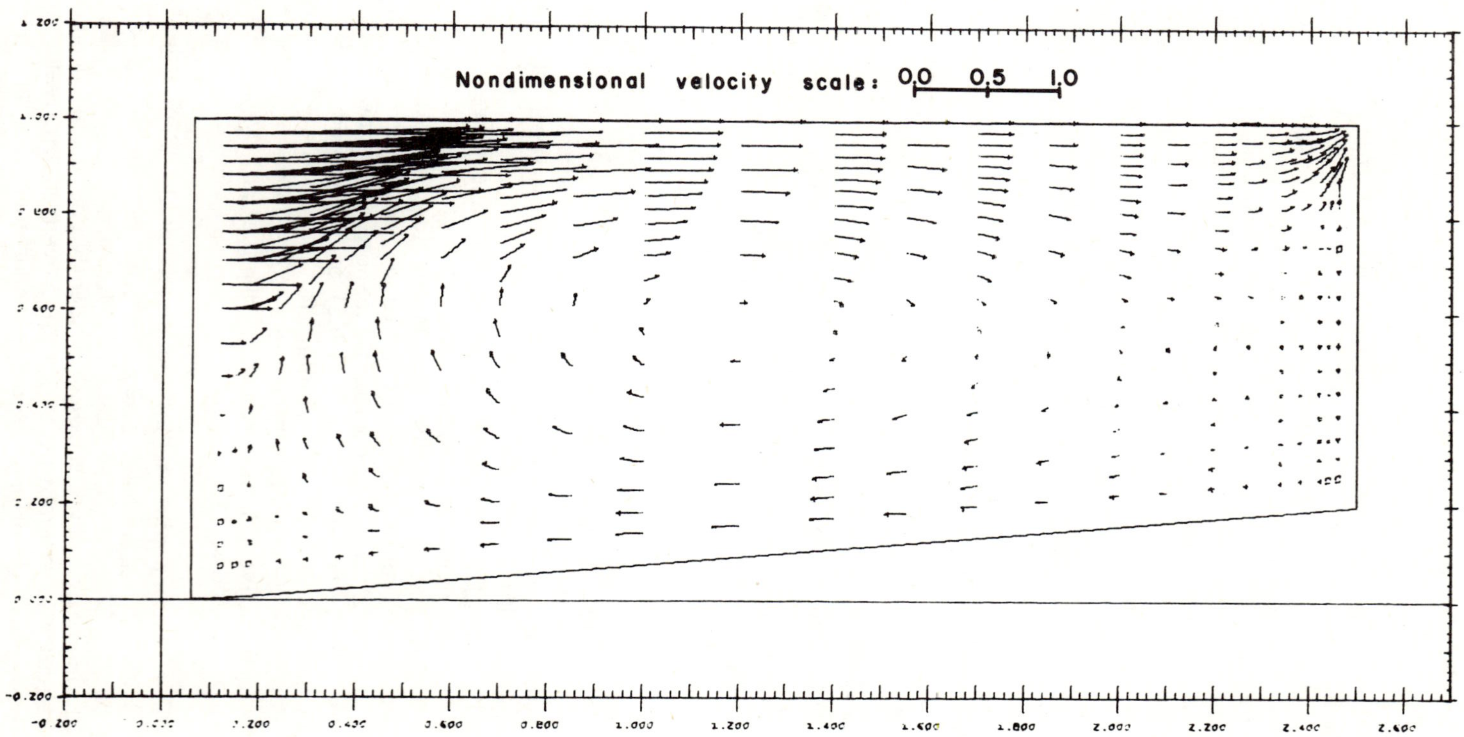

Figure 2. Velocity field in circular basin.

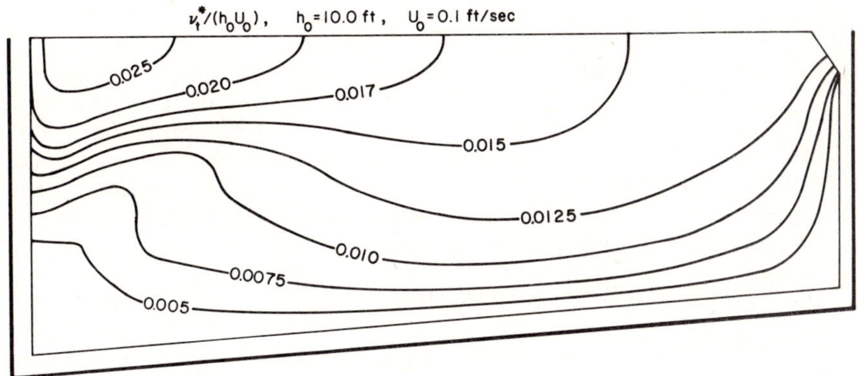

Figure 3. Contour plot of turbulent kinematic viscosity.

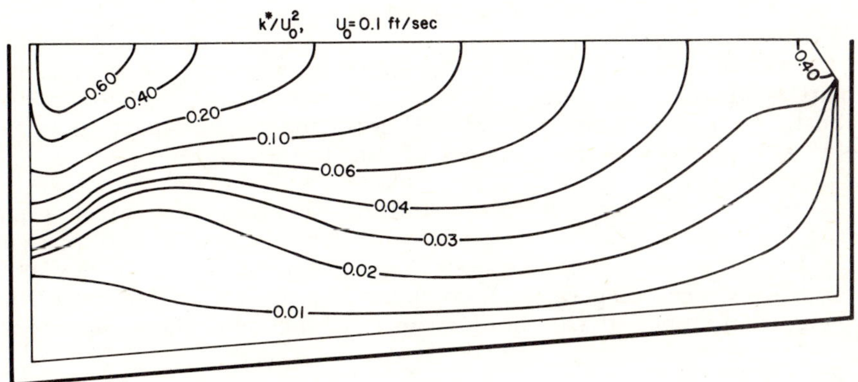

Figure 4. Contour plot of turbulent kinetic energy.

seen in Figure 4 which is a contour plot of the nondimensional kinetic energy k.

The cost of obtaining a solution of the mean flow equations for a constant eddy viscosity is roughly one tenth of the cost of obtaining a complete solution with variable k and ε. The obvious difficulty with constant viscosity simulations is choosing the appropriate value of ν_t (cf. Schamber and Larock, 1980b). The k-ε model, of course, determines this factor directly from the structure of the flow field and boundary conditions. Without the k-ε computations, only experience and inspection of experimental data, if they exist, can be used in selecting ν_t. It is doubtful that these sources could adequately and dependably replace the computed information that the turbulence model supplies, however.

ALTERNATIVE SOLUTION SCHEMES

The present solution scheme has been successfully used to solve a few complex, turbulent, recirculating flow problems, including the circular basin example, but the method nevertheless has some significant disadvantages. The performance of Newton's method, in combination with the flip-flop method, is sensitive to the closeness of the initialization vector to the ultimate solution vector; this behavior is undoubtedly exacerbated by the strong nonlinearity of the turbulence model and the fact that k and ε by definition must everywhere be non-negative quantities. As progressively more complex turbulent flows are studied via numerical methods, it is doubtful that investigators can regularly conjure up suitably good initialization vectors so that this method will be successful and computationally efficient. In its place one would like to have a robust, efficient technique whose performance, in particular, is relatively insensitive to goodness of the initialization.

This section will outline three solution schemes that offer some intriguing possibilities as replacement candidates for the present scheme. The first two methods are similar in form to Newton's method; by beginning with a compact but general presentation of Newton's method, these similarities will become more apparent.

Newton's Method

Any system of n nonlinear algebraic equations, including those obtained from Equations 11-15, may be written as

$$f_m(x_i) = 0 \qquad (30)$$

where $1 \leq m \leq n$, x_i is the set of n unknowns, and each m denotes one of the equations. The set x_i could represent the set of unknown nodal values of U, W, P, k and ε. It is convenient to revert to vector notation and express Equation (30) as

$$f(x) = 0 \qquad (31)$$

where $\mathbf{f}$ and $\mathbf{x}$ represent the sets f_i and x_i respectively. A linear Taylor series expansion of $\mathbf{f}$ about a particular solution β leads to

$$\mathbf{x}^{\ell+1} = \mathbf{x}^{\ell} - \lambda \mathbf{J}^{-1}(\mathbf{x}^{\ell}) f(\mathbf{x}^{\ell}) \qquad (32)$$

for the new estimate $\mathbf{x}^{\ell+1}$ of β, wherein ℓ = iteration number. For Newton's method $\lambda = 1$ and the Jacobian matrix $\mathbf{J}$ is

$$\mathbf{J}^{\ell} = \left[\frac{\partial f_j}{\partial x_k} \right]_{\mathbf{x}^{\ell}} \qquad 1 \le j, \, k \le n \qquad (33)$$

One begins with an initial estimate $\mathbf{z}^{0}$ to β and computes a sequence of iterates until $\|\mathbf{x}^{\ell}* - \beta\| < \delta$, where δ is a small chosen number and ℓ_* is the last iteration. In all these methods it is understood that Jacobian terms are evaluated at the old iterate ℓ. Newton's method is characterized by a quadratic convergence rate. However, if $\mathbf{x}^{0}$ is not sufficiently close to β, the method will fail to converge. The method described in the previous section modifies some of the elements in Equation 33 and uses $\lambda < 1$ to produce a convergent algorithm. Such modifications usually produce an algorithm which converges more slowly and thus requires greater computational effort.

Method of Steepest Descent

This method is attractive because it will apparently converge from any initial approximation if the Jacobian is everywhere nonsingular, although the convergence rate is only linear. At the least it may be effective in suitably obtaining an initialization for Newton's method.

The method will be developed with the aid of the "least squares" minimization function

$$F(x) = \mathbf{f}^{T}(x) f(x) = \sum_{m=1}^{n} f_m^2 \qquad (34)$$

When $\mathbf{F}$ assumes its absolute minimum $\mathbf{F(x)} = 0$, one also obtains the solution to Equation 30. One can seek the absolute minimum of $\mathbf{F}$ with the aid of the gradient operator

$$\nabla = \frac{\partial}{\partial x_m} e_m \qquad m = 1, n \qquad (35)$$

which, when applied to $\mathbf{F}$, yields

$$\nabla F(\mathbf{x}) = 2\mathbf{J}^T(\mathbf{x})\mathbf{f}(\mathbf{x}) \tag{36}$$

where e_i are unit vectors in n-dimensional space and $\mathbf{J}^T$ is the transpose of the Jacobian matrix. A sequence of points $\mathbf{x}^{\ell+1}$ is now sought that causes $|F(\mathbf{x}^{\ell+1})| < |F(\mathbf{x}^{\ell})|$. If $\mathbf{x}^{\ell}$ is not a stationary point so that $\nabla F(\mathbf{x}^{\ell}) \neq 0$, then an interval $A^{\ell} = (0, \phi^{\ell})$ exists so that

$$|F(\mathbf{x}^{\ell+1})| = |F[\mathbf{x}^{\ell} - \phi^{\ell}\nabla F(\mathbf{x}^{\ell})]| \tag{37}$$

if $\phi^{\ell} \in A^{\ell}$. Hence one chooses $\mathbf{x}^{\ell+1} = \mathbf{x}^{\ell} - \phi^{\ell} \nabla F(\mathbf{x}^{\ell})$ and applies Equation 36 to find

$$\mathbf{x}^{\ell+1} = \mathbf{x}^{\ell} - 2\phi^{\ell}\mathbf{J}^T(\mathbf{x}^{\ell})\mathbf{f}(\mathbf{x}^{\ell}) \tag{38}$$

Unfortunately the interval A^{ℓ} depends on both the function $\mathbf{F}$ and point $\mathbf{x}^{\ell}$ and generally is unknown a priori. The steepest descent iterative scheme, Equation 38, is identical to Newton's method, Equation 32, if $\mathbf{J}^T$ is replaced by $\mathbf{J}^{-1}$. A standard finite element code which computes $\mathbf{J}$ may therefore be easily modified to obtain $\mathbf{J}^T$ and create the steepest descent algorithm. Of course, if $\mathbf{J}$ is singular for any $\mathbf{x}^{\ell}$, then the method converges to a local minimum at which $\mathbf{f}(\mathbf{x}^{\ell}) \neq 0$. However, Newton's method also fails if $\mathbf{J}$ becomes singular.

A trial application of this method to the one-dimensional plane channel flow problem described in the next section was inconclusive in determining whether the method would ultimately succeed or fail. Several numerical difficulties are worth noting at this writing, however. Via Newton's method the corrections $\delta \mathbf{x}^{\ell}$ for this simple example are symmetric for nodes at the same elevation in the streamwise direction. The form of the functional described by Equation 34 produced unsymmetric corrections with respect to the midside nodes in the elements. In addition, the appropriate value of ϕ^{ℓ} which would enhance convergence is difficult to specify a priori. (Booth (1966) describes an approximate method for determining ϕ^{ℓ}.) Finally, the convergence rate was found to be very slow after the first few iterations, as Marquardt (1963) also noted. It is therefore natural to consider applying this method to create a good initial approximation with which to begin use of

Newton's method with its quadratic convergence rate. The difficulty is in deciding when one can switch to Newton's method; a premature switch could lead to nonconvergence. The next method systematically addresses this question.

Levenberg-Marquardt Method

The Levenberg-Marquardt algorithm (Ralston and Rabinowitz, 1978) combines Newton's method and the method of steepest descent in the expression

$$x^{\ell+1} = x^{\ell} - \phi^{\ell}(J^TJ + \lambda^{\ell}I)^{-1}J^Tf(x^{\ell}) \tag{39}$$

where $\lambda^{\ell} \geq 0$, $\phi^{\ell} = 1$ and I = identity matrix. When $\lambda = 0$, $(J^TJ)^{-1}J^T = J^{-1}$ and Newton's method is recovered. The method of steepest descent is obtained by allowing $\lambda^{\ell} \to \infty$, which causes the Jacobian terms to approach $(\lambda^{\ell})^{-1}J^T$ and leads to Equation 38. With this method computations begin with an initially large value of λ^{ℓ}. As the solution is approached, progressively smaller values of λ^{ℓ} (tending to zero) are used, and the quadratic convergence rate of Newton's method is approached.

Newton's method, the method of steepest descent, and the Levenberg-Marquardt algorithm all require the computation of the Jacobian. For large equation sets this can become expensive. One way of reducing this cost is to compute the Jacobian by finite-difference approximations (Ralston and Rabinowitz, 1978). Another promising possibility is to consider use of a tensor matrix product formulation of the Jacobian (Baker, 1980).

Continuation Method

The continuation method imbeds the problem one would like to solve within a system of equations whose solution is known. The problem is formulated in terms of a parameter, say t, such that $t = 0$ corresponds to the system of equations with a known solution, while $t = 1$ corresponds to the system of equations one would like to solve. By incrementing t towards unity, the solution at the previous t-value may be used to initialize the solution being sought. The desired solution ($t = 1$) is therefore approached in a controlled manner, and divergence of the computations is avoided.

As shown in the previous sections, applications of the finite element method to the k-ε turbulence model produces a system of nonlinear algebraic equations represented by Equation 30. The continuation method imbeds this problem in a family of related problems which have the form

$$H(x, t) = 0 \tag{40}$$

with the parameter t defined on $[0,1]$. The imbedding $\mathbf{H}$ is chosen such that $t = 0$ corresponds to a problem where the solution $\mathbf{x}^0$ is known. The value $t = 1$ corresponds to the desired solution $\mathbf{x} = \mathbf{b}$ of Equation 40. For example, $\mathbf{H}$ might have the form (Avila, 1974)

$$H(\mathbf{x},\ t) = f(\mathbf{x}) + (t-1)f(\mathbf{x}^0) \tag{41}$$

The solution for various values of t is then traced out along the continuation branch

$$\mathbf{x} = \mathbf{x}(t) \tag{42}$$

starting from the initial point $\mathbf{x}^0 = \mathbf{x}(0)$ to the end point $\beta = \mathbf{x}(1)$.

The continuation branch given by Equation 42 may have simple or multiple bifurcation points (i.e., points which branch in several directions) and limit points (i.e., points where the slope of the curve represented by Equation 42 changes sign). Keller (1977) and Rheinboldt (1977) present various techniques for handling these bifurcation and turning points.

Previous experience has shown that the global form of Equations 11-13 will generally converge, using Newton's method with initial values of U, W, and P being constant throughout the solution domain, for constant values $k = k_o$ and $\varepsilon = \varepsilon_o$ throughout the solution domain. This specification for k and ε corresponds to a constant kinematic eddy viscosity simulation (cf. Equation 4). In this instance the nonlinear equations for k and ε (Equations 14 and 15) are eliminated. These results suggest the following possible form for Equation 40 (other forms may also be appropriate):

$$H(\mathbf{x},\ t) = \begin{bmatrix} g(\mathbf{x}) \\ h'(\mathbf{x},\ t) \end{bmatrix} = 0 \tag{43}$$

in which $\mathbf{g}$ represents the nonlinear, algebraic, assembled finite element form of Equations 11, 12 and 13, and h' is

$$h'(\mathbf{x},\ t) = t\ h(\mathbf{x}) + (t-1)\begin{bmatrix} k-k_o \\ \varepsilon-\varepsilon_o \end{bmatrix} \tag{44}$$

Here h(x) represents the global nonlinear, algebraic, finite element form of Equations 14 and 15. For $t = 0$ the constant eddy viscosity simulation is recovered. This solution may thus be used as the starting vector x^0. As $t \to 1$ in Equation 44, the k and ε equations are gradually brought into a more dominant role. For $t = 1$ the desired solution is obtained.

As an example of the finite element application of this method, the continuation method described by Equation 44 has been used to compute the fully developed turbulent flow occurring between parallel walls. Previous attempts to solve this relatively simple example via Newton's method failed unless x^0 was sufficiently close to the solution β (Schamber and Larock, 1980a). For a specified pressure gradient $\Delta p^*/\Delta x^*$ (starred variables are dimensional) and appropriate boundary conditions, the unknowns are U, k and ε per computational node with $W = 0$ for one-dimensional flow. The computational mesh consists of a single column of six elements which begins near the wall at $z = \delta$ and extends to the channel centerline $z = 1.0$. Here the reference length $h_o = d$ in which d is the channel halfwidth. Near-wall values of U, k and ε are determined as boundary conditions from Equations 22 and 23, and the symmetry conditions $\partial U/\partial z = \partial k/\partial z = \partial \varepsilon/\partial z = 0$ are specified at the channel centerline. With the reference velocity chosen to be $U_o = u_\tau$, Equations 22 and 23 at $y = \delta$ reduce to the following: $U = U^*/u_\tau = \kappa^{-1} \ln[Ey^+]$, $k = k^*/u_\tau^2 = c_\mu^{-\frac{1}{2}}$, and $\varepsilon = \varepsilon^* d/u_\tau^3 = (\kappa\delta)^{-1}$ in which $y^+ = \delta R_\tau$ and $R_\tau = u_\tau d/\nu$. The pressure gradient (nondimensionalized on u_τ, d and ρ) is then $\Delta p/\Delta x = -1.0$ since a force balance on a control volume enclosing the flow yields a wall shear stress $\tau_{wall} \equiv \rho u_\tau^2 = -(\Delta p^*/\Delta x^*)d$. Specification of R_τ and δ yields a velocity profile which corresponds to a unique channel Reynolds number $R = R_\tau U_m/u_\tau$; the maximum nondimensional channel velocity U_m/u_τ is obtained as part of the solution. For the results shown in Figures 5 and 6 the values $\delta = 0.03$ and $R_\tau = 2275$ have been used; they yield a computed channel Reynolds number of $R = 57,000$. The wall boundary conditions are then $k = 3.333$, $\varepsilon = 83.33$ and $U = 16.05$ so that $y^+ = 68$, which falls within the inertial subrange as required.

The computational schedule used to generate a convergent solution is presented in Table I. Newton's method with $\lambda = 1$ was used for

Table I. Computation Schedule for Continuation Method

j	1	2	3	4	5	6	7	8	9	10	11	12
t_j	0	0.2	0.4	0.6	0.8	0.9	0.94	0.96	0.98	0.99	0.998	1.0
n_j	2	5	4	4	4	5	5	4	5	4	4	3

each t_j to obtain a convergent solution. The parameter n_j in Table I is the number of iterations for a fixed t_j which were required to

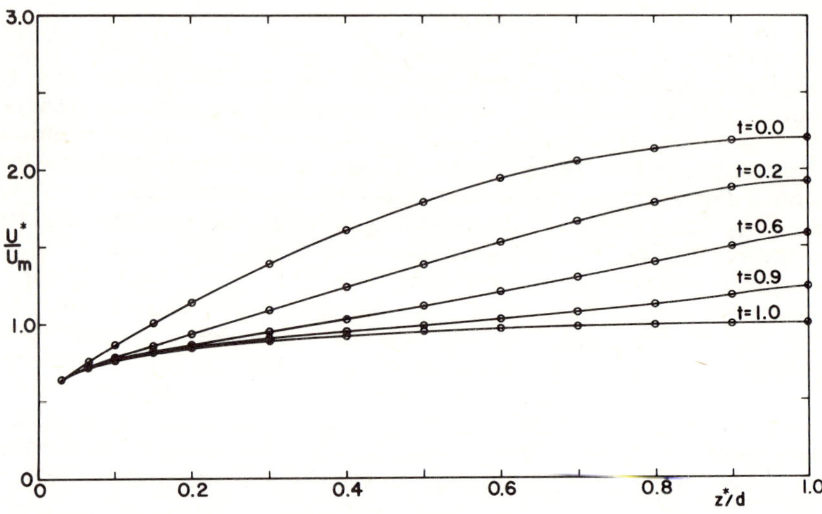

Figure 5. Continuation method for channel flow, mean velocity profile.

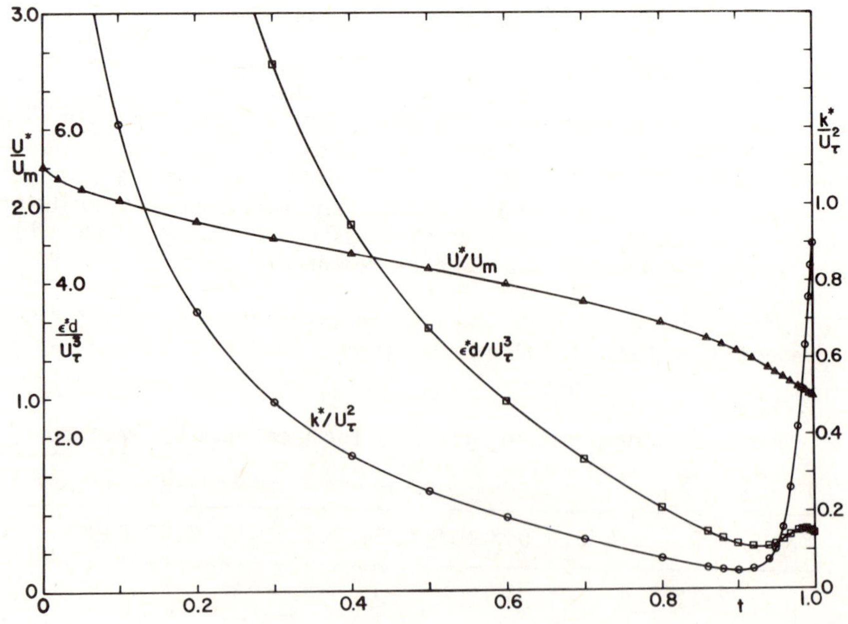

Figure 6. Channel flow continuation curves for z*/d = 1.0.

achieve a relative error at all nodes in the domain below 0.05% (a tolerance of 1% would probably have been sufficient; a corresponding decrease in n_j at each t_j cycle would also have resulted). The converged solution for t_j served as the initialization $\mathbf{x}^0$ for t_{j+1}. After several converged t_j-solutions have been obtained, an extrapolation scheme could be formulated to obtain a better starting vector $\mathbf{x}^0$ (Rheinboldt, 1977) for the next t_j-solution, but this technique was not applied here.

In Equation 44 the wall boundary values $k_o = 3.333$ and $\varepsilon_o = 83.33$ were used to start the procedure. For $t_1 = 0$ a constant eddy viscosity solution was generated which, as shown in Figure 5, yielded a parabolic mean velocity profile. The t_1 initial vector of unknowns for velocity used the wall boundary value $U = 16.05$ for all nodes in the domain. Several intermediate converged velocity profiles are also shown in Figure 5 for various values of t_j.

The continuation branches for nodal values of k, ε, and U at the channel centerline $z = 1.0$ are shown in Figure 6. Additional t_j points were used in a run subsequent to that tabulated in Table I to define accurately the continuation curves. The continuation curves for U and ε are relatively smooth and well behaved. The continuation branch for k changes rapidly in the range $0.9 < t \leq 1.0$, however, requiring the use of a small increment in t in this region to obtain a converged solution.

Large Eddy Simulation

The finite element implementation of this method has been developed primarily by Findikakis and Street (1979); they spatially filter, or average, the original conservation equations and then solve the resulting transient equations via the Galerkin finite element approach. With this method the steady-state solution of any problem is sought by beginning with some known initial flow state and integrating, via a finite difference time-step treatment, forward in time until a steady flow state is obtained. Clearly the amount of computation required to reach a steady state is problem dependent and could in some instances exceed one's computing resources; primarily for this reason the writers prefer to seek a steady solution via a direct solution of the time-averaged mean flow equations and an appropriate turbulence model. The Large Eddy Simulation method does have two attributes that deserve attention, however: 1. The choice of a suitable initialization does not usually appear to be a problem. 2. The algebraic model proposed by Findikakis and Street contains only velocity and pressure as unknowns in a non-stratified two-dimensional flow, whereas the present approach also must compute k and ε.

SUMMARY AND CONCLUDING COMMENTS

This article presented a Galerkin finite element method for the representation of turbulent flows which might reasonably be approximated as two-dimensional. The mean-flow equations were closed by use of the k-ε turbulence model. The implementation of suitable boundary conditions was described. Solution of the resulting equation set was achieved by partitioning the global equation set into mean-flow and turbulence-model subsets and applying Newton's method in an alternating fashion. As an application of these methods, the flow pattern in a circular sedimentation basin was determined. Although the solution was successfully obtained, the relatively slow convergence of the procedure and its sensitivity to a suitable initialization greatly limit the general utility of the method. Thus the writers attempted to identify some other potentially promising solution schemes.

At least two alternative solution strategies deserve further attention to ascertain whether they can be used to develop a robust and efficient solution procedure for the current finite element turbulent flow formulation. The method of steepest descent is a candidate procedure, even though it is known to have a slow rate of convergence, because of its relative insensitivity to choice of an initial estimate of the solution vector; in combination with the Levenberg-Marquardt algorithm the slow convergence rate might not be particularly disadvantageous. The continuation method currently appears even more promising, however, and it is for this reason that a relatively simple example of its use has been presented herein.

The greatest initial attraction of the continuation method is the opportunity to use a systematic approach to the solution while avoiding initialization difficulties. However, more experience in applying this approach to both one- and two-dimensional turbulent flow problems is certainly needed before one can clearly assess the true worth of the approach. The example solved simultaneously for both the mean-flow and turbulence variables; can the partitioning scheme be productively combined with this method to balance core storage needs and solution speed? Can the method be applied equally well to much larger problems? What are the relative merits of using a frontal solver or an in-core solver? As Figure 6 illustrates, the continuation method also gives one some understanding of the behavior of the nonlinearities in the model equation set. Further experience with this method is needed to learn whether its perceived advantages are real or illusory.

In the event that neither the continuation method nor the method of steepest descent offers any substantial improvement over the current approach, then it would be appropriate to consider further the seeking of steady-flow solutions via time-stepping or pseudo-time-stepping procedures, including the large eddy simulation method.

ACKNOWLEDGEMENTS

The work reported herein, and the writing of this article, have been supported in part by National Science Foundation Grants ENG76-18846, CME-7914762 and CME-8006743.

REFERENCES

Avila, J. H. (1974) The Feasibility of Continuation Methods for Nonlinear Equations. SIAM J. Numerical Analysis, 11, 1, 102-122.

Baker, A. J. (1980) Research on Numerical Algorithms for the Three-Dimensional Navier-Stokes Equations, I. Accuracy, Convergence and Efficiency. USAF Report AFFDL-TR-79-3141.

Booth, A. D. (1966) Numerical Methods, Butterworths, London, 172-177.

Chieng, C. C. and B. E. Launder (1980) On the Calculation of Turbulent Heat Transport Downstream From an Abrupt Pipe Expansion. Numerical Heat Transfer, 3, 189-207.

Clements, M. S. and A. F. M. Khattab (1968) Research into Time Ratio in Radial Flow Sedimentation Tanks. Instn. Civ. Engrs., 40, Aug., 471-494.

Findikakis, A. N. and R. L. Street (1979) An Algebraic Model for Sub-Grid-Scale Turbulence in Stratified Flows. J. Atm. Sci., 36, 10, 1934-1949.

Gosman, A. D. and F. J. K. Ideriah (1976) Teach-T: a General Computer Program for Two-Dimensional, Turbulent, Recirculating Flows. Report, Imperial College, London.

Gresho, P. M. and R. L. Lee (1979) Don't Suppress the Wiggles-- They're Telling You Something, in Finite Element Methods for Convection Dominated Flows, Hughes, T. J. R. (ed.), ASME AMD, 34, 37-61.

Huebner, K. H. (1975) The Finite Element Method for Engineers, John Wiley, New York.

Keller, H. B. (1977) Numerical Solution of Bifurcation and Nonlinear Eigenvalue Problems, in Applications of Bifurcation Theory, Rabinowitz, P. H. (ed.), Academic, New York.

Kolmogorov, A. N. (1942) Equations of Turbulent Motion of an Incompressible Turbulent Fluid. Izv. Akad. Nauk SSSR Ser. Phys. VI, no. 1-2, 56.

Larock, B. E. and D. R. Schamber (1980) Finite Element Computation of Turbulent Flows, in Finite Elements in Water Resources, Wang, S. Y. et al. (eds.), Univ. of Mississippi, University, MS, 4.31-4.47.

Larsen, P. (1977) On the Hydraulics of Rectangular Settling Basins. Report 1001, Dept. of Water Resources Engineering, Univ. of Lund, Sweden.

Launder, B. E. and D. B. Spalding (1974) The Numerical Computation of Turbulent Flows. Comp. Meth. Appl. Mech. and Engng., 3, 269-289.

Marquardt, D. W. (1963) An Algorithm for Least-Squares Estimation of Nonlinear Parameters. J. Soc. Indust. Appl. Math., 11, 2, 431-441.

Montens, A. (1954) The Use of Radioactive Isotopes for Water Flow and Velocity Measurements, in Radioisotope Conference 1954, Johnston, J. E. (ed.), Butterworths, London, 169-180.

Oden, J. T. and J. N. Reddy (1976) Mathematical Theory of Finite Elements, Wiley, New York, 323-326.

Oliver, A. J. (1980) The Prediction of Turbulent Flow and Heat Transfer Over Backward Facing Steps, in Computer Methods in Fluids, Morgan, K. et al. (eds.), Pentech, London, 309-338.

Patankar, S. V. and D. B. Spalding (1970) Heat and Mass Transfer in Boundary Layers, 2nd Ed., Intertext, London.

Ralston, A. and P. Rabinowitz (1978) A First Course in Numerical Analysis, McGraw-Hill, New York.

Reynolds, W. C. (1976) Computation of Turbulent Flows, in Annual Review of Fluid Mechanics, Van Dyke, M. et al. (eds.), Annual Reviews, Inc., Palo Alto, CA, 183-208.

Rheinboldt, W. C. (1977) Numerical Continuation Methods for Finite Element Applications, in Formulations and Computational Algorithms in Finite Element Analysis: U.S.-Germany Symposium, Bathe, K. J., Oden, J. T. and W. Wunderlich (eds.), M.I.T. Press, 599-631.

Rodi, W. (1980) Turbulence Models and Their Application in Hydraulics - A State-of-the-Art Review, IAHR.

Sani, R. L., Gresho, P. M., Lee, R. L. and D. F. Griffiths (1981) The Cause and Cure (?) of the Spurious Pressures Generated by Certain FEM Solutions of the Incompressible Navier-Stokes Equations. Int. J. Num. Meth. in Fluids, 1, two parts.

Schamber, D. R. (1979) Finite Element Analysis of Flow in Sedimentation Basins. Ph.D. Thesis, University of California, Davis.

Schamber, D. R. and B. E. Larock (1980a) Computational Aspects of Modeling Turbulent Flows by Finite Elements, in Computer Methods in Fluids, Morgan, K. et al. (eds.), Pentech, London, 339-361.

Schamber, D. R. and B. E. Larock (1980b) Constant and Variable Eddy Viscosity Flow Simulations in Settling Tanks, in Computer and Physical Modeling in Hydraulic Engineering, Ashton, G. (ed.) ASCE, New York, 345-355.

APPENDIX I. MATRIX COEFFICIENTS

This appendix presents the expanded form of the individual Jacobian coefficients which appear in Equations 28 and 29.

The mean flow coefficients are

$$\frac{\partial f_U}{\partial U_j} = \int_{\Omega_e} N_i \left[r^m \left(N_j \frac{\partial U}{\partial r} + U \frac{\partial N_j}{\partial r} + W \frac{\partial N_j}{\partial z} \right) + 2m\nu_t N_j \right] d\Omega$$

$$+ \int_{\Omega_e} r^m \nu_t \left[2 \frac{\partial N_i}{\partial r} \frac{\partial N_j}{\partial r} + \frac{\partial N_i}{\partial z} \frac{\partial N_j}{\partial z} \right] d\Omega$$

$$+ \int_{\Gamma_e} r^m N_i \left[-2\nu_t \frac{\partial N_j}{\partial r} \ell_r + \frac{\partial \overline{uw}}{\partial U_j} \ell_z \right] d\Gamma \qquad (A\text{-}1)$$

$$\frac{\partial f_U}{\partial W_j} = \int_{\Omega_e} r^m N_i N_j \frac{\partial U}{\partial z} d\Omega + \int_{\Omega_e} r^m \nu_t \frac{\partial N_i}{\partial z} \frac{\partial N_j}{\partial r} d\Omega$$

$$- \int_{\Gamma_e} r^m N_i \nu_t \frac{\partial N_j}{\partial r} \ell_z d\Gamma \qquad (A\text{-}2)$$

$$\frac{\partial f_U}{\partial P_j} = - \int_{\Omega_e} m N_i M_j d\Omega - \int_{\Omega_e} r^m M_j \frac{\partial N_i}{\partial r} d\Omega$$

$$+ \int_{\Gamma_e} r^m N_i M_j \ell_r d\Gamma \qquad (A\text{-}3)$$

$$\frac{\partial f_W}{\partial U_j} = \int_{\Omega_e} r^m N_i N_j \frac{\partial W}{\partial r} d\Omega + \int_{\Omega_e} r^m \nu_t \frac{\partial N_i}{\partial r} \frac{\partial N_j}{\partial z} d\Omega$$

$$+ \int_{\Gamma_e} r^m N_i \nu_t \frac{\partial N_j}{\partial z} \ell_r d\Gamma \qquad (A\text{-}4)$$

K

$$\frac{\partial f_W}{\partial W_j} = \int_{\Omega_e} r^m N_i \left[U\frac{\partial N_j}{\partial r} + N_j\frac{\partial W}{\partial z} + W\frac{\partial N_j}{\partial z} \right] d\Omega$$

$$+ \int_{\Omega_e} r^m \nu_t \left[\frac{\partial N_i}{\partial r}\frac{\partial N_j}{\partial r} + 2\frac{\partial N_i}{\partial z}\frac{\partial N_j}{\partial z} \right] d\Omega$$

$$- \int_{\Gamma_e} r^m N_i \left[\frac{\partial \overline{uw}}{\partial W_j} \ell_r + 2\nu_t \frac{\partial N_j}{\partial z} \ell_z \right] d\Gamma \qquad (A\text{-}5)$$

$$\frac{\partial f_W}{\partial P_j} = - \int_{\Omega_e} r^m M_j \frac{\partial N_i}{\partial z} d\Omega + \int_{\Gamma_e} r^m N_i M_j \ell_z d\Omega \qquad (A\text{-}6)$$

$$\frac{\partial f_P}{\partial U_j} = \int_{\Omega_e} M_i \frac{\partial}{\partial r} \left[r^m N_j \right] d\Omega \qquad (A\text{-}7)$$

and

$$\frac{\partial f_P}{\partial W_j} = \int_{\Omega_e} M_i \frac{\partial}{\partial z} \left[r^m N_j \right] d\Omega \qquad (A\text{-}8)$$

The line integrals in these expressions are formed only when Γ_e coincides with a global boundary on which natural, i.e. derivative, boundary conditions are specified.

The modified k-ϵ coefficients, for use in the first iterations, are

$$\frac{\partial f_k}{\partial k_j} = \int_{\Omega_e} r^m N_i \left[U\frac{\partial N_j}{\partial r} + W\frac{\partial N_j}{\partial z} \right] d\Omega$$

$$+ \int_{\Omega_e} r^m \frac{\nu_t^*}{\sigma_k} \left[\frac{\partial N_j}{\partial r}\frac{\partial N_i}{\partial r} + \frac{\partial N_j}{\partial z}\frac{\partial N_i}{\partial z} \right] d\Omega \qquad (A\text{-}9)$$

$$\frac{\partial f_k}{\partial \epsilon_j} = \int_{\Omega_e} r^m N_i N_j d\Omega \qquad (A\text{-}10)$$

$$\frac{\partial f_\epsilon}{\partial k_j} = - \int_{\Omega_e} r^m N_i N_j c_{\epsilon_1} c_\mu \, \Phi \, d\Omega \qquad \text{(A-11)}$$

with

$$\Phi = 2\left[\frac{\partial U}{\partial r}\right]^2 + \left[\frac{\partial U}{\partial z} + \frac{\partial W}{\partial r}\right]^2 + 2\left[\frac{\partial W}{\partial z}\right]^2 + 2m\left[\frac{U}{r}\right]^2 \qquad \text{(A-12)}$$

and

$$\frac{\partial f_\epsilon}{\partial \epsilon_j} = \int_{\Omega_e} r^m N_i \left[U\frac{\partial N_j}{\partial r} + W\frac{\partial N_j}{\partial z} + c_{\epsilon_2} \frac{\epsilon^*}{k^*} N_j \right] d\Omega$$

$$+ \int_{\Omega_e} r^m \frac{\nu_t^*}{\sigma_\epsilon} \left[\frac{\partial N_j}{\partial r} \frac{\partial N_i}{\partial r} + \frac{\partial N_j}{\partial z} \frac{\partial N_i}{\partial z} \right] d\Omega \qquad \text{(A-13)}$$

Equations A-9 to A-13 assume no nonzero natural boundary conditions exist. The stars indicate quantities which remain constant while the derivatives are formed.

The k-ϵ coefficients, to be used for <u>unmodified</u> Newton computations, follow:

$$\frac{\partial f_k}{\partial k_j} = \int_{\Omega_e} r^m N_i \left[U\frac{\partial N_j}{\partial r} + W\frac{\partial N_j}{\partial z} - 2c_\mu \frac{k}{\epsilon} N_j \, \Phi \right] d\Omega$$

$$+ \int_{\Omega_e} r^m \frac{\nu_t}{\sigma_k} \left[\frac{\partial N_j}{\partial r} \frac{\partial N_i}{\partial r} + \frac{\partial N_j}{\partial z} \frac{\partial N_i}{\partial z} \right] d\Omega$$

$$+ \int_{\Omega_e} 2r^m \frac{c_\mu}{\sigma_k} \frac{k}{\epsilon} N_j \left[\frac{\partial k}{\partial r} \frac{\partial N_i}{\partial r} + \frac{\partial k}{\partial z} \frac{\partial N_i}{\partial z} \right] d\Omega \qquad \text{(A-14)}$$

$$\frac{\partial f_k}{\partial \epsilon_j} = \int_{\Omega_e} r^m N_i N_j \left[-c_\mu \frac{k^2}{\epsilon^2} \, \Phi + 1 \right] d\Omega$$

$$- \int_{\Omega_e} r^m \frac{c_\mu}{\sigma_k} \frac{k^2}{\epsilon^2} N_j \left[\frac{\partial k}{\partial r} \frac{\partial N_i}{\partial r} + \frac{\partial k}{\partial z} \frac{\partial N_i}{\partial z} \right] d\Omega \qquad \text{(A-15)}$$

282

$$\frac{\partial f_\varepsilon}{\partial k_j} = - \int_{\Omega_e} r^m N_i N_j \left[c_{\varepsilon_1} c_\mu \Phi + c_{\varepsilon_2} \frac{\varepsilon^2}{k^2} \right] d\Omega$$

$$+ \int_{\Omega_e} 2r^m \frac{c_\mu}{\sigma_\varepsilon} \frac{k}{\varepsilon} N_j \left[\frac{\partial \varepsilon}{\partial r} \frac{\partial N_i}{\partial r} + \frac{\partial \varepsilon}{\partial z} \frac{\partial N_i}{\partial z} \right] d\Omega \qquad (A-16)$$

and

$$\frac{\partial f_\varepsilon}{\partial \varepsilon_j} = \int_{\Omega_e} r^m N_i \left[U \frac{\partial N_j}{\partial r} + W \frac{\partial N_j}{\partial z} + c_{\varepsilon_2} \frac{2\varepsilon}{k} N_j \right] d\Omega$$

$$+ \int_{\Omega_e} r^m \frac{\nu_t}{\sigma_\varepsilon} \left[\frac{\partial N_j}{\partial r} \frac{\partial N_i}{\partial r} + \frac{\partial N_j}{\partial z} \frac{\partial N_i}{\partial z} \right] d\Omega$$

$$- \int_{\Omega_e} r^m \frac{c_\mu}{\sigma_\varepsilon} \frac{k^2}{\varepsilon^2} N_j \left[\frac{\partial \varepsilon}{\partial r} \frac{\partial N_i}{\partial r} + \frac{\partial \varepsilon}{\partial z} \frac{\partial N_i}{\partial z} \right] d\Omega \qquad (A-17)$$

The expression for Φ, Equation A-12, remains valid in Equations A-14 through A-17.

CHAPTER 10

ANALYSIS OF TURBULENT FLOW WITH SEPARATION USING THE
FINITE ELEMENT METHOD

C. TAYLOR, C.E. THOMAS and K. MORGAN

Department of Civil Engineering, University College, Singleton
Park, Swansea SA2 8PP, U.K.

INTRODUCTION

The present chapter deals exclusively with flow problems where
abrupt changes in domain geometry results in, for the Reynolds'
numbers considered, separated flow.   In such cases, whether
considering laminar or turbulent flow, the advent of separated
regions has usually necessitated the use of numerical techniques to obtain a solution.   The more dominant techniques
currently being employed are the Finite Difference (F.D.),
Gosman et al. (1969), and Finite Element (F.E.), Hinton and
Owen (1980) and Taylor and Hughes (1981), methods.   At significant values of Reynolds numbers specialised techniques have
been introduced to suppress oscillations or 'wriggles' which
are apparent in converged solutions, Launder and Spalding (1974)
and Gresho et al. (1979).   These techniques commonly known as
'upwinding', have been used quite extensively when either F.D.
or F.E. technique has been used.   Discussion and controversy
regarding their use are currently topical and the reader is
referred to the cited literature, Gosman et al. (1969) and
Hughes and Brooks (1979).   In a F.E. context the advocates
range from rejection, Gresho et al. (1980) to schemes with no
crosswind difference, Hughes and Brooks (1979).   Each has its
merit but a general observation is that the introduction of
such techniques can lead to an underprediction of such
quantities as turbulence kinetic energy and re-attachment
lengths, Atkins et al. (1980) and Thomas et al. (1981), by the
introduction of damping.   This is amply demonstrated when a
comparison is made between numerical results obtained with and
without upwinding are compared for flow over a backward facing
step.

Within the last decade the F.E.M. has been used, with measurable
success, to investigate the flow of viscous fluids governed by
the Navier-Stokes equations.   Earlier publications, Oden (1972)
and Taylor and Hood (1973) were limited to low values of
Reynolds numbers.   In these earlier investigations inaccuracies

K2

in the converged pressures, Taylor and Hood (1974), was largely
obviated by using mixed interpolation techniques. The introduction of such techniques was later rationalised by Olson
and Tuann (1978) and Gresho and Lee (1981). For most cases
mesh refinement and a judicious choice of boundary conditions
was used to avoid converged oscillatory, chequerboard, nature
of the pressure.

More generalised applications of the F.E.M. to solve turbulent
flow problems were then introduced. Notable amongst the early
publications are those by Baker (1978) and Taylor et al. (1977).
Initially, algebraic models were used to depict local variations
in turbulence viscosity. Later, however both one and two
equation models were utilised, Baker and Soliman (1980) and
Thomas et al. (1981). One important feature of the solution
technique when the F.E.M. is used for solving turbulent flow
is that no upwinding is required. This is an observation
common to all investigators.

The particular problem chosen for demonstrating the applicability of the F.E.M. to solve problems where a recirculation
zone is apparent at very low values of Reynolds number, is flow
over a backward facing step. The fluid is assumed to be
steady, incompressible, two dimensional and can be depicted by
utilising the Navier-Stokes equations. The effective viscosity
is assumed constant and equal to the molecular viscosity when
considering laminar flow and varies spatially when turbulent.
The spatial variation in effective viscosity in turbulent
flows is usually written in terms of the local values of turbulence kinetic energy, k, and the Prandtl mixing length, ℓ.
In the one equation model of turbulence, Wolfshtein (1970), the
local values of k are obtained from a transport equation in k
and an empirical relationship defines ℓ. A further transport
equation defining ℓ is used in the two equation model, Launder
and Spalding (1972), and all equations are solved simultaneously.

When laminar flow is considered upwinding techniques are employed to avoid excessive local mesh refinement. As stated
earlier no such technique is required when considering turbulent
flow and the damping effect of introducing upwinding is
demonstrated.

The values obtained using the F.E.M. for laminar flow are
compared with other numerically obtained values, Hutton and
Smith (1979) and experimental results, Atkins et al. (1980).
Similarly, those for turbulent flow are compared with those
of Atkins et al. (1980) and Denham et al. (1975).

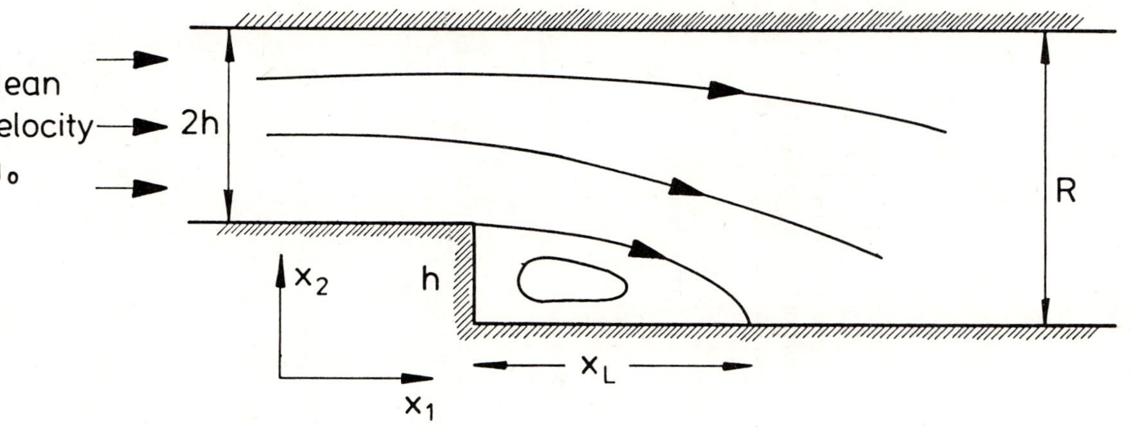

FIGURE 1   GENERAL LAYOUT DIAGRAM

MATHEMATICAL CONCEPTS

Equations which are used to predict the steady state, two dimensional flow of an incompressible viscous Newtonian fluid are,

Momentum

$$u_j \frac{\partial u_i}{\partial x_j} = - \frac{1}{\rho} \frac{\partial}{\partial x_j} (\tau_{ij}) \qquad i,j = 1,2 \tag{1}$$

and

Continuity

$$\frac{\partial u_i}{\partial x_i} = 0 \tag{2}$$

in which ρ is the fluid density, u_i the local velocity and τ_{ij} the mean viscous shear stress. Following the notation adopted in the text, as well as previous chapters in the current series, (1980), u_i denotes the component of velocity in the x_ith coordinate direction in a cartesian orthogonal coordinate system x_1, x_2.

The mean shear stress in (1) can be expanded to,

$$\tau_{ij} = - \tilde{p} \, \delta_{ij} + \mu \frac{\partial}{\partial x} \left(\frac{\partial u_i}{\partial x_j} + \frac{\partial u_j}{\partial x_i} \right) \tag{3}$$

where p is the local pressure, μ the molecular viscosity and δ_{ij} the Kronecker delta.

If sufficient and appropriate boundary conditions can be defined then equations (1) to (3) can be solved simultaneously to give the spatial distribution of the primitive variables u_i and $\tilde{p}$.

When considering turbulent flow a time averaged form of the Navier-Stokes equations are usually employed.

Momentum

$$\rho u_j \frac{\partial u_i}{\partial x_j} = \frac{\partial}{\partial x_j} (\tau_{ij} - \rho \, \overline{u_i' u_j'}) \tag{4}$$

and the mass conservation equation (2). In (4), $\rho \, \overline{u_i' u_j'}$ represents the Reynolds shear stress, Davies (1972), the primes instantaneous fluctation from the mean velocity and the overbar signifies time averaging. This Reynolds stress is usually interpreted in the following manner,

$$\rho \, \overline{u_i' u_j'} = \mu_T \left[\frac{\partial u_i}{\partial x_j} + \frac{\partial u_j}{\partial x_i} \right] - \frac{2}{3} \rho k \, \delta_{ij} \qquad (5)$$

in which μ_T is the turbulent dynamic viscosity and k the mean turbulence kinetic energy,

$$k = \frac{1}{2} \overline{u_i'^2} \qquad (6)$$

Introducing the new variable,

$$p = \tilde{p} + \frac{2}{3} \rho k \qquad (7)$$

and employing (5), equation (4) can be rewritten,

$$u_j \frac{\partial u_i}{\partial x_j} = - \frac{1}{\rho} \frac{\partial p}{\partial x_i} + \frac{1}{\rho} \frac{\partial}{\partial x_j} \left[\mu_e \left(\frac{\partial u_i}{\partial x_j} + \frac{\partial u_j}{\partial x_i} \right) \right] \qquad (8)$$

where μ_e is the so-called 'effective' viscosity, $\mu_e = \mu_T + \mu$.

It is evident that once μ_T can be defined then equation (8) degenerates into a similar form to (1) and a solution can again be effected.   The two models one and two equation turbulence models, differ in the manner in which the additional variables, ℓ and k, defining spatial variations in μ_T are evaluated. In both, the Prandtl (1945) and Kolmogorov (1942) relationship,

$$v_T = \frac{\mu_T}{\rho} = C_\mu \, k^{\frac{1}{2}} \ell \qquad (9)$$

is used.   Here C_μ is considered to be constant.

One Equation Model

An additional 'transport' differential equation depicting the spatial variation in the turbulence kinetic energy has been derived, Wolfshtein (1970), and appendix 1,

$$\rho u_j \frac{\partial k}{\partial x_j} = \frac{\partial}{\partial x_j} \left[(\frac{\mu_T}{\sigma_k} + \mu) \frac{\partial k}{\partial x_j} \right] + \mu_T \frac{\partial u_i}{\partial x_j} \left(\frac{\partial u_i}{\partial x_j} + \frac{\partial u_j}{\partial x_i} \right)$$

CONVECTION   DIFFUSION   PRODUCTION

$$- C_D \frac{k^{3/2}}{\ell} \qquad (10)$$

DISSIPATION

in which $\frac{\mu_T}{\sigma_k}$ is commensurate with a turbulent diffusion coefficient for the turbulence kinetic energy.   Usually, σ_k and C_D are considered to be constant.

Equations (2), (8), (9) and (10) form the basis for calculations associated with the one equation model. Clearly, in order that computations can proceed the mixing length ℓ, needs to be defined spatially. For the present example this can be defined in a similar manner to that advocated by Runchal (1969) and Wolfshtein (1967) in the form,

and

$$\ell = \begin{cases} \lambda & \text{the smaller of which is} \\ 0.045R \dfrac{x_1}{h} & \text{used for } x_1 \le 5.5555h, \end{cases}$$

and

$$\ell = \begin{cases} \lambda & \text{the smaller of which is} \\ 0.09R \; x_1/h - 0.25R & \text{used for } x_1 \ge 5.5555h. \end{cases} \tag{11}$$

where λ is the distance measured normal from the nearest wall, h the step height and R the distance between the walls, figure 1.

Two Equation Model

In this case both the turbulence kinetic energy and length scale are evaluated from 'transport' type equations. Some of the advantages of using such a model have been presented by Launder and Spalding (1972), Ng and Spalding (1972) and Taylor et al. (1980). A form of transport equation relating to the distribution of the length scale has been derived, Wolfshtein (1970) and see Appendix 1, in the form,

$$u_j \frac{\partial \varepsilon}{\partial x_j} = \frac{\partial}{\partial x_j} \left| \left(\mu + \frac{\mu_T}{\sigma_\varepsilon} \right) \frac{\partial \varepsilon}{\partial x_j} \right| + C_1 \; \rho k \frac{\partial u_i}{\partial x_j} \left(\frac{\partial u_i}{\partial x_j} + \frac{\partial u_j}{\partial x_i} \right)$$

$\qquad$ CONVECTION $\qquad\qquad$ DIFFUSION $\qquad\qquad$ PRODUCTION

$$- \; C_2 \; \rho \; \frac{\varepsilon^2}{k} \tag{12}$$

$\qquad$ DISSIPATION

where $\varepsilon = k^{3/2}/\ell$ and C_1, C_2 and σ_ε are again assumed to be constant. Again μ/σ_ε can be interpreted as a diffusion co-efficient for ε.

The values attributed to the above constants by Atkins (1974), are used for the problems analysed. These are defined when specific numerical examples are considered.

The two equation model comprises, therefore, a simultaneous solution of equations (2), (8), (9), (10) and (12).

Having outlined the equations required for solution, each model will be considered in sequence, starting with laminar flow. It is taken for granted that the reader is familiar with the F.E.M. Galerkin weighted residual approach. For those not familiar with the utilisation of the technique a brief outline and further references is given in Appendix 2.

SOLUTION PROCEDURE

Laminar Flow

For both laminar and turbulent flow the application of the F.E.M. results in a matrix equation of the form,

$$[H]\,\{\beta\} = \{f\} \tag{13}$$

in which [H] is a non-symmetric matrix and, for laminar flow,

$$\beta_{(k)} = \begin{Bmatrix} u_{1(k)} \\ p_{(k)} \\ u_{2(k)} \end{Bmatrix} \qquad \text{for corner nodes} \tag{14}$$

and $\beta_{(k)} = \begin{Bmatrix} u_{1(k)} \\ u_{2(k)} \end{Bmatrix}$ for midside nodes.

The coefficients in the matrix [H] are detailed in Appendix 2. The existence of such terms as $u_j \dfrac{\partial u_i}{\partial x_i}$ in (1) clearly make any solution procedure iterative when retaining the present form of the equations. These are usually handled by replacing such terms by $\tilde{u}_j \dfrac{\partial u_i}{\partial x_j}$ in which $\tilde{u}_j$ represent previous iteration or, for the first iteration, initial values. These are assumed constant for a current iteration solution to the equations and the procedure is deemed converged when $\tilde{u}_j \simeq u_j$. Further details can be obtained, for example, from Hood (1974).

Even when a suitable algorithm has been adopted for evaluating the primitive variables, care must be exercised with regard to the imposition of boundary conditions, Sani et al. (1981) and Thomas et al. (1981). In the event that suitable pre-cautions are not taken then spurious pressure modes are generated. These often result in ostensibly acceptable velocities but exhibit a 'chequerboard syndrome' regarding pressure, Sani et al. (1981) where oscillations appear in the converged pressures. In other cases, Thomas (1981) the velocities are also affected to a marked extent.

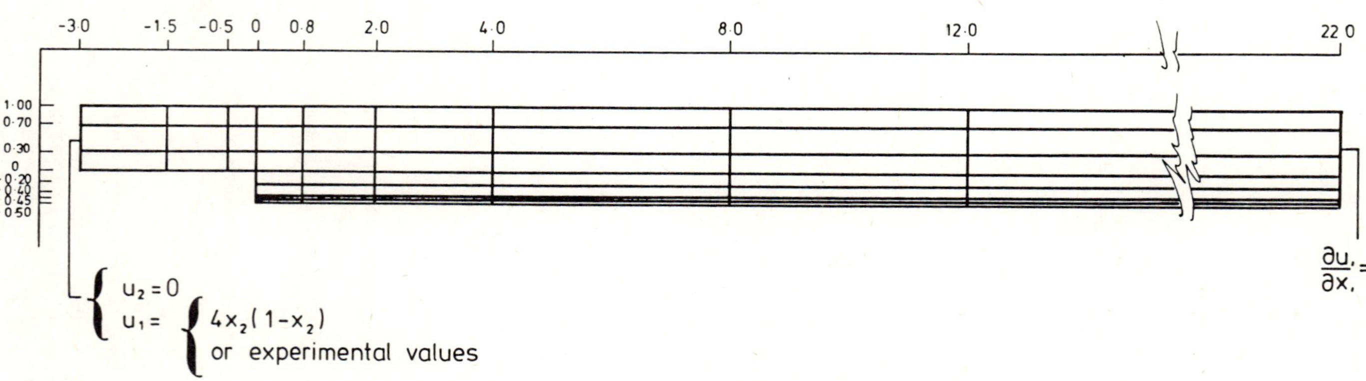

FIGURE 2   F.E. MESH FOR LAMINAR FLOW PROBLEMS

Boundary Conditions

i) Upstream

$$\left.\begin{array}{l} u_1 - \text{specified} \\[2ex] u_2 = 0 \end{array}\right\} \quad x_1 = 0, \quad 0 < x_2 < R$$

u_1 - specified by utilising experimental results, Denham and Patrick (1974) or a fully developed profile assumed.

ii) Downstream

$$\left.\begin{array}{l} \dfrac{\partial u_1}{\partial x_1} = 0 \\[3ex] \dfrac{\partial u_2}{\partial x_1} = 0 \end{array}\right\} \quad 0 < x_2 < R$$

which simulates fully developed flow and therefore the boundary should be chosen sufficiently far downstream for this condition to be acceptable.

iii) On Walls

$$u_1 = u_2 = 0$$

Pressure was specified as zero at one node point on the down-stream boundary.   All calculated values are, therefore, relative to this point.

The finite element mesh used when analysing laminar flow problems is as shown in Figure 2.   It was found that upwinding techniques, Heinrich (1977), are necessary to obtain oscillation free results with the mesh utilised.

The velocity profiles at two chosen Reynolds numbers, 73 and 191, and two types of inlet boundary conditions are compared with experiment and other numerically determined values on Figures 3 and 4.   Recirculation zone lengths are compared with experiment and other numerical results, Figures 5, 6 and 7.

Turbulent Flow

The large near wall velocity gradients which exist in turbulent flow merits special consideration when discretising the flow domain.   Such gradients could be accommodated by mesh refinement in the near wall zone. However, this would result in a

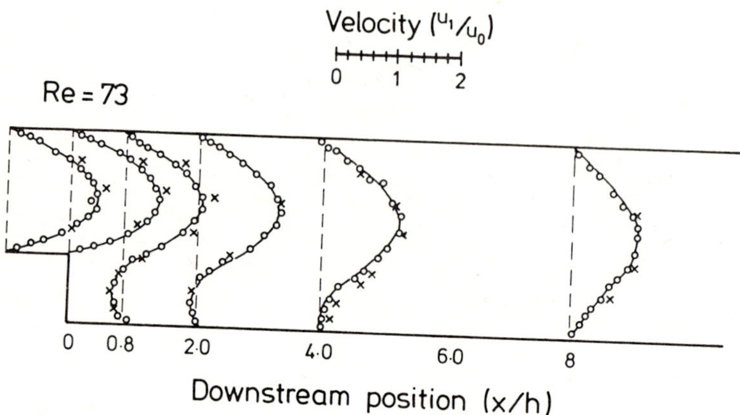

FIGURE 3   VELOCITY PROFILES - LAMINAR FLOW RN 73

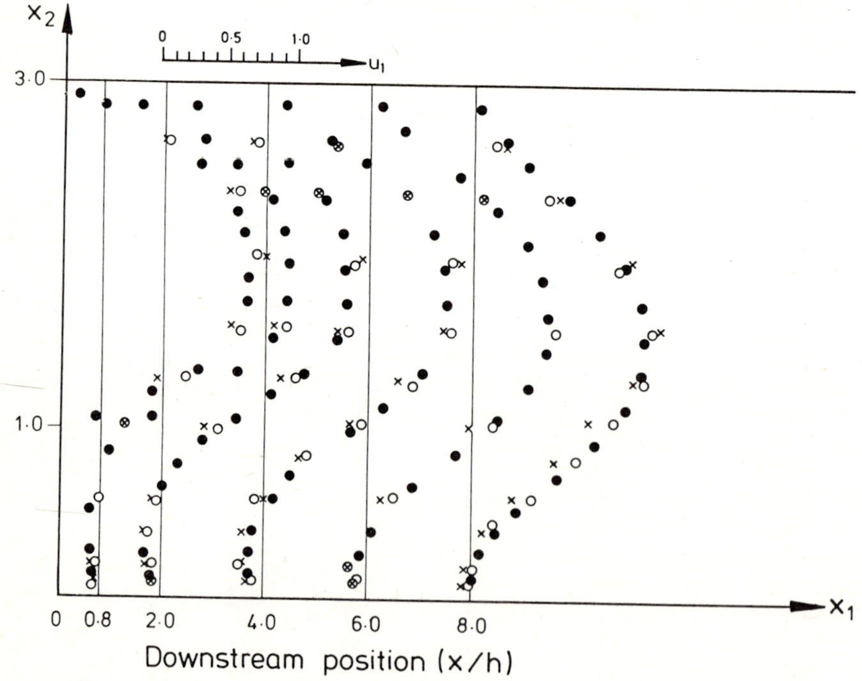

FIGURE 4   VELOCITY PROFILES - LAMINAR FLOW RN 191

Predictions

× Fully developed inlet profile with upwinding
⊙ Experimental inlet profiles with upwinding
✳ Finite elements, fully developed inlet profiles
○ Finite elements, experimental inlet profiles

} A.G.Hutton
} J.R.M.Smith
   (1978)

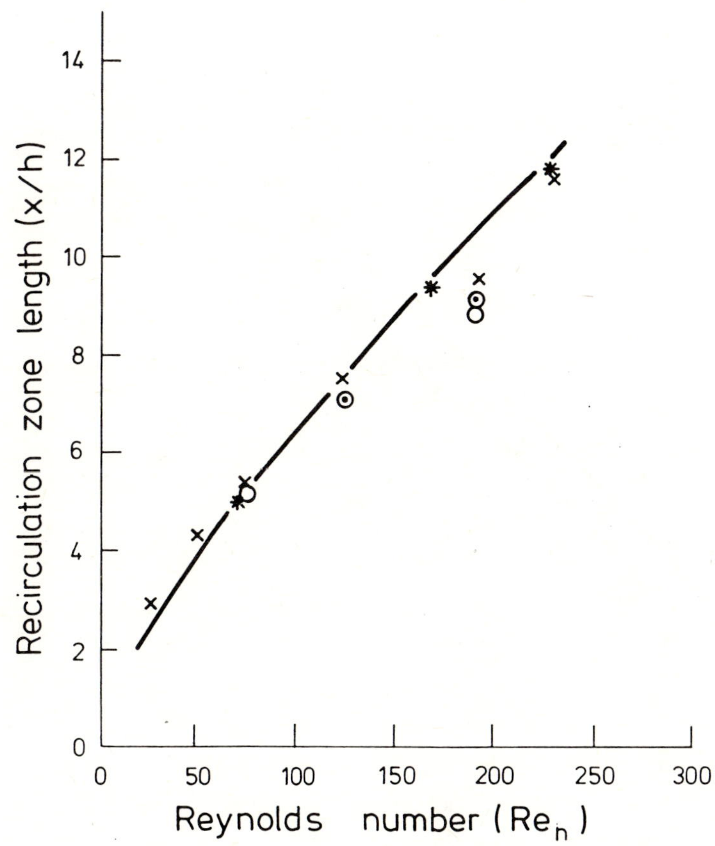

FIGURE 5   RECIRCULATION ZONE LENGTH – COMPARISON WITH OTHER NUMERICAL RESULTS

<u>Predictions</u>

×   Fully developed inlet profiles with upwinding   } Finite
⊙   Experimental inlet profiles with upwinding   } elements

<u>Experiment</u>

●   Dye - tracer   } Denham and
○   Lazer anemometer   } Patrick

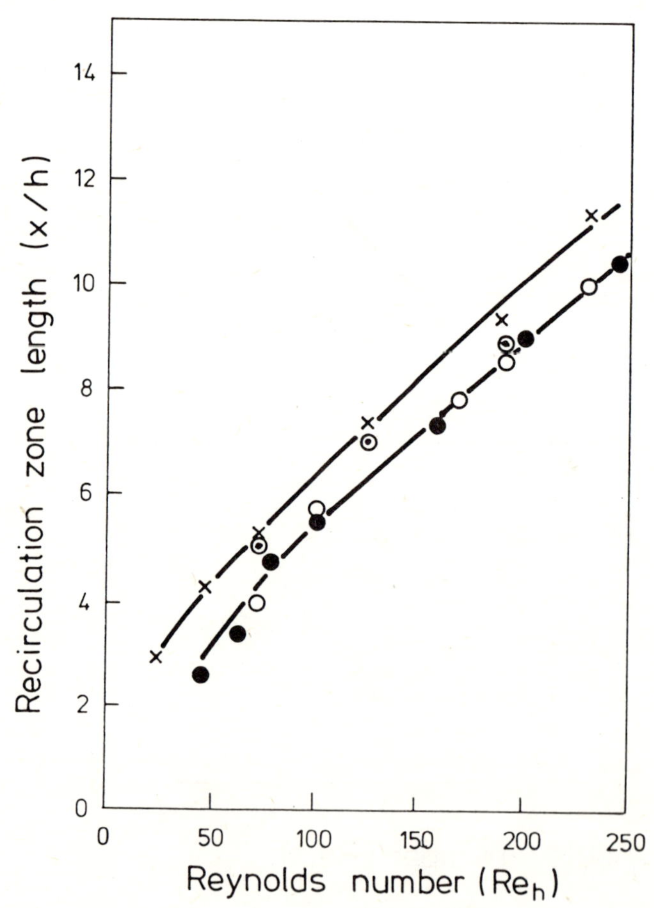

FIGURE 6   RECIRCULATION ZONE LENGTH - COMPARISON WITH EXPERIMENTAL RESULTS

<u>Predictions</u>

×   Fully developed inlet profiles - with upwinding

⊙   Experimental inlet profiles - with upwinding

✳   Fully developed inlet profiles - without upwinding

} Finite elements

∗   Fully developed inlet profiles - finite differences

+   Experimental inlet profiles - finite differences

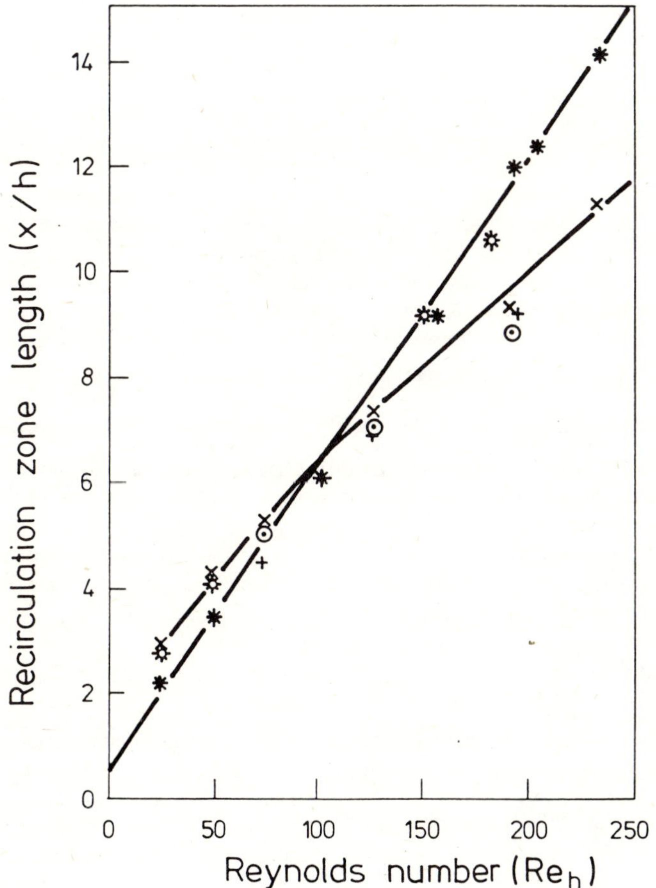

FIGURE 7   RECIRCULATION ZONE LENGTH – FURTHER COMPARISON WITH EXPERIMENTAL AND F.D. RESULTS

marked increase in both computer core and execution time. One
solution is to use special wall elements, Taylor et al. (1977)
or utilise the concept of 'universal laws' Davis (1972) over
the solid boundaries, Figure 8, where

$$
\left.
\begin{aligned}
u_i^* &= \lambda^* & 0 \le \lambda^* \le 5 \\
u_i^* &= (-3.05+5.0 \log \lambda^*) (\tau_w/|\tau_w|) & 5 \le \lambda^* \le 30 \\
u_i^* &= (-5.5+2.5 \log \lambda^*) (\tau_w/|\tau_w|) & \lambda^* \ge 30
\end{aligned}
\right\}
\qquad (15)
$$

in which,

$$
u_i^* = u_i/\sqrt{|\tau_w/\rho|}; \quad \lambda^* = \left(\frac{\lambda}{\mu}\right) \sqrt{|\tau_w/\rho|}
$$

$$
i = 1,2 \text{ for boundaries parallel to } x_1, x_2 \text{ axes, respectively.}
$$

The shear stress at the wall is given by,

$$
\tau_w = \mu \frac{\partial u_i}{\partial \lambda}
$$

and is assumed to be constant within the small distance λ'
from the wall where the finite element mesh begins, Figure 8.
For regions such as that shown on Figure 9, some problems may
arise when attempting to evaluate shear stresses near the
reattachment points.   For the node 'i' considered the value
of $\frac{\partial u_1}{\partial \lambda}$ is positive and thus gives rise to a positive shear
stress at the wall, $\mu \frac{\partial u_1}{\partial \lambda}$.   In reality however, this should
have been negative with a corresponding negative u^*.   This
would, in effect, move the re-attachment point upstream if
no compensation were made.   Such a movement would be exag-
gerated by each successive iteration of the solution procedure
leading, eventually, to a zero recirculating length.   This
can be overcome, once understood, by simply calculating,

$$
\tau_w = \mu \left| \frac{\partial u_i}{\partial \lambda} \right| \left(\frac{u_i}{|u_i|} \right)
$$

along the walls.   This requires that the re-attachment point
lies between two pre-selected nodes.   An incorrect initial
guess at these nodes is, however, soon realised due to the
advent of large/small streamwise velocity gradients on opposite
sides of one of the nodes.   A better approximation to the
correct location can then be made and the solution procedure
continued.   In practice this converges very rapidly and is
not a limiting factor on the overall solution procedure.

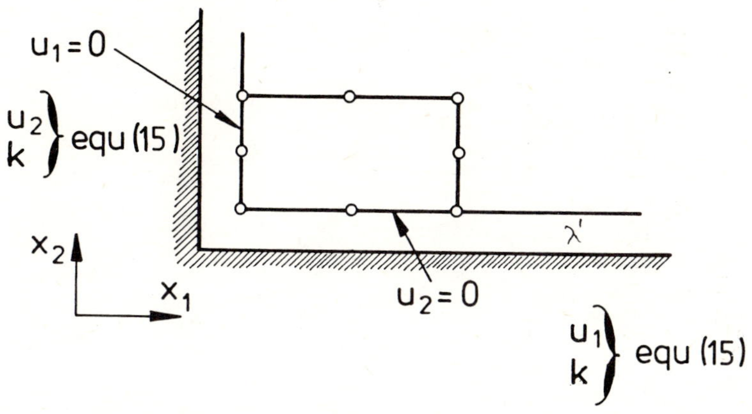

FIGURE 8   CONCEPT OF DISCRETISATION INITIATED INSIDE
DOMAIN

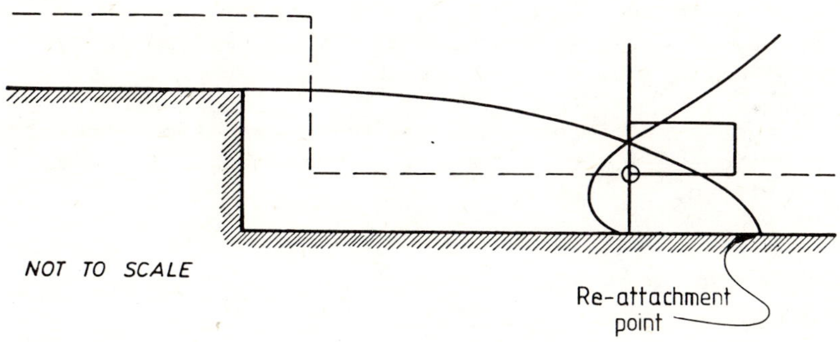

FIGURE 9   DETAILED RE-ATTACHMENT ZONE

One Equation Model

The boundary conditions used in the one equation model are as follows,

i) Upstream

$$u_1 \text{ - specified}$$
$$u_2 \text{ - 0}$$
$$k \text{ - specified}$$

$$x_1 = 0, \ 0 < x_2 < R$$
$$u_1 \text{ - experimental values}$$
$$k \text{ - experimental values}$$
$$\text{Denham et al. (1979)}$$

ii) Downstream

$$\frac{\partial u_1}{\partial x_1} = 0$$

$$\frac{\partial u_2}{\partial x_1} = 0$$

$$\frac{\partial k}{\partial x_1} = 0$$

$$0 < x_2 < R$$

again effectively simulating fully developed conditions. On the solid boundary, again

$$u_1 = u_2 = 0$$

A rather finer mesh, Figure 10, was used in this case, which proved satisfactory for Reynolds numbers in the range 3025 - 200,000 which is the same as that investigated by Atkins et al. (1980).

As in the case of laminar flow the matrix equation resulting from equations (2), (8) and (10), see Appendix 2 for details, can be written,

$$[H] \ \{\beta\} = \ \{f\} \tag{16}$$

in which

$$\beta_{(k)} = \begin{Bmatrix} u_{1(k)} \\ P_{(k)} \\ u_{2(k)} \\ k_{(k)} \end{Bmatrix} \qquad \text{for corner nodes}$$

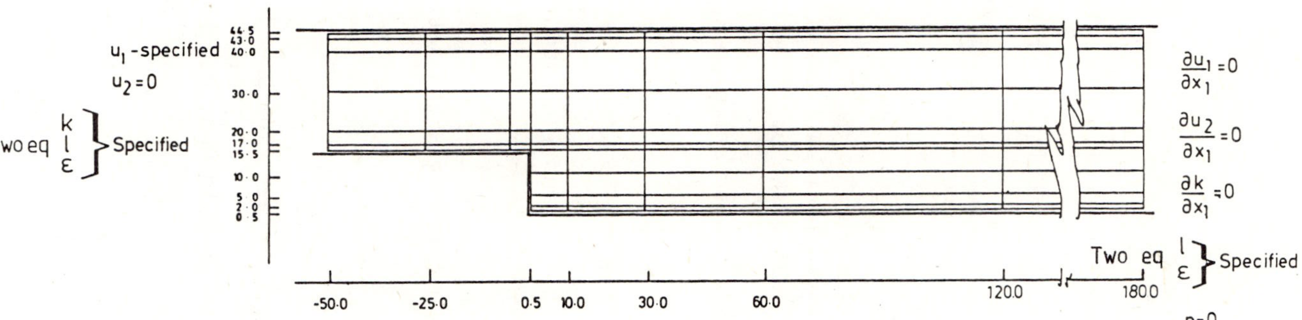

FIGURE 10   F.E. MESH FOR ONE AND TWO EQUATION MODEL

300

and

$$\beta_{(k)} = \begin{Bmatrix} u_{1(k)} \\ u_{2(k)} \\ k_{(k)} \end{Bmatrix} \qquad \text{for a midside node.}$$

Solution Procedure

Again non-linear terms such as $u_j \dfrac{\partial u_i}{\partial x_j}$ are replaced by $\tilde{u}_j \dfrac{\partial u_i}{\partial x_j}$ and the following iterative procedure adopted.

i) impose initial conditions on velocity.

ii) evaluate the resulting distribution of wall shear stress using the relationship

$$\tau_w = \left| \mu \frac{\partial u_i}{\partial \lambda} \right| \frac{u_i^*}{|u_i|} \tag{17}$$

iii)   use equation (15) and re-evaluate initial values of velocity along the boundary of the near wall region.

iv)   determine the variation in k in the near wall region from,

$$k = |\tau_w| / \sqrt{C_D}$$

v)   solve (2), (8) and (10) simultaneously to obtain the current iteration values for u_i, p and k.

vi)   repeat the process from (ii), until the required convergence criteria are satisfied.

For both laminar and turbulent flow a relaxation procedure,

$$\underset{\sim}{\beta} = \alpha \underset{\sim}{\beta} + (1-\alpha) \underset{\sim}{\beta}$$

was used.   For laminar flow $\alpha = 0.5$ proved adequate whereas for turbulent flow $\alpha = 0.8$ proved more appropriate. No up-winding was required to obtain oscillation free solutions for the range of Reynolds numbers considered.

Using the values $C_D = 0.22$, $\sigma_k = 1.53$, $C_1 = 1.45$ and $C_2 = 0.18$, a comparison was made with the results of Denham et al. (1975) and the one equation finite difference model of Atkins (1980) on Figure 11.   Although the numerical results for velocity distribution as obtained by both numerical methods agreed quite well, marked differences in the turbulence kinetic energy profiles are apparent.   The finite element model results are in closer agreement with experimental measurements than those obtained using the Finite Difference technique. Similar plots for a Reynolds number of 200,000 is shown in Figure 12.
The streamline plot corresponding to the one equation model is presented on Figure 13.   This exhibits a re-attachment length of approximately 5.6h.

Two Equation Model

In this case the imposed boundary conditions are,

 i) Upstream

u_1 - specified

$u_2 = 0$

k   - specified

ℓ, ε - fully developed values of the one equation model

These were obtained by analysing flow in a channel of constant width equal to the width upstream of the step. Fully developed flow was simulated by imposing the downstream boundary condition

$$\left. \begin{array}{l} \dfrac{\partial u_i}{\partial x_1} = 0 \\[2em] \dfrac{\partial u_2}{\partial x_1} = 0 \\[2em] \dfrac{\partial k}{\partial x_1} = 0 \\[2em] p = 0 \end{array} \right\} \quad 0 < x_2 < R$$

ii) Downstream

These are again compatible with fully developed flow,

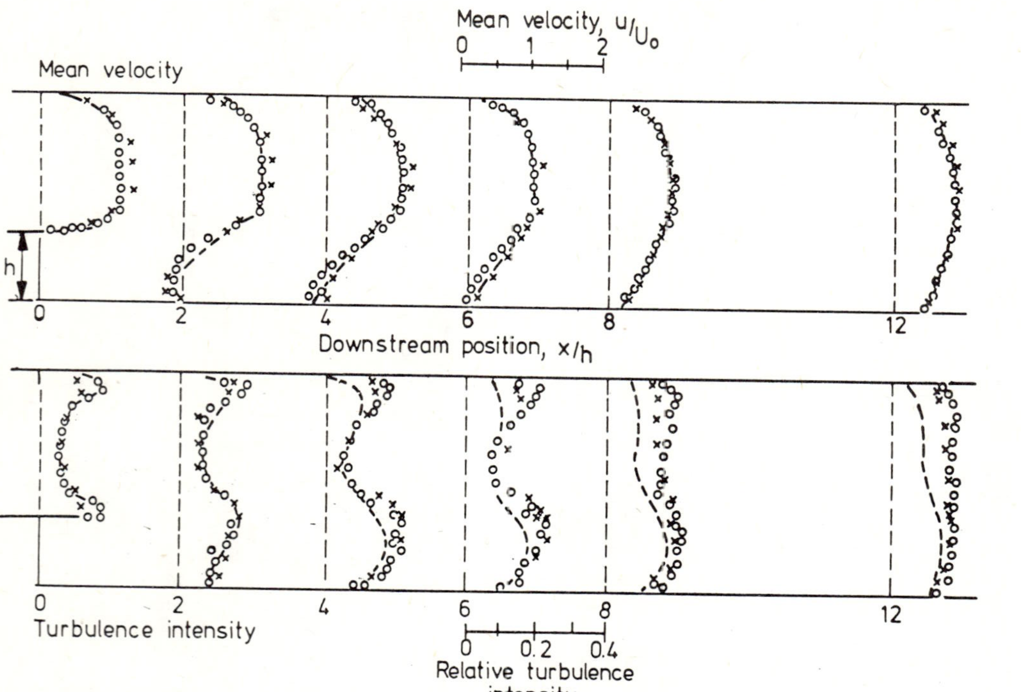

FIGURE 11   VELOCITY AND TURBULENCE INTENSITY PLOTS FOR RE = 3025 – ONE EQUATION MODEL

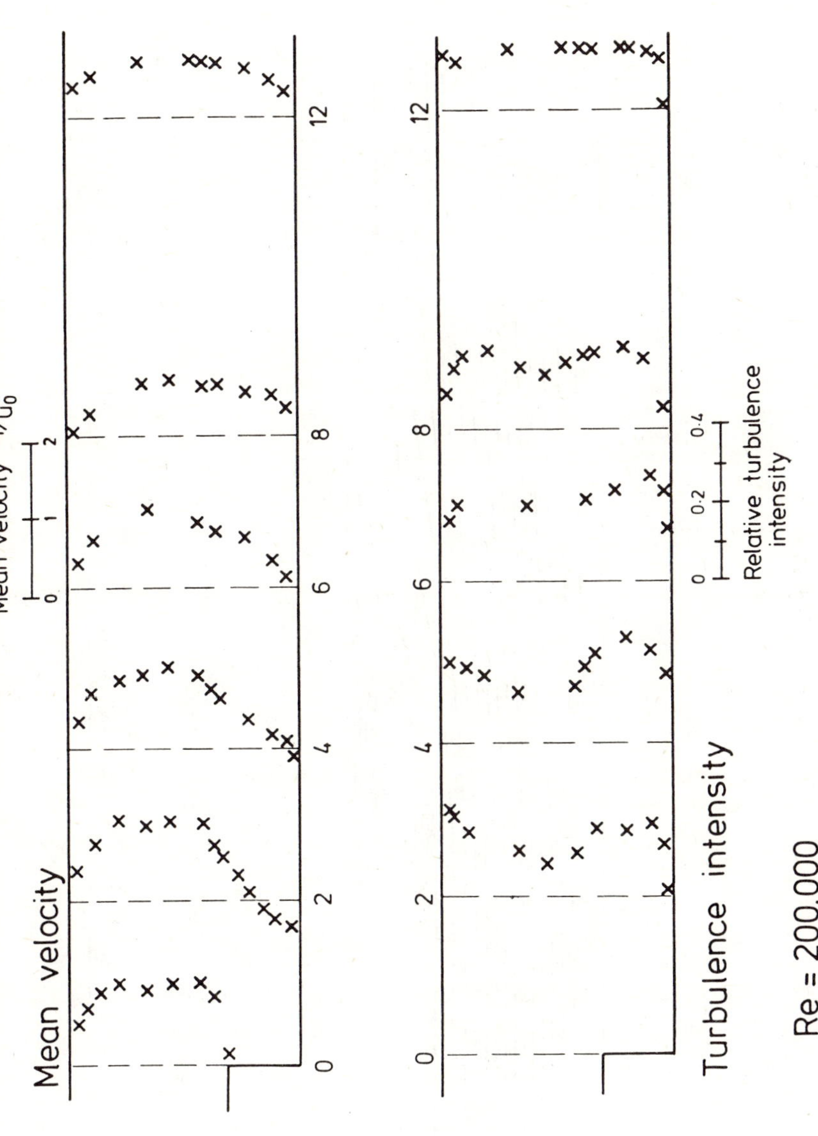

FIGURE 12   VELOCITY AND TURBULENCE INTENSITY PLOTS RE = 200,000 – ONE
EQUATION MODEL

L

FIGURE 13   STREAMLINE PLOTS FOR THE ONE EQUATION MODEL RE = 3025

$$\frac{\partial u_1}{\partial x_1} = 0$$

$$\frac{\partial u_2}{\partial x_1} = 0$$

$$\frac{\partial k}{\partial x_1} = 0$$

ℓ - fully developed values from one equation model

$$\frac{\partial \varepsilon}{\partial x_1} = 0$$

$$p = 0$$

iii) Wall

The universal profile is again utilised.

The one equation model discretisation, Figure 10, was found to be suitable.   In this case,

$$\beta_{(k)} = \begin{Bmatrix} u_{1(k)} \\ p_{(k)} \\ u_{2(k)} \\ k_{(k)} \\ \varepsilon_{(k)} \end{Bmatrix} \qquad \text{for a corner node}$$

and

$$\beta_{(k)} = \begin{Bmatrix} u_{1(k)} \\ u_{2(k)} \\ k_{(k)} \\ \varepsilon_{(k)} \end{Bmatrix} \qquad \text{for a midside node}$$

when all equations were incorporated into the global matrix [H].A slight modification, however, resulted in a measurable saving of both core space and execution time. The modification consisted, simply, of uncoupling the ε equation and solving separately. The overall solution technique employed was, when starting from zero initial conditions,

   i) Set all initial values to zero and assume that the effective viscosity corresponds to the molecular viscosity.

  ii) Solve for u_i, p and k for a fixed distribution of ℓ (3 iterations on 1st entry to problem proved advantageous).

iii) Estimate near wall and boundary conditions for ε using,

$$\varepsilon = C_\mu \frac{k^{3/2}}{\ell}$$

  iv) Solve for ε using fixed values of u_i, p and k.

   v) Test on convergence of ε, (i) not converged repeat from (ii).

  vi) Update ℓ using $\ell = C_\mu \dfrac{k^{3/2}}{\varepsilon}$

 vii) Update μ_ε using the new k and ℓ values.

viii) Using equation (15) re-estimate the wall shear stress and therefore boundary conditions on u_i and k.

  ix) Repeat from (ii) until convergence criteria satisfied.

Relaxation factors, for the particular mesh used, of 0.8 on all variables except μ_t where a value of 0.5 was used. If negative values of k and ε were encountered, k was set to a small positive value and if ε continued to be negative then ℓ_n was set to $\frac{1}{2}(\ell_{n-1}+\ell_{n+1})$. Following such a procedure and incorporating the particular technique for dealing with the re-attachment point resulted in a stable convergent algorithm.

The mesh used for evaluating fully developed profiles in the constant width channel is shown on Figure 14. The fully developed values of k were then used as upstream boundary conditions for the step problem.

The coefficients, assumed constant, used in the analyses of the two equation mode are,

$$C_\mu = 0.22, \ \sigma_k = 1.00, \ C_1 = 1.45, \ C_2 = 0.18, \ \sigma_\varepsilon = 1.3 \text{ and}$$
$$C_D = 0.092$$

which correspond to those used by Atkins (1974). The results are compared, Figure 15, with the experimental values of Denham et al. (1975) and the numerical results of Atkins (1974). The overall distribution of k is as shown on Figure 16 and a stream function plot is presented on Figure 17. The two

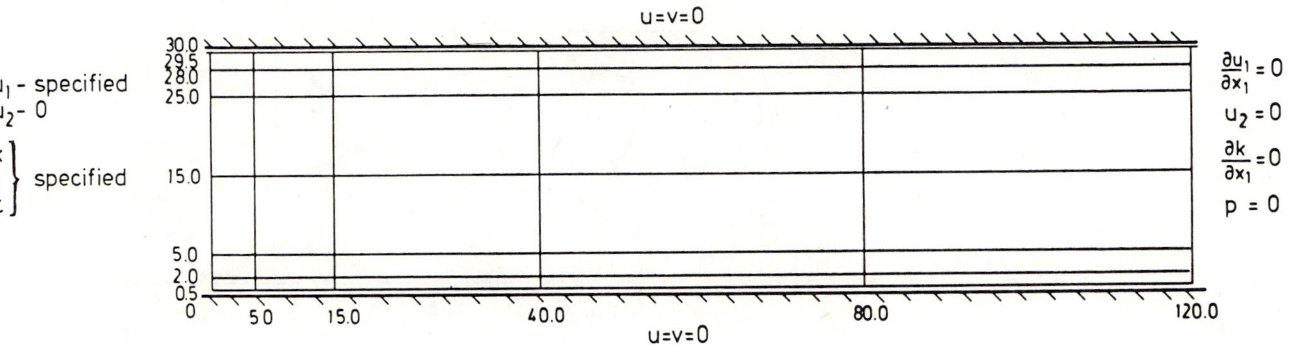

FIGURE 14   STRAIGHT CHANNEL MESH AND BOUNDARY CONDITIONS

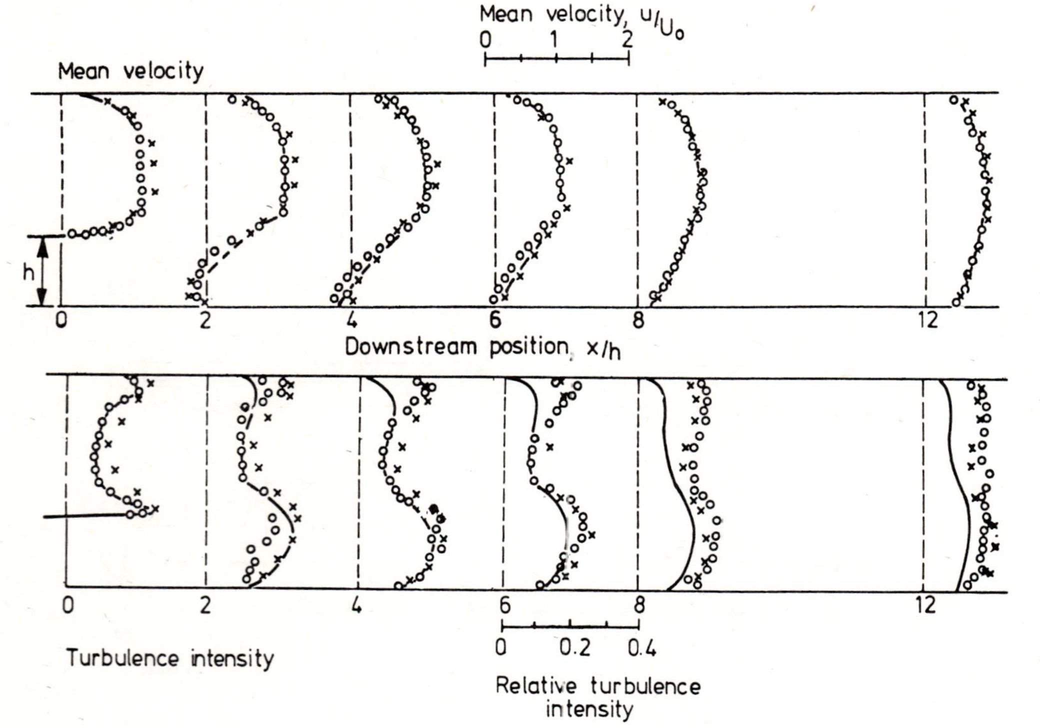

FIGURE 15   VELOCITY AND TURBULENCE INTENSITY PLOTS FOR RE = 3025 TWO EQUATION MODEL

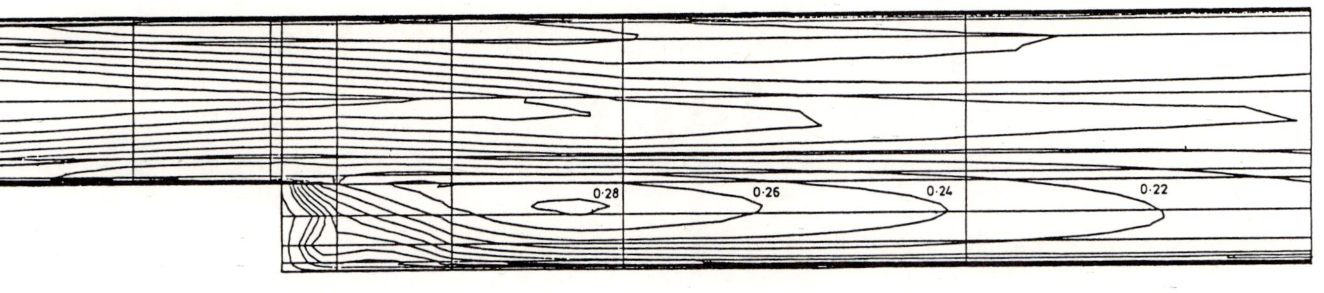

FIGURE 16   CONTOUR PLOTS OF TURBULENT KINETIC ENERGY

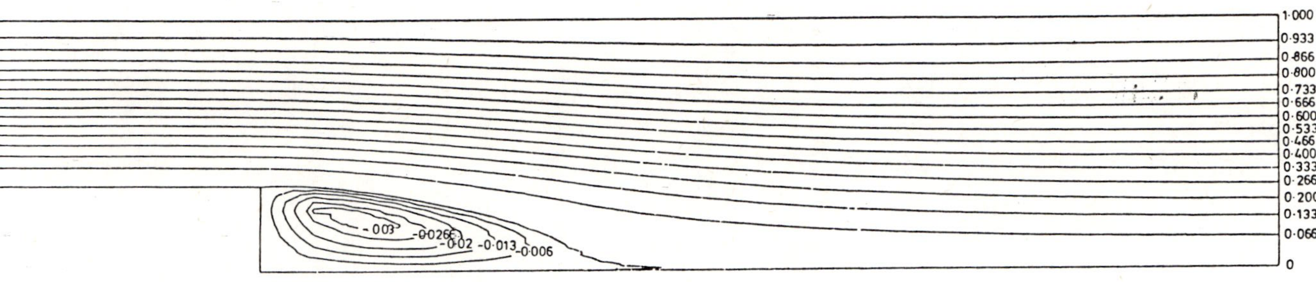

FIGURE 17   STREAMLINE PLOT FOR THE TWO EQUATION MODEL RE = 3025

equation model converged after 125 iterations compared with 30 iterations for the one equation model.

An estimate of the re-attachmeng length is 4.5 step heights, compared with 5.6 step heights obtained in the one equation model. Again, marked differences are apparent in the turbulence kinetic energy values depending on whether finite difference or finite element models are used.

Generally, it can be stated that the F.E.M. is a viable technique for solving laminar and turbulent flow problems where circulation is a principal feature of the flow pattern. It appears, however, that for the particular problem analysed nothing seems to have been gained in introducing the more complex two equation model. Comparison of velocity with measured quantities are equally good whilst the distribution of turbulence kinetic energy seems slightly better with the one equation model.

In both cases, however, the prediction of turbulence kinetic energy is better than that obtained using the finite difference models, which seem to overdamp. This is probably due to the introduction of upwinding which proved unnecessary when using the F.E.M. for turbulent flow prediction. The re-attachment length was also estimated rather better with the one equation model although both were less than that measured,

	F.D.		F.E.	
	ONE EQ.	TWO EQ.	ONE EQ.	TWO EQ.
Re-attachment length	5.2h	4.2h	5.6h	4.5h

TABLE 1

It appears that both numerical techniques result in some over-damping although some experimentation with the coefficients utilised could be a worthwhile exercise, particularly in the F.E.M.

REFERENCES

ATKINS,D.J. (1974) Numerical Studies of Separated Flows, Ph.D. Thesis, Univ. Exeter.

ATKINS,D.J., MASKELL,S.J. and PATRICK,M.A. (1980), Numerical Prediction of Separated Flows, Int. J. Num. Meth. in Engng., 15, 1, 129-144.

BAKER,A.J. (1978), Finite Element Analysis of Turbulent Flows, Proc. Int. Conf. Num. Meth. in Laminar and Turbulent Flow, Swansea, 203-209.

BAKER,A.J. and SOLIMAN,M.O. (1980), Analysis of a Finite Element Algorithm for Numerical Predictions in Water Resources Research, Proc. 3rd Int. Conf. on F.E. in Water Resources, Mississippi, 1980.

CRANDAL,S.H., (1966), Engineering Analysis, McGraw-Hill.

DENHAM,M.K., BRIARD,P. and PATRICK,M.A. (1975), A Directionally Sensitive Laser Anemometer for Velocity Measurements in Highly Turbulent Flow, J. of Phys. E. Scientific Inst., 8, 681-683.

GRESHO,P.M. and LEE,R.L. (1979), Don't Suppress the Wiggles - They're Telling You Something, Finite Element Methods for Convection Dominated Flows, Hughes,T.J.R. (Ed.) A.S.M.E., AMD, 34, 37-61.

GRESHO,P.M. and LEE,R.L. (1980), On the Time-dependent Solution of the Incompressible Navier-Stokes Equations in Two and Three Dimensions, Recent Advances in Numerical Methods in Fluids, 1, Pineridge Press, Swansea, 27-79.

GOSMAN,A.D., PUN,W.M., RUNCHAL,A.K., SPALDING,D.B. and WOLFSHTEIN,M. (1969), Heat and Mass Transfer in Recirculating Flows, Academic Press, London.

HINTON,E. and OWEN,D.R.J. (1980), An Introduction to F.E. Computations, Pineridge Press, U.K.

HINTON,E. and OWEN,D.R.J. (1980), A Simple Guide to F.E. Pineridge Press, U.K.

HINTON,E. and OWEN,D.R.J. (1980), An Introduction to Finite Element Computations, Pineridge Press, Swansea, U.K.

HUGHES,T.J.R. and BROOKS,A. (1979), A Multi-dimensional Upwind Scheme with no Crosswind Diffusion, Finite Element Methods for Convection Dominated Flows, Hughes, T.J.R. (ed), A.S.M.E., AMD 34, 19-36.

HUTTON,A.G. and SMITH,R.M. (1979), The Prediction of Laminar Flow over a Downstream Facing Step, C.E.G.B. Report RD/B/H3660.

LAUNDER,B.E. and SPALDING,D.B. (1972), Lectures in Mathematical Models of Turbulence, Academic Press, London.

LAUNDER,B.E. and SPALDING,D.B. (1974), The Numerical Computation of Turbulent Flows, Computer Methods in App. Mech. and Engng., 3, 269.

NG,K.H. and SPALDING,D.B. (1972), Turbulence Model for Boundary Layers Near Walls, Phys. of Fluids, 15, 1:20.

ODEN,J.T. (1972), Finite Elements of Non-linear Continua, McGraw-Hill.

OLSON,M.D. and TUANN,S.Y. (1978) Primitive Variables versus Stream Function Finite Elements for the Solution of the Navier-Stokes Equations, F.E. In Fluids, Vol.3, Wiley.

RUNCHAL,A.K. (1969), Transfer Processes in Two Dimensional Steady Separated Flows, Ph.D. Thesis, Univ. of London.

TAYLOR,C. and HOOD,P. (1973), A Numerical Solution of the Navier-Stokes Equations using the Finite Element Technique, Int. J. Computers and Fluids, 1, 1, 73-100.

TAYLOR,C. and HOOD,P. (1974), Navier-Stokes Equation using Mixed Interpolation, Proc. Int. Conf. F.E.M. in Flow Problems, Pentech Press, London, 121-132.

TAYLOR,C., HUGHES,T.G. and MORGAN,K. (1977), A Numerical Analysis of Turbulent Flow in Pipes, J. Comp. and Fluids, 5, 191-204.

THOMAS,C.E., MORGAN,K. and TAYLOR,C. (1981), A Finite Element Analysis of Flow over a Backward Facing Step - to be published, J. Comp. and Fluids.

WOLFSHTEIN,M. (1967), Convection Process in Turbulent Impinging Jet, Ph.D. Thesis, Univ. of London.

WOLFSHTEIN,M. (1970), Some Solutions of the Plane Turbulent Impinging Jet, Trans. A.S.M.E., J. Basic Engng., 92, 915-922.

APPENDIX 1

TRANSPORT EQUATIONS

Transport equations for both k and ε are presented below.

The Transport Equation for k

The Navier Stokes and mean momentum equations are

$$\rho\{\frac{\partial u_i}{\partial t} + u_j \frac{\partial u_i}{\partial x_j}\} = - \frac{\partial P}{\partial x_i} + \frac{\partial}{\partial x_j} \{\mu(\frac{\partial u_i}{\partial x_j} + \frac{\partial u_j}{\partial x_i})\} \tag{A.1.1}$$

and

$$\rho\{\frac{\partial \bar{u}_i}{\partial t} + \bar{u}_j \frac{\partial \bar{u}_i}{\partial x_j}\} = - \frac{\partial \bar{P}}{\partial x_i} + \frac{\partial}{\partial x_j} \{\mu(\frac{\partial \bar{u}_i}{\partial x_j} + \frac{\partial \bar{u}_j}{\partial x_i}) - \rho\overline{u_i'u_j'}\} \tag{A.1.2}$$

Subtracting (A.2.2) from (A.2.1) we have,

$$\rho\{\frac{\partial u_i'}{\partial t} + \bar{u}_j \frac{\partial u_i'}{\partial x_j}\} = - \frac{\partial P'}{\partial x_i} + \frac{\partial}{\partial x_j} \{\mu(\frac{\partial u_i'}{\partial x_j} + \frac{\partial u_j'}{\partial x_i})\}$$

$$+ \frac{\partial}{\partial x_j} (\rho\overline{u_i'u_j'}) - \rho u_j' \frac{\partial u_i}{\partial x_j} \tag{A.1.3}$$

Multiply (A.1.3) by u_j' and add to the product of u_i' and equation (A.1.3) for u_j

$$\rho \frac{\partial}{\partial t} (u_i'u_j') + \rho \bar{u}_k \frac{\partial}{\partial x_k} (u_i'u_j') = - \{u_i' \frac{\partial P'}{\partial x_i} + u_j' \frac{\partial P'}{\partial x_j}\}$$

$$+ \mu\{u_i' \frac{\partial}{\partial x_k} (\frac{\partial u_i'}{\partial x_k} + \frac{\partial u_k'}{\partial x_i})\} + u_i' \frac{\partial}{\partial x_k} (\frac{\partial u_j'}{\partial x_k} + \frac{\partial u_k'}{x_j})\} - \rho u_i'u_k' \frac{\partial u_j}{\partial x_k}$$

$$+ \rho u_j' \frac{\partial}{\partial x_k} (\rho\overline{u_i'u_k'}) + \rho u_i' \frac{\partial}{\partial x_k} (\rho\overline{u_j'u_k'}) - \rho u_j' u_k' \frac{\partial u_i}{\partial x_k} \tag{A.1.5}$$

Taking the mean of (A.2.5)

$$\rho \frac{\partial}{\partial t} (\overline{u_i'u_j'}) + \rho \bar{u}_k \frac{\partial}{\partial x_k} (\overline{u_i'u_j'}) = - \overline{\{u_i' \frac{\partial P'}{\partial x_i} + u_j' \frac{\partial P'}{\partial x_j}\}}$$

$$+ \overline{\mu\{u_j' \frac{\partial}{\partial x_k} (\frac{\partial u_i'}{\partial x_k} + \frac{\partial u_k'}{\partial x_i}) + u_i' \frac{\partial}{\partial x_k} (\frac{\partial u_j'}{\partial x_k} + \frac{\partial u_k'}{\partial x_j}\}} - \rho \frac{\partial}{\partial x_k}$$

$$\overline{(u_i'u_j'u_k')} - \overline{\rho u_i'u_k'} \frac{\partial \bar{u}_j}{\partial x_k} - \overline{\rho u_j'u_k'} \frac{\partial \bar{u}_i}{\partial x_k} \tag{A.1.6}$$

314

Since $\rho u_i \dfrac{\partial}{\partial x_k} \overline{(\rho u'_j u'_k)} = \rho \overline{u}_i \dfrac{\partial}{\partial} \overline{(\rho u'_j u'_k)} = 0$

and $- \overline{\rho u'_i u'_k \dfrac{\partial u'_j}{\partial x_k}} - \overline{\rho u'_j u'_k \dfrac{\partial u'_i}{\partial x_k}} = - \rho \dfrac{\partial}{\partial x_k} \overline{(u'_i u'_j u'_k)}$

$+ \overline{\rho u'_j u'_k \dfrac{\partial u'_i}{\partial x_k}} - \overline{\rho u'_j u'_k \dfrac{\partial u'_i}{\partial x_k}}$

Multiply equation (A.1.6) by δ_{ij}:

$$\rho \frac{\partial}{\partial t} (\overline{u'_i{}^2}) + \rho \overline{u}_k \frac{\partial}{\partial x_k} (\overline{u'_i{}^2}) = - 2 \overline{u'_i \frac{\partial P'}{\partial x_i}}$$

$$+ 2\mu \, \overline{u_i \frac{\partial}{\partial x_k} \left(\frac{\partial u'_i}{\partial x_k} + \frac{\partial u'_k}{\partial x_i} \right)} - 2 \overline{\rho u'_i u'_k} \frac{\partial u'_i}{\partial x_k} - \rho \frac{\partial}{\partial x_k} \overline{(u'_i{}^2 u'_k)}$$

$$(A.1.7)$$

i.e. $\rho \dfrac{\partial k}{\partial t} + \rho \overline{u}_k \dfrac{\partial k}{\partial x_k} = - \dfrac{\partial}{\partial x_i} \overline{(u'_i P')} + \mu \dfrac{\partial}{\partial x_k} \overline{\left\{ u'_i \left(\frac{\partial u'_i}{\partial x_k} + \frac{\partial u'_k}{\partial x_i} \right) \right\}}$

$$- \mu \overline{\left(\frac{\partial u'_i}{\partial x_k} + \frac{\partial u'_k}{\partial x_i} \right) \frac{\partial u'_i}{\partial x_k}} - \rho \overline{u'_i u'_k} \frac{\partial \overline{u}_i}{\partial x_k} - \frac{\rho}{2} \frac{\partial}{\partial x_k} \overline{(u'_i{}^2 u'_k)}$$

$$= - \frac{\partial}{\partial x_i} \overline{(u'_i P')} + \mu \frac{\partial^2 k}{\partial x_k{}^2} - \mu \overline{\left(\frac{\partial u'_i}{\partial x_k} \right)^2} - \frac{\rho}{2} \frac{\partial}{\partial x_k} \overline{(u'_i{}^2 u'_k)}$$

$$- \rho \overline{u'_i u'_k} \frac{\partial \overline{u}_i}{\partial x_k} = - \frac{\partial}{\partial x_i} \overline{\left(u'_i P' + \frac{\rho}{2} u'_k{}^2 u'_i \right)} + \mu \frac{\partial^2 k}{\partial x_k{}^2}$$

$$- \mu \overline{\left(\frac{\partial \overline{u}_i}{\partial x_k} \right)^2} - \rho \overline{u'_i u'_k} \frac{\partial \overline{u}_i}{\partial x_k}$$

$$(A.1.8)$$

We now use the Reynolds stress hypothesis,

$$-\rho \, \overline{u'_i u'_k} \frac{\partial \overline{u}_i}{\partial x_k} = \rho \, \mu_T \left(\frac{\partial \overline{u}_i}{\partial x_k} + \frac{\partial \overline{u}_k}{\partial x_i} \right) \frac{\partial \overline{u}_i}{\partial x_k} - \frac{1}{3} \overline{u'_i{}^2} \, \delta_{ik} \frac{\partial \overline{u}_i}{\partial x_k}$$

$$= \rho \, \mu_T \frac{\partial \overline{u}_i}{\partial x_k} \left(\frac{\partial \overline{u}_i}{\partial x_k} + \frac{\partial \overline{u}_k}{\partial x_i} \right)$$

$$(A.1.9)$$

Also, since $- \dfrac{\partial}{\partial x_k} \overline{\{u_i' P' + \dfrac{\rho}{2} u_k'^2 u_i\}}$ is analogous to

$- \dfrac{\partial}{\partial x_i} (\overline{u_i' u_k'})$ in the mean momentum equation, it is reasonable to assume that

$$- \overline{u_i' P'} + \dfrac{\rho}{2} \overline{u_k'^2 u_i'} = \dfrac{\mu_T}{\sigma_k} \dfrac{\partial k}{\partial x_i} \qquad (A.1.10)$$

where μ_T / σ_k is the turbulent diffusion coefficient, and σ_k is the turbulent Prandtl or Schmidt number.

Substituting from equations (A.1.9) and (A.1.10) in (A.1.8)

$$\rho \dfrac{\partial k}{\partial t} + \rho \bar{u}_k \dfrac{\partial k}{\partial x_k} = \mu \dfrac{\partial^2 k}{\partial x_k^2} + \rho\, \mu_T \dfrac{\partial \bar{u}_i}{\partial x_k} \left(\dfrac{\partial \bar{u}_i}{\partial x_k} + \dfrac{\partial \bar{u}_k}{\partial x_i} \right)$$

$$+ \dfrac{\partial}{\partial x_k} \left(\dfrac{\mu_T}{\sigma_k} \dfrac{\partial k}{\partial x_k} \right) - \mu \overline{\left(\dfrac{\partial u_i'}{\partial x_k} \right)^2} \qquad (A.1.11)$$

The last term represents physically viscous dissipation of energy; for which the simplest combination of k and ℓ giving dimensional consistency is

$$\mu \overline{\left(\dfrac{\partial u_i}{\partial x_k} \right)^2} = \rho\, C_D \dfrac{k^{3/2}}{\ell} \qquad (A.1.12)$$

so that the transport equation for k is

$$\rho \dfrac{\partial k}{\partial t} + \rho \bar{u}_k \dfrac{\partial k}{\partial x_k} = \dfrac{\partial}{\partial x_k} \left\{ \left(\mu + \dfrac{\mu_T}{\sigma_k} \right) \dfrac{\partial k}{\partial x_k} \right\} + \mu_T \dfrac{\partial \bar{u}_i}{\partial x_k} \left(\dfrac{\partial \bar{u}_i}{\partial x_k} + \dfrac{\partial \bar{u}_k}{\partial x_i} \right)$$

$$\qquad\qquad \text{CONVECTION} \qquad\qquad \text{DIFFUSION} \qquad\qquad\qquad \text{GENERATION}$$

$$- C_D\, \rho \dfrac{k^{3/2}}{\ell} \qquad (A.1.13)$$

$$\text{DISSIPATION}$$

The Transport Equation for Epilson

We start by differentiation of equation (A.1.3) w.r.t. x_k;

$$\frac{\partial}{\partial t}\left(\frac{\partial u_i'}{\partial x_k}\right) + \bar{u}_j \frac{\partial^2 u_i'}{\partial x_k \partial x_j} + \frac{\partial \bar{u}_j}{\partial x_k}\frac{\partial u_i'}{\partial x_j} = -\frac{1}{\rho}\frac{\partial^2 P'}{\partial x_i \partial x_k} + \nu \frac{\partial^3 u_i'}{\partial x_k \partial x_j}$$

$$+ \frac{\partial^2}{\partial x_k \partial x_j}\overline{(u_i' u_j')} - u_j'\frac{\partial}{\partial x_k}\frac{\partial}{\partial x_j}(\bar{u}_i + u_i') - \frac{\partial u_j'}{\partial x_k}\frac{\partial}{\partial x_j}(\bar{u}_i + u_i')$$

$$(A.1.14)$$

Multiply throughout by $2\nu\,\dfrac{\partial u_i'}{\partial x_k}$ and take the mean.

$$\frac{\partial}{\partial t}\overline{\left|\left(\frac{\partial u_i'}{\partial x_k}\right)^2 \nu\right|} + 2\nu \overline{\frac{\partial u_i'}{\partial x_k}\frac{\partial}{\partial x_k}\left(\bar{u}_j\frac{\partial u_i'}{\partial x_j}\right)} = -\frac{2\nu}{\rho}\overline{\frac{\partial u_i'}{\partial x_k}\frac{\partial^2 P'}{\partial x_i \partial x_k}}$$

$$+ 2\nu \overline{\frac{\partial u_i'}{\partial x_k}\frac{\partial^2}{\partial x_k \partial x_j}(u_i' u_j')} + 2\nu^2 \overline{\frac{\partial u_i'}{\partial x_k}\frac{\partial^3}{\partial x_j{}^2}\left(\frac{\partial u_i'}{\partial x_k}\right)}$$

$$- 2\nu \overline{\frac{\partial u_i'}{\partial x_k}\frac{\partial}{\partial x_k}\left\{u_j'\frac{\partial u_i'}{\partial x_j} + u_j'\frac{\partial u_i}{\partial x_j}\right\}} \qquad (A.1.15)$$

The terms in this equation may be interpreted as follows:-

$$\overline{\frac{\partial u_i'}{\partial x_k}\frac{\partial}{\partial x_k}\left(\bar{u}_j\frac{\partial u_i'}{\partial x_j}\right)} = \overline{\frac{\partial u_i'}{\partial x_k}\left(\frac{\partial u_j}{\partial x_k}\frac{\partial u_i'}{\partial x_j} + \bar{u}_j\frac{\partial^2 u_i'}{\partial x_j \partial x_k}\right)} = \overline{\frac{\partial \bar{u}_j}{\partial x_k}\frac{\partial u_i'}{\partial x_k}\frac{\partial u_i'}{\partial x_j}}$$

$$+ \bar{u}_j\frac{\partial}{\partial x_j}\left\{\frac{1}{2}\left(\frac{\partial u_i}{\partial x_k}\right)^2\right\} \qquad (A.16.a)$$

$$\overline{\frac{\partial u_i'}{\partial x_k}\frac{\partial^2 P'}{\partial x_i \partial x_k}} = \frac{\partial}{\partial x_i}\overline{\left\{\frac{\partial u_i}{\partial x_k}\frac{\partial P'}{\partial x_k}\right\}} - \overline{\frac{\partial P'}{\partial x}\frac{\partial^2 u_i'}{\partial x_i \partial x_k}}$$

$$= \frac{\partial}{\partial x_i}\overline{\left\{\frac{\partial u_i'}{\partial x_k}\frac{\partial P'}{\partial x_k}\right\}} \qquad (A.16.b)$$

$$\overline{\frac{\partial u_i'}{\partial x_k}\frac{\partial^2}{\partial x_k \partial x_j}(u_i' u_j')} = \overline{\frac{\partial u_i'}{\partial x_k}\frac{\partial^2}{\partial x_k \partial x_j}(u_i' u_j')} = \overline{\frac{\partial \bar{u}_i'}{\partial x_k}\frac{\partial^2}{\partial x_k \partial x_j}(u_i' u_j')} = 0$$

$$(A.16.c)$$

$$\overline{\frac{\partial u_i'}{\partial x_k}\frac{\partial}{\partial x_k}\left(u_j'\frac{\partial \bar{u}_i}{\partial x_j}\right)} = \overline{\frac{\partial u_i'}{\partial x_k}\frac{\partial u_j'}{\partial x_k}\frac{\partial \bar{u}_i}{\partial x_j}} + \overline{u_j'\frac{\partial u_i'}{\partial x_k}\frac{\partial^2 \bar{u}_i}{\partial x_j \partial x_k}} \qquad (A.16.d)$$

$$\overline{\frac{\partial u_i'}{\partial x_k}\frac{\partial}{\partial x_k}\left(u_j'\frac{\partial u_i'}{\partial x_j}\right)} = \overline{\frac{\partial u_i'}{\partial x_k}\frac{\partial u_j'}{\partial x_k}\frac{\partial u_i'}{\partial x_j}} + \overline{u_j'\frac{\partial}{\partial x_j}\left(\frac{1}{2}\frac{\partial u_i'}{\partial x_k}\right)^2} \qquad (A.16.e)$$

Defining $\varepsilon' = \dfrac{\mu}{\rho C_D}\left(\dfrac{\partial u_j}{\partial x_k}\right)^2$

then $2\nu\, u_j'\dfrac{\partial}{\partial x_j}\left\{\dfrac{1}{2}\left(\dfrac{\partial u_i'}{\partial x_k}\right)^2\right\} = 2\nu\, u_j'\dfrac{C_D \rho}{\partial \mu}\dfrac{\partial \varepsilon'}{\partial x_j} = u_j\, C_D\dfrac{\partial \varepsilon'}{\partial x_j}$

$$\qquad (A.16.f)$$

$$\overline{\frac{\partial u_i'}{\partial x_k}\frac{\partial^2}{\partial x_j^2}\left(\frac{\partial u_i'}{\partial x_k}\right)} = \frac{\partial}{\partial x_j}\overline{\left[\frac{\partial u_i'}{\partial x_k}\frac{\partial}{\partial x_j}\left(\frac{\partial u_i'}{\partial x_k}\right)\right]} - \overline{\left\{\frac{\partial}{\partial x_k}\left(\frac{\partial u_i'}{\partial x_k}\right)\right\}^2}$$

$$= \frac{\partial^2}{\partial x_j^2}\overline{\left\{\frac{1}{2}\left(\frac{\partial u_i'}{\partial x_k}\right)^2\right\}} - \overline{\left|\frac{\partial^2 u_i'}{\partial x_j \partial x_k}\right|^2} \qquad (A.16.g)$$

and substituting from equation (A.1.16,a,b,c,d,e,f,g) in
equation (A.1.15) gives

$$\frac{\partial}{\partial t}(C_D \varepsilon) + 2\nu\frac{\partial \bar{u}_j}{\partial x_k}\overline{\frac{\partial u_i'}{\partial x_k}\frac{\partial u_i'}{\partial x_j}} + 2\nu\,\bar{u}_j\frac{\partial}{\partial x_j}\overline{\left\{\frac{1}{2}\left(\frac{\partial u_i}{\partial x_k}\right)^2\right\}}$$

$$= -\frac{2\nu}{\rho}\frac{\partial}{\partial x_i}\overline{\left(\frac{\partial u_i'}{\partial x_k}\frac{\partial P'}{\partial x_k}\right)} + 2\nu^2\frac{\partial^2}{\partial x_j^2}\overline{\left\{\frac{1}{2}\left(\frac{\partial u_i'}{\partial x_k}\right)^2\right\}} - 2\nu^2\overline{\left\{\frac{\partial^2 u_i'}{\partial x_j \partial x_k}\right\}}$$

$$- 2\nu\frac{\partial^2 \bar{u}_i}{\partial x_k \partial x_j}\overline{u_j'\frac{\partial u_i'}{\partial x_k}} - C_D\,\overline{u_j'\frac{\partial \varepsilon'}{\partial x_j}} - 2\nu\overline{\frac{\partial u_i'}{\partial x_k}\frac{\partial u_j'}{\partial x_k}\frac{\partial u_i'}{\partial x_j}}$$

i.e.

$$C_D\frac{\partial \varepsilon}{\partial t} + C_D\,\bar{u}_j\frac{\partial \varepsilon}{\partial x_j} = -2\nu\frac{\partial \bar{u}_i}{\partial x_k}\left\{\overline{\frac{\partial u_j'}{\partial x_k}\frac{\partial u_j'}{\partial x_i}} + \overline{\frac{\partial u_k'}{\partial x_j}\frac{\partial u_i'}{\partial x_j}}\right\}$$

$$- \frac{2\nu}{\rho}\frac{\partial}{\partial x_i}\overline{\left\{\frac{\partial u_i'}{\partial x_k}\frac{\partial P}{\partial x_k}\right\}} + \nu\, C_D\frac{\partial^2 \varepsilon}{\partial x_j^2} - 2\nu^2\overline{\left(\frac{\partial^2 u_i'}{\partial x_j \partial x_k}\right)^2}$$

$$- 2\nu\frac{\partial^2 \bar{u}_i}{\partial x_k \partial x_j}\overline{u_j'\frac{\partial u_i'}{\partial x_k}} - C_D\,\overline{u_j'\frac{\partial \varepsilon'}{\partial x_j}} - 2\nu\overline{\frac{\partial u_i'}{\partial x_k}\frac{\partial u_j'}{\partial x_k}\frac{\partial u_i'}{\partial x_j}} \qquad (A.1.17)$$

This is the exact transport equation for ε.   The following assumptions will now be made to express unknown correlations in terms of mean quantities:-

$$-2\nu\,\frac{\partial \bar{u}_i}{\partial x_k}\,\overline{\left\{\frac{\partial u'_j}{\partial x_k}\frac{\partial u'_j}{\partial x_i} + \frac{\partial u'_k}{\partial x_j}\frac{\partial u'_i}{\partial x_j}\right\}} \qquad \text{is a production term.}$$

$$\delta_{ik}\left\{\overline{\frac{\partial u'_j}{\partial x_k}\frac{\partial u'_j}{\partial x_i}} + \overline{\frac{\partial u'_k}{\partial x_j}\frac{\partial u'_i}{\partial x_k}}\right\} \qquad \text{consists of components of}$$

$$\text{factors of } \frac{C_D\,\varepsilon}{\nu} \quad . \text{ (eq.A.1.12)}.$$

Consequently, and for dimensional consistency (Hanjalic and Launder, 1972)

$$-2\nu\,\frac{\partial \bar{u}_i}{\partial x_k}\left\{\overline{\frac{\partial u'_j}{\partial x_k}\frac{\partial u'_j}{\partial x_i}} + \overline{\frac{\partial u'_i}{\partial x_j}\frac{\partial u'_k}{\partial x_j}}\right\} = -\frac{\partial \bar{u}'_i}{\partial x_k}\left\{-C_1 C_D \varepsilon\,\frac{\overline{u'_i u'_k}}{k}\right.$$

$$\left. + C_3 C_D \varepsilon\,\delta_{ik}\right\} \qquad \text{where } C_1, C_3 \text{ are constants}$$

$$= C_1 C_D \frac{\varepsilon}{k}\frac{\partial \bar{u}_i}{\partial x_k}\left\{\frac{\mu_T}{\rho}\left(\frac{\partial \bar{u}_i}{\partial x_k} + \frac{\partial \bar{u}_k}{\partial x_i}\right) - \frac{2}{3}\,k\delta_{ik}\right\}$$

$$= C_1 C_D \frac{\varepsilon}{k}\frac{\mu_T}{\rho}\frac{\partial \bar{u}_i}{\partial x_k}\left(\frac{\partial \bar{u}_i}{\partial x_k} + \frac{\partial \bar{u}_k}{\partial x_i}\right) \qquad\qquad \text{(A.1.18a)}$$

$-\dfrac{\partial}{\partial x_j}\left(\overline{u'_j \varepsilon'}\right)$ is analogous to $-\dfrac{\partial}{\partial x_i}\left(\overline{u'_i u'_k}\right)$ in the mean momentum equations, representing diffusion of ε through velocity fluctuations, so that,

$$-\frac{\partial}{\partial x_j}\left(\overline{u'_j \varepsilon'}\right) = \frac{1}{\rho}\frac{\partial}{\partial x_j}\left(\frac{\mu_T}{\sigma_\varepsilon}\frac{\partial \varepsilon}{\partial x_j}\right) \qquad\qquad \text{(A.1.18b)}$$

where μ_T/σ_ε is a diffusion coefficient and σ_ε is the corresponding Prandtl or Schmidt number.

$-\dfrac{\nu}{\rho}\dfrac{\partial}{\partial x_j}\left(\overline{\dfrac{\partial P'}{\partial x_i}\dfrac{\partial u'_j}{\partial x_i}}\right)$ represents diffusion of ε through pressure fluctuations.

Chan (1945) derived an exact expression for this consisting of two terms, both of which contain a higher order of mean flow derivatives than appears in any other terms in the transport equation for ε.   For this

reason, this term is considered negligible relative to the other terms in (A.1.17)

i.e.

$$- \frac{\nu}{\rho} \frac{\partial}{\partial x_j} \overline{\left\{ \frac{\partial P'}{\partial x_i} \frac{\partial u'_j}{\partial x_i} \right\}} = 0 \qquad\qquad (A.1.18)$$

Consider the term, $- \frac{\partial^2 \bar{u}_i}{\partial x_k \partial x_j} \overline{u'_j \frac{\partial u'_i}{\partial x_k}}$

We approximate

$$\overline{\rho u'_j \frac{\partial u'_i}{\partial x_k}} = - \mu_T \frac{\partial^2 \bar{u}_i}{\partial x_j \partial x_k} \quad A_i \, \delta_{jk} + B_j \, \delta_{ki} + C_k \, \delta_{ij} + D\varepsilon_{ijk}$$

where A_i, B_j, C_k and D are undetermined functions and ε_{ijk} is the alternating tensor. Now

$$-\rho \frac{\partial^2 \bar{u}_i}{\partial x_k \partial x_j} \overline{u'_j \frac{\partial u'_i}{\partial x_k}} = \mu_T \left(\frac{\partial^2 \bar{u}_i}{\partial x_j \partial x_k} \right)^2 - A_i \frac{\partial^2 \bar{u}_i}{\partial x_j^2} - B_j \frac{\partial^2 \bar{u}_i}{\partial x_k \partial x_i}$$

$$- C_k \frac{\partial^2 \bar{u}_i}{\partial x_k \partial x_i} + D\varepsilon_{ijk} \frac{\partial^2 \bar{u}_i}{\partial x_k \partial x_i} = \mu_T \left(\frac{\partial^2 \bar{u}_i}{\partial x_j \partial x_k} \right)^2 - A_i \frac{\partial^2 \bar{u}_i}{\partial x_j^2} .$$

Now, setting

$$i = j \qquad \overline{\rho u'_i \frac{\partial u'_i}{\partial x_k}} = - \mu_T \frac{\partial^2 \bar{u}_i}{\partial x_i \partial x_k} + A_i \delta_{ik} + B_i \delta_{ki} + 3C_k + D\varepsilon_{iik}$$

$$\rho \frac{\partial k}{\partial x_k} = A_k + B_k + 3C_k$$

$$j = k \qquad \overline{\rho u'_k \frac{\partial u'_i}{\partial x_k}} = - \mu_T \frac{\partial^2 \bar{u}_i}{\partial x_k^2} + 3A_i + B_k \delta_{ki} + C_k\,_{ik} + D\varepsilon_{ikk}$$

$$- \frac{\partial}{\partial x_k} \left\{ \mu_T \left(\frac{\partial \bar{u}_i}{\partial x_k} + \frac{\partial \bar{u}_k}{\partial x_i} \right) - \frac{2}{3} \rho \, k^2 \, \delta_{ik} \right\} = 3A_k + B_k + C_k - \mu_T \frac{\partial^2 \bar{u}_i}{\partial x_k^2}$$

$$\frac{2}{3} \rho \frac{\partial k}{\partial x_k} - \frac{\partial \mu_T}{\partial x_k} \left(\frac{\partial \bar{u}_i}{\partial x_k} + \frac{\partial \bar{u}_k}{\partial x_i} \right) = 3A_k + B_k + C_k .$$

$$k = i \neq 0 = A_k \delta_{jk} + B_j \delta_{kk} + C_k \delta_{kj} + D \varepsilon_{kjk}$$

$$= A_k \delta_{jk} + 3B_j + C_k \delta_{kj}$$

$$= A_k + 3B_k + C_k .$$

Solving,

$$A_j = \frac{1}{6} \rho \frac{\partial k}{\partial x_j} - \frac{2}{5} \frac{\partial \mu_T}{\partial x_k} \left(\frac{\partial \bar{u}_i}{\partial x_k} + \frac{\partial \bar{u}_k}{\partial x_i} \right)$$

i.e.

$$- \rho \frac{\partial^2 \bar{u}_i}{\partial x_k \partial x_j} \overline{u'_j \frac{\partial u'_i}{\partial x_k}} = \mu_T \left(\frac{\partial^2 \bar{u}_i}{\partial x_j \partial x_k} \right)^2 - \frac{1}{6} \rho \frac{\partial k}{\partial x_j} \frac{\partial^2 \bar{u}_i}{\partial x_j^2}$$

$$+ \frac{2}{5} \frac{\partial \mu_T}{\partial x_k} \left(\frac{\partial \bar{u}_i}{\partial x_k} + \frac{\partial \bar{u}_k}{\partial x_i} \right) \frac{\partial^2 \bar{u}_i}{\partial x_j^2}$$

As in the term $- \dfrac{\nu}{\rho} \dfrac{\partial}{\partial x_j} \overline{\left\{ \dfrac{\partial P'}{\partial x_i} \dfrac{\partial u'_j}{\partial x_i} \right\}}$; products of first and
second derivatives of mean flow quantities are negligible
w.r.t. remaining terms in (A.1.17).

Hence $- \rho \dfrac{\partial^2 \bar{u}_i}{\partial x_k \partial x_j} \overline{u'_j \dfrac{\partial u'_i}{\partial x_k}} = 0$
(A.1.18c)

Finally, the terms $-2\nu \overline{\dfrac{\partial u'_i}{\partial x_k} \dfrac{\partial u'_j}{\partial x_k} \dfrac{\partial u'_i}{\partial x_j}}$; $- 2\nu^2 \overline{\left(\dfrac{\partial^2 u'_i}{\partial x_j \partial x_k} \right)^2}$ represent,

respectively the generation rate of vorticity fluctuations,
and the decay of dissipation rate as a consequence of viscosity.
For high Reynolds numbers the sum of these terms is independent
of viscosity, since it is controlled by the process of trans-
porting energy from low to high wave numbers.   A simple approximation ensuring dimensional consistency is to set

$$+2\nu \overline{\frac{\partial u'_i}{\partial x_k} \frac{\partial u'_j}{\partial x_k} \frac{\partial u'_i}{\partial x_j}} + 2\nu^2 \overline{\left(\frac{\partial^2 u'_i}{\partial x_j \partial x_k} \right)^2} = \frac{C_2}{C_\mu} C_D \frac{\varepsilon^2}{k} \qquad \text{(A.1.18d)}$$

where C_2 is a constant to be determined.

Substituting from (A.1.18a,b,c,d) in equation (A.1.17)
yields

$$\rho \, \frac{\partial \varepsilon}{\partial t} + \rho \bar{u}_j \, \frac{\partial \varepsilon}{\partial x_j} = C_1 \, \rho \, \frac{\varepsilon}{k} \, \frac{\mu_T}{\rho} \, \frac{\partial \bar{u}_i}{\partial x_k} \left(\frac{\partial \bar{u}_i}{\partial x_k} + \frac{\partial \bar{u}_k}{\partial x_i} \right)$$

$$- \frac{C_2}{C_\mu} \, \rho \, \frac{\varepsilon^2}{k} + \frac{\partial}{\partial x_j} \left\{ \left(\mu + \frac{\mu_T}{\sigma_\varepsilon} \right) \frac{\partial \varepsilon}{\partial x_j} \right\} \, \rho \left\{ \frac{\partial \varepsilon}{\partial t} + \bar{u}_j \, \frac{\partial \varepsilon}{\partial x_j} \right\}$$

$$= \rho \, C_\mu \, C_1 \, k \, \frac{\partial \bar{u}_i}{\partial x_k} \left(\frac{\partial \bar{u}_i}{\partial x_k} + \frac{\partial \bar{u}_k}{\partial x_i} \right) - \frac{C_2}{C_\mu} \, \rho \, \frac{\varepsilon^2}{k} + \frac{\partial}{\partial x_k} \left\{ \left(\mu + \frac{\mu_T}{\sigma_\varepsilon} \right) \frac{\partial \varepsilon}{\partial x_j} \right\}$$

$$(A.1.19)$$

    Equations (1), (2), (A.1.13), (A.1.19) together
with the relationships (6) and (9) form the two
equation model of turbulence.   In the derivation of these
equations we have assumed that C_1, C_2, σ_ε, C_D, σ_k and C_μ are
constants.

APPENDIX 2

Method of Weighted Residuals

Weighted residual methods are, in essence, numerical techniques which can be used to solve a single or set of partial differential equations. Consider such a set, say representative of (1) and (2)

$$L(\underline{u}) = p \qquad\qquad (A.2.1)$$

in a domain Ω, where $\underline{u}$ is the exact solution and may represent a single variable or a column vector of variables. The prevalent type boundary conditions are,

 i) prescribed value

ii) prescribed gradient
and
iii) traction

in which

$$\Gamma_{uo} + \Gamma_q = \Gamma \qquad\qquad (A.2.2)$$

the total length of the boundary.

The first step in the application of the weighted residual procedure is to assume that $\underline{u}$ can be approximated over the whole domain by,

$$\hat{\underline{u}} = \sum_{L=1}^{n} \alpha^L \underline{\beta}^L \qquad\qquad (A.2.3)$$

where α^L are functions described in terms of independent variables, such as spatial coordinates (x,y), and $\underline{\beta}$ are undetermined parameters, and L represents a prime and not a power.

Utilising this approximation and incorporating (A.2.3) in (A.2.1) results in an error or residual, ε, such that

$$\varepsilon = L(u) - p \neq 0 \qquad\qquad (A.2.3)$$

where ε is exactly zero when $\hat{u} = u$ i.e. an exact solution is possible.

In order to make ε identically zero a set of 'arbitrary' weighting functions, W, are employed such that over the whole domain, Ω,

$$\int_{\Omega} W\varepsilon \; d\Omega = 0 \qquad\qquad (A.2.4)$$

If the number of unknown parameters is s then there are s linearly independent weighting functions and (A.2.4) can be re-written,

$$\int_{\Omega} W^k \; \varepsilon \; d\Omega = \int_{\Omega} W^k \; (L(\hat{u}) - p) \; d\Omega = 0 \qquad\qquad (A.2.5)$$
$$k = 1,2,3...s$$

in which W^k must be, positive, single valued and finite.

There are a number of ways in which the above concepts can be utilised to transform the differential equations into a form where finite element techniques can be adopted with effect. These have been expounded in various texts, Hinton and Owen (1980,1980), Taylor and Morgan (1980), Taylor and Hughes (1981). The present text will be confined to one method only, the Galerkin, Crandall (1966), method.

The Galerkin Weighted Residual Method

Before embarking on the main objective of this section a further brief introduction must be given to the commonly adopted concept of trial or shape functions in a finite element context.

The technique of defining approximated values of the required variable via a discrete summation was introduced in (A.2.3). The approximate values were defined in terms of some functions α and discrete values β.   This applied over the whole domain under consideration in which s refers to the total number of discrete values.   If we now refine this concept and subdivide the domain into elements, the variable value within that sub-region can now be defined in terms of discrete values on the boundary of, or within that region,

$$\hat{u} = \sum_{L=1}^{n} \underline{N}^L \; \underline{\beta}^L \qquad\qquad (A.2.6)$$

where $\underline{N}$ are a set of trial functions written in terms of local coordinates associated with n discrete values within or on the boundary of an element.   Each element will, normally, possess a unique set of equations and β^L is now confined to each element.

The residual now becomes,

$$\varepsilon = L\left(\sum_{L=1}^{n} (\underline{N}^L \; \underline{\beta}^L) \right) - p \qquad\qquad (A.2.7)$$

324

such that (A.2.5) can be rewritten,

$$\int_{\Omega} W^k \left[L\left(\sum_{i=1}^{n} \underline{N}^L \underline{\beta}^L \right) - P \right] d\Omega = 0 \qquad (A.2.8)$$

$$k = 1,2,\ldots s$$

In the Galerkin technique the same approximating functions are used for the weighting and trial functions i.e. $W_k = N_k$ and the generalised equation becomes,

$$\int_{\Omega} N^k \left[L\left(\sum_{i=1}^{n} \underline{N}^L \underline{\beta}^L \right) - P \right] d\Omega = 0 \qquad (A.2.9)$$

in which orthogonalisation has been effected with the same functions.

Finite Element Model

Considering the continuity equation, (2), as a typical example,

$$\frac{\partial u_i}{\partial x_i} = 0 \qquad\qquad i = 1,2 \qquad (A.2.10)$$

would be written, after the application of the Galerkin weighted residual approach,

$$\iint_{\Omega} W^k \left(\frac{\partial u_i}{\partial x_i} \right) d\Omega = 0 \quad k = 1\ldots s \qquad (A.2.11)$$

If the variables variation across an element are defined in accordance with,

$$\{\lambda\} = N^J \lambda^J$$

and $\quad p = M^J p^J$

equation (A.2.11) can be re-written,

$$\int_{\Omega} M^k \left(\frac{\partial N^L}{\partial x_1} u_i^L \right) d\Omega = 0 \qquad (A.2.12)$$

where the order of M is one lower than that for N, Taylor and Hood (1974).   If each element is considered in sequence then (A.2.12) can be simplified,

$$\sum_{e=1}^{ne} \iint_{\Omega_e} M^k \left(\frac{\partial N^L}{\partial x_i} u_i^L \right) d\Omega = 0 \qquad (A.2.13)$$

in which ne is the total number of elements and the integration is over each element.   A similar procedure can be adopted when considering the momentum or transport equations.   If Green's theorem is utilised to reduce second order terms, equation (8) can be written,

$$\sum_{e=1}^{ne} \iint_{\Omega_e} N^K \left(N^J u_j^J \frac{\partial N^L}{\partial x_j} u_i^L + \frac{1}{\rho} \frac{\partial M^J}{\partial x_i} p^J \right)$$
$$+ N^J v_e \left(\frac{\partial N^K}{\partial x_j} \frac{\partial N^L}{\partial x_j} u_i^L \right) - \int_{\Gamma_e} N^K N^J v_e \left(\frac{\partial u_i}{\partial x_j} \ell_j \right) ds_e = 0$$

$$(A.2.14)$$

in which Γ_e is the element boundary and ℓ_j represents the outward normal to the boundary.

Both transport equations, for k and ε, can be treated in exactly the same manner.   Fuller details on this can be found in the previous book in this series, Taylor and Morgan (1980) and in the chapter by Larock and Schamber in the current book.

SUBJECT INDEX

Adams Bashforth integration 99,115,118,119,120

Advection, 111,117,133,138,147, 152,161

Advection-Diffusion, 108,113, 115,118

Aeronautical, 37,39,47

Airfoil, 47,49

Airplane, 47

Airy function, 78

Alternating Direction Method 34

Amplification Factor, 114

Amplification Matrix, 117

Boundary Layers, 1,14,15,24,139, 232

Buoyancy effect, 246

Cavity, 224,225

Cavity driven, 121,122,123,124, 125,126,127

Cell, computational, 63

Centrifugal force, 204,215,219

Central Difference, 1,2,4,6,8, 11,13,14,15,16,17,19,23,33

Characteristics, 39,40,42,45,51

Circulation, 48

Clarifier, 264

Collocation, 149

Combustion Chamber, 225,226

Contact Discontinuity, 60,74, 75,77,79,92

Contact Strip, 76,77,79,87,89, 90,92

Convection, 1,2,3,4,6,7,8,9,10, 11,12,13,15,18,25,33,34,191, 200,204,206,207,209,229,234, 235,244,256

Convergence, 43,47,49

Cooling Tower, 227

Coriolis, 135,245

Courant Number, 66

Critical Flow, 167

Cyclon Separator, 216

Cylinder, 129

Damping, 3,4,5,33,47,63,109, 110,112,113,114,115,119,120, 145,193

Neutral, 3

Inherent, 3,5,60,139

Dams, 187

Density, 49,66,99

Diffusion, 1,2,4,5,6,7,8,9,14, 13,17,19,33,98,104,110,117, 136,191,200,203,229,230,231, 135,236,256,287

   Artificial, 2,4,9,10,20,136

Dirac Functions, 39,44,45

Dirichlet Boundary Conditions, 106

Divergence Form, 67,238

Eddy Viscosity, 254,265,272

Eddies, Secondary, 127

Eigenvalue, 72,110,137,148,149, 150

Einstein, 245

Elastic Flexure, 33

Energy Dissipation Term, 193,209

Enthalpy, 230,247

Error Analyses, 134,139

Error
   Discretisation, 6,9

   Evolutionary, 141

   Residual, 256

   Truncation, 7,141,143,144, 145,146,151

Euler
   Equations, 37,38,44,45,51,61

   Integration, 99,115

   Scheme, 114

Eularian
   Grid, 73,74

   Variables, 70

Expansion
   Sudden, 128

Extremum, 65

Feedback Sensitivity, 5,6,9, 11,12

   Negative, 4,5,6,7,9,11

   Neutral, 5,7,8

   Positive, 5,6

Flux, 41

Forecasting, Weather, 133,134

Fourier
   Component, 108

   Transform, 144,145,146

Fractional Step Method 98,100,102,105,107

Free Flow, 192

Free Surface, 165,169,171,172, 173,183,187,188,265

Functional, 169,172

Galerkin
   Formulation, 108,142,143,144, 145,146,148,149,152,154,161, 228,254,256,257,259,264,265, 275,289

   Spline, 149,151

Gas Flow, 59,60,70,81

Gaussian
   Elimination, 107,199,261

   Integration, 262

Generator, Steam, 225,226,245, 247,248

Geopotential, 135

Geostrophere, 138,157

Godunov Scheme, 69

Gravity Flow, 165,167,187

Green's Formula, 100

Halow's Method, 70,71,72

Harmonic, Spherical, 133,154,
    155,156,161

Heat
    Exchangers, 227

    Transfer, 223,247,251

Hilbert Space, 98,100,140

Hugoniot Conditions, 69

Hybrid Techniques, 17,18,33

Impact, 73

Implicit Schemes, 47

Incompressible Flow, 1,37

Incompressible Transport
    Equation, 30

Inviscid, 37

Irrotational Flow, 168

Jacobian, 178

Jets, 225,253

Joukowski Airfoil, 49

Kinetic Energy of Turbulence,
    193,230,245

Kutta-Joukowski Condition, 48

Lagrange Elements, 98

Laplacian, 168

Latent Heat, 136

Latitude, 135,157

Lax Scheme, 69,71,85

Lax-Wendroff, 66,71,79,80,82,
    83,88,91,92

Least Squares, 143,144,148,149,
    150,161,269

Leuenberg-Marquardt Method, 271

Lift Factor, 49,54,55

Locus, Geometric, 77

Longitude, 135

Lumping, 3

Map Projection, 152

Mode
    Fundamental, 121

    Normal, 136

    Spurious, 121

Modal Frequency, 115

Nappe, 181,182

Newtonian Fluid, 99,286

Newton Scheme, 261,263,268,269,
    270,271,272,273,281

Newton-Raphson Method, 178

Nuclear Plant, 225,245

Nusselt No. 244,246

Operator
    Identity, 67,82

    Non-linear, 140

    Projection, 98,104,106,111,
    113

330

Operator
    Translation, 81

Oscillations, Numerical, 121,129

Particle-in-Cell, 74

Peclet No. 7,8,10,13,14,15,16,17
    18,19,20,21,22,24,25,26,34,
    236,237

Phase
    Difference, 155

    Speed, 115

Plume, Chimney, 225

Porous Media, 165

Potential, Velocity, 62

Prandtl, 77,284

Projection Method, 97

Rank Deficiency, 106

Recirculation Zone, 284,291,
    295,296

Relaxation Procedure, 238,261,
    300

Reservoir, 167,173,178

Reynolds
    Approach, 254

    No. 121,128,129,130,139,256,
    273,283,291,298,300,301

    Stress, 191,192,196,200,204,
    206,207,215,219,254,286

Riemann Problem, 74

Rotating
    Disc, 212
    Duct, 245

Runge-Kutta Technique, 41,46

Saturation Temperature, 249

Secondary Flow, 191,192,211,
    224,244

Sedimentation Basin, 264

Separated Flow, 283

Separation, 265

Shock, Wave/Front, 44,59,60,
    62,63,64,65,66,69,72,73,74,
    138

Shallow Water Equations, 136,
    154

Shrouded Fin, 244

Singularities, 59

Sobolov Space, 39,100

Source Term, 3,203,229,230,
    240

Specific Heat, 136

Spectral Components, 155,156

Spillway, 165,166,167,171,172,
    178,179,180,183,184,187,188

Spline Functions, 87,146

Stability, 78,92,113,114,117

Staggered Grid, 240,241

Stagnation Level, 168,169,171,
    172,180,183,187

State, Equation of, 62,84,85,
    86,87

Stationary Points, 68,270

Steepest Descent Method,
    269,270

Stratosphere, 139

Streamlines, 49,50,124,126,128,
    129,130,167,172,173,179,187,
    244,246

Supersonic Flow, 86

Stream Function, 38,42,48,121,
    157,169,171,188,195,201,204,
    205,206,208,213,219

Swirl, 191,192,195,204,205,208,
    209,212,213,215

Tainter Gates, 183

Tangent Function, 19

Taylor Series, 3,4,6,11

Time Integration
    Explicit, 115,116,118

    Splitting, 34

Traction, 99

Trailing Edge, 48

Transient Flow, 37

Transition Zone, 74

Transport Equations
    k - 313
    ε - 316

Trotosphere, 139

Truncation Error, 4

Turbulence, 47

    Atmospheric, 49

    Diffusion, 194,2,45,254,
    263,264

    Kinetic Energy, 193,254,263,
    284,287,310

Two-Phase Flow, 247

Universal Laws, 296,305

Upwinding, 1,2,8,11,17,202,283,
    284,295,310

    Optimal, 1,2,3,4,18,19,20,
    21,22,23,33

    1st Order, 9,10,11,15,23
    2nd Order, 3,10
    3rd Order, 1,11,33

Wake, 48,79,253

Waves
    Gravity, 138

    Progressive, 62,63,64,68,69

Wave Motion, 33

    Number, 109,112,154,155

    Shock, 60,63

    Short, 110,115,145

    Sound, 138

Weather Forecasting, 133,134,
    136,139,151,156,161

Weighting Procedure, 29,30,33,
    34,254,256

Weighted Residuals, 149,228,254

Weir, Submerged, 171

Wiggles,4,5,8,11,13,14,15,18,24,
    25,33,34,283

Wind Component, 135

Variational Formulation,
    99,100,165,166,167,169,171,
    172

Viscosity
    38,63,64,70,99,121,227,230,
    272,287

    Artificial, 63,64,65,66,68,
    72

332

Viscosity,
  Effective, 284, 287

Vortex, 38,45,49,52,127,210,
  216

  Method, 37,39,40,44,126,
  127

Vorticity, 40,41,154,195,203,
  205,207,211,219